MATHEMATICS

For

B.Sc. BRANCH–I

[VOLUME-IV]

MATHEMATICS

For

B.Sc. BRANCH–I

FOURTH SEMESTER

Containing

Vector Calculus, Fourier Series and Fourier Transforms

VOLUME-IV

P. KANDASAMY
M.A., Ph.D.
Professor of Mathematics (Retd.)
P.S.G. College of Technology
Coimbatore
&
Visiting Professor
Amrita Institute of Technology
Deemed University
COIMBATORE–641105

K. THILAGAVATHI
M.Sc., M.Phil, Ph.D.
Reader in Mathematics
Kongunadu Arts and Science College
COIMBATORE–641021

S. CHAND
PUBLISHING

S Chand And Company Limited

(ISO 9001 Certified Company)

S Chand And Company Limited

(ISO 9001 Certified Company)

Head Office: Block B-1, House No. D-1, Ground Floor, Mohan Co-operative Industrial Estate, New Delhi – 110 044 | Phone: 011-66672000

Registered Office: A-27, 2nd Floor, Mohan Co-operative Industrial Estate, New Delhi – 110 044 | Phone: 011-49731800

www.**schandpublishing.com**; e-mail: **info@schandpublishing.com**

Branches

Chennai : Ph: 23632120; chennai@schandpublishing.com
Guwahati : Ph: 2738811, 2735640; guwahati@schandpublishing.com
Hyderabad : Ph: 40186018; hyderabad@schandpublishing.com
Jalandhar : Ph: 4645630; jalandhar@schandpublishing.com
Kolkata : Ph: 23357458, 23353914; kolkata@schandpublishing.com
Lucknow : Ph: 4003633; lucknow@schandpublishing.com
Mumbai : Ph: 25000297; mumbai@schandpublishing.com
Patna : Ph: 2260011; patna@schandpublishing.com

First Edition 2004
Reprint 2015

Reprint 2022

ISBN: 978-81-219-2473-3 **Product Code:** H6MBF61MATH10ENAA05S

PRINTED IN INDIA

By Vikas Publishing House Private Limited, Plot 20/4, Site-IV, Industrial Area Sahibabad, Ghaziabad – 201 010 and Published by S Chand And Company Limited, A-27, 2nd Floor, Mohan Co-operative Industrial Estate, New Delhi – 110 044.

Dedicated
to
Students
Past and the Present

Authors

PREFACE TO THE FIRST EDITION

We are happy to bring out this volume "Mathematics for B.Sc.-Branch I" for fourth Semester containing the syllabus of Mathematics namely, Vector calculus, Fourier series and Fourier Transform to suit the syllabus of Madras University, under the semester pattern, from the year 2003 onwards.

As our books on Engineering Mathematics, Numerical methods and our books on Mathematics for B.Sc. Branch-I and Allied Mathematics have been well received and appreciated by the students as well as the facility, we are inspired to bring out this volume IV also for students of Arts and Science Colleges. Theoretical concepts have been presented exhaustively with clear and lucid explanations. Very many illustrative examples have been fully worked out for better understanding. An extensive collection of problems is given at the end of each chapter.

Critical evaluation and suggestions for improvement of the book will always be most welcome and will be acknowledged most gratefully.

We are also thankful how publishers, S. Chand & Company Ltd., New Delhi and in particular to the Managing Director, Shri Ravindra Kumar Gupta and Advisor, Shri R.S. Saxena for the support readily extended to for publishing this book.

AUTHORS

CONTENTS

CHAPTER 1

Differentiation of Vectors

(*The Scalar and Vector Fields*)

A physical quantity expressible as a continuous function and which can assume one or more definite values at *each* point of a region of space, is called a *point function* in the region and the region concerned is called a *field.* Point functions are classified as *scalar point function* and *vector point function* according as the nature of the quantity concerned is a scalar or a vector. At each point P of the field if the function denoted by $f(P)$ is a scalar, it is known as scalar point function while if $\mathbf{f}(P)$ is a vector, then the function $\mathbf{f}(P)$ is called a vector point function. The concerned field is called a scalar field or a vector field respectively. The examples of scalar fields are the temperature distribution in a medium, the gravitational potential of a system of masses and the electrostatic potential of a system of charges. The examples of vector fields are the velocity of a moving particle, the electrostatic, the magnetostatic and gravitational fields.

Derivative of a vector. Let $\mathbf{r}$ be a continuous single valued function of a scalar variable t. Then $\mathbf{r}$ is said to be a vector function of the scalar variable t.

Let $\mathbf{r} = \mathbf{f}(t)$ be a continuous and single valued function of t. As t changes to $t + \delta t$, $\mathbf{r}$ changes to $\mathbf{r} + \delta\mathbf{r}$

$$\therefore \qquad \mathbf{r} + \delta\mathbf{r} = \mathbf{f}(t + \delta t)$$

$$\delta\mathbf{r} = \mathbf{f}(t + \delta t) - \mathbf{f}(t); \quad \frac{\delta\mathbf{r}}{\delta t} = \frac{\mathbf{f}(t + \delta t) - \mathbf{f}(t)}{\delta t}$$

As $\delta t \to 0$, $\frac{\delta\mathbf{r}}{\delta t}$ is defined as the derivative of $\mathbf{r}$ w.r.t. t.

Thus, $$\underset{\delta t \to 0}{\text{Lt}} \frac{\delta\mathbf{r}}{\delta t} = \underset{\delta t \to 0}{\text{Lt}} \frac{\mathbf{f}(t + \delta t) - \mathbf{f}(t)}{\delta t} \quad \therefore \frac{d\mathbf{r}}{dt} = \underset{\delta t \to 0}{\text{Lt}} \frac{\mathbf{f}(t + \delta t) - \mathbf{f}(t)}{\delta t}$$

$\frac{d\mathbf{r}}{dt}$ is itself a vector point function of t. If its derivative exists, it is denoted by $\frac{d^2\mathbf{r}}{dt^2}$.

Suppose a particle P moves on a curve.

At time t, it is at P.

(Fig. 1.)

At time $(t + \delta t)$, it is at P'.

If $\overrightarrow{OP} = \mathbf{r}$, $\overrightarrow{OP'} = \mathbf{r} + \delta\mathbf{r}$

Then, $\overrightarrow{PP'} = \overrightarrow{OP'} - \overrightarrow{OP} = \delta\mathbf{r}$,

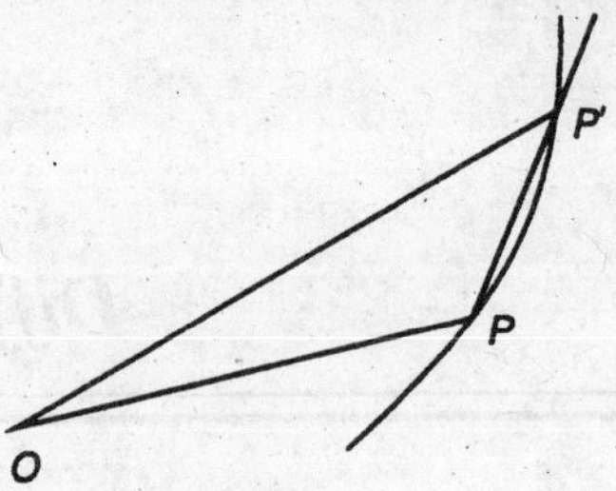

Fig. 1

$\therefore \quad \underset{\delta t \to 0}{\text{Lt}} \dfrac{\delta \mathbf{r}}{\delta t} = \dfrac{d\mathbf{r}}{dt} =$ velocity of P

relative to O and is in the direction of the tangent to the path of P, in the sense of t increasing.

Also $\dfrac{d^2\mathbf{r}}{dt^2}$ is the acceleration of P relative to O.

If arc $PP' = \delta s$, as $P' \to P$, the ratio of the lengths of the chord $P'P$ and arc $P'P$ tends to unity. Hence $\dfrac{d\mathbf{r}}{ds}$ is a unit vector along the tangent at P to the path, in the sense of t increasing.

$\therefore \quad \dfrac{d\mathbf{r}}{ds} = \mathbf{t}$ where $\mathbf{t}$ is the unit tangent vector at P.

1. Derivative of a constant vector is zero

Let **c** be a constant vector. As t changes from t to $t + \delta t$, there is no change in **c**.

Therefore, $\quad \delta c = 0$; *i.e.*, $\underset{\delta t \to 0}{\text{Lt}} \dfrac{\delta \mathbf{c}}{\delta t} = 0$; *i.e.*, $\dfrac{d\mathbf{c}}{dt} = 0$

2. If **u** and **v** are both vector point functions of the scalar t then $\dfrac{d}{dt}(\mathbf{u} \pm \mathbf{v}) = \dfrac{d\mathbf{u}}{dt} \pm \dfrac{d\mathbf{v}}{dt}$

Let Δt be the increment in t. Let the corresponding increments in **u** and **v** be $\delta\mathbf{u}$ and $\delta\mathbf{v}$ respectively.

$$\frac{\delta(\mathbf{u} \pm \mathbf{v})}{\delta t} = \frac{\delta \mathbf{u}}{\delta t} \pm \frac{\delta \mathbf{v}}{\delta t}$$

$$\therefore \underset{\delta t \to 0}{\text{Lt}} \frac{\delta(\mathbf{u} \pm \mathbf{v})}{\delta t} = \underset{\delta t \to 0}{\text{Lt}} \frac{\delta \mathbf{u}}{\delta t} \pm \underset{\delta t \to 0}{\text{Lt}} \frac{\delta \mathbf{v}}{\delta t} \quad \textit{i.e.,} \quad \frac{d}{dt}(\mathbf{u} \pm \mathbf{v}) = \frac{d\mathbf{u}}{dt} \pm \frac{d\mathbf{v}}{dt}.$$

Note : The result can be extended to any number of vectors.

3. If $\mathbf{v} = u\mathbf{r}$ where u is a scalar differentiable function of the scalar variable t and **r** is a vector differentiable function of t, then

$$\frac{d\mathbf{v}}{dt} = \frac{du}{dt}\mathbf{r} + u\frac{d\mathbf{r}}{dt}$$

Let t change from t to $t + \delta t$. Then,

$$\delta\mathbf{v} = (u + \delta u)(\mathbf{r} + \delta\mathbf{r}) - u\mathbf{r} = u\delta\mathbf{r} + \delta u\mathbf{r} + \delta u\delta\mathbf{r}$$

$$\therefore \qquad \frac{\delta\mathbf{v}}{\delta t} = u\frac{\delta\mathbf{r}}{\delta t} + \frac{\delta u}{\delta t}\mathbf{r} + \frac{\delta u}{\delta t}\frac{\delta\mathbf{r}}{\delta t}.\delta t$$

Taking limit as $\delta t \to 0$, $\dfrac{d\mathbf{v}}{dt} = u\dfrac{d\mathbf{r}}{dt} + \dfrac{du}{dt}\mathbf{r}$

4. Derivative of a function of a function

If $\mathbf{r}$ is a continuous function of a scalar variable s and s is a continuous function of another scalar variable t, then $\dfrac{\delta\mathbf{r}}{\delta t} = \dfrac{\delta\mathbf{r}}{\delta s}.\dfrac{\delta s}{\delta t}$

Taking limit as $\delta t \to 0$, we have $\dfrac{d\mathbf{r}}{dt} = \dfrac{d\mathbf{r}}{ds}.\dfrac{ds}{dt}$

5. Derivative of a dot and cross product of two vectors

(*i*) Let $\mathbf{f} = \mathbf{u} \,.\, \mathbf{v}$ and $\mathbf{w} = \mathbf{u} \times \mathbf{v}$ where $\mathbf{u}$ and $\mathbf{v}$ are vector point functions of the scalar variable t. Let δt be the increment in t.

Let $\delta u, \delta v, \delta f, \delta w$ be the corresponding increments in **u, v, f, w** respectively.

$$\mathbf{f} + \delta\mathbf{f} = (\mathbf{u} + \delta\mathbf{u}) \,.\, (\mathbf{v} + \delta\mathbf{v}) = \mathbf{u}.\mathbf{v} + \mathbf{u} \,.\, \delta\mathbf{v} + \delta\mathbf{u}.\mathbf{v} + \delta\mathbf{u}.\delta\mathbf{v}$$

$$\delta\mathbf{f} = \mathbf{u} \,.\, \delta\mathbf{v} + \delta\mathbf{u} \,.\, v + \delta\mathbf{u} \,.\, \delta\mathbf{v}$$

$$\frac{\delta\mathbf{f}}{\delta t} = \mathbf{u} \,.\, \frac{\delta\mathbf{v}}{\delta t} + \frac{\delta\mathbf{u}}{\delta t} \,.\, \mathbf{v} + \left(\frac{\delta\mathbf{u}}{\delta t} \,.\, \frac{\delta\mathbf{v}}{\delta t}\right)\delta t$$

As $\delta t \to 0$, the last equation becomes $\dfrac{d\mathbf{f}}{dt} = \mathbf{u} \,.\, \dfrac{d\mathbf{v}}{dt} + \dfrac{d\mathbf{u}}{dt} \,.\, \mathbf{v} + \left(\dfrac{d\mathbf{u}}{dt} \,.\, \dfrac{d\mathbf{v}}{dt}\right)(0)$

$$\therefore \qquad \frac{d}{dt}(\mathbf{u} \,.\, \mathbf{v}) = \mathbf{u} \cdot \frac{d\mathbf{v}}{dt} + \frac{d\mathbf{u}}{dt} \cdot \mathbf{v}$$

(*ii*) Similarly, $\delta\mathbf{w} = \mathbf{u} \times \delta\mathbf{v} + \delta\mathbf{u} \times \mathbf{v} + \delta\mathbf{u} \times \delta\mathbf{v}$

$$\frac{\delta\mathbf{w}}{\delta t} = \mathbf{u} \times \frac{\delta\mathbf{v}}{\delta t} + \frac{\delta\mathbf{u}}{dt} \times \mathbf{v} + \left(\frac{\delta\mathbf{u}}{\delta t} \times \frac{\delta\mathbf{v}}{\delta t}\right)\delta t$$

Taking limits as $\delta t \to 0$, $\dfrac{d\mathbf{w}}{dt} = \mathbf{u} \times \dfrac{d\mathbf{v}}{dt} + \dfrac{d\mathbf{u}}{dt} \times \mathbf{v} + \left(\dfrac{d\mathbf{u}}{dt} \times \dfrac{d\mathbf{v}}{dt}\right)(0)$

$$\therefore \qquad \frac{d}{dt}(\mathbf{u} \times \mathbf{v}) = \mathbf{u} \times \frac{d\mathbf{v}}{dt} + \frac{d\mathbf{u}}{dt} \times \mathbf{v}.$$

Cor 1. $\qquad \dfrac{d}{dt}(\mathbf{r} \,.\, \mathbf{r}) = 2\mathbf{r} \,.\, \dfrac{d\mathbf{r}}{dt}$

But $\mathbf{r} \,.\, \mathbf{r} = r^2$ where $r = |\mathbf{r}|$ and $\dfrac{d}{dt}(r^2) = 2r\dfrac{dr}{dt}$

$$\therefore \qquad \mathbf{r}\cdot\frac{d\mathbf{r}}{dt} = r\frac{dr}{dt}$$

Cor 2. If **a** is a constant vector, then

$$\frac{d}{dt}(\mathbf{a}\,.\,\mathbf{u}) = \mathbf{a}\,.\,\frac{d\mathbf{u}}{dt} + \frac{d\mathbf{a}}{dt}\,.\,\mathbf{u} = \mathbf{a}\,.\,\frac{d\mathbf{u}}{dt} \quad \left(\because \frac{d\mathbf{a}}{dt} = 0\right)$$

Also $\frac{d}{dt}(\mathbf{a}\times\mathbf{u}) = \mathbf{a}\times\frac{d\mathbf{u}}{dt} + \frac{d\mathbf{a}}{dt}\times\mathbf{u} = \mathbf{a}\times\frac{d\mathbf{u}}{dt} \quad \left(\because \frac{d\mathbf{a}}{dt} = 0\right)$

6.
$$\frac{d}{dt}[\mathbf{p},\mathbf{q},\mathbf{r}] = \frac{d}{dt}\{(\mathbf{p}\,.\,(\mathbf{q}\times\mathbf{r})\}$$
$$= \left[\frac{d\mathbf{p}}{dt},\mathbf{q},\mathbf{r}\right] + \left[\mathbf{p},\frac{d\mathbf{q}}{dt},\mathbf{r}\right] + \left[\mathbf{p},\mathbf{q},\frac{d\mathbf{r}}{dt}\right]$$

7. $\frac{d}{dt}[\mathbf{p}\times(\mathbf{q}\times\mathbf{r})] = \frac{d\mathbf{p}}{dt}\times(\mathbf{q}\times\mathbf{r}) + \mathbf{p}\times\left(\frac{d\mathbf{q}}{dt}\times\mathbf{r}\right) + \mathbf{p}\times\left(\mathbf{q}\times\frac{d\mathbf{r}}{dt}\right)$

The proofs of these are left as exercises to the reader.

Example 1. *If* **r** *is a vector of constant magnitude, prove*

$$\mathbf{r}\,.\,\frac{d\mathbf{r}}{dt} = 0 \text{ and}$$ **(*B.Sc. M.U.*)**

if **r** *is vector of constant direction, prove that* $\mathbf{r}\times\frac{d\mathbf{r}}{dt} = \mathbf{0}$.

Solution. Let **r** be a vector of constant magnitude.

Then $|\mathbf{r}| = r =$ constant ; $\mathbf{r}\,.\,\mathbf{r} = r^2 =$ constant

$$\therefore \qquad \frac{d}{dt}(\mathbf{r}\,.\,\mathbf{r}) = \frac{d}{dt}(\text{constant}) = 0$$

$$2\,\mathbf{r}\,.\,\frac{d\mathbf{r}}{dt} = 0 \;\therefore\; \mathbf{r}\,.\,\frac{d\mathbf{r}}{dt} = 0\,;\ \text{but } \mathbf{r}\neq\mathbf{0}$$

$\therefore$ either $\frac{d\mathbf{r}}{dt} = \mathbf{0}$ or $\frac{d\mathbf{r}}{dt}$ is perpendicular to **r**

Suppose **r** is a vector of constant direction

Let $\mathbf{r} = r\mathbf{n}$ where **n** is a unit vector in the constant direction

$$\frac{d\mathbf{r}}{dt} = \frac{d}{dt}(r\mathbf{n}) = \frac{dr}{dt}\mathbf{n}$$

$$\therefore \qquad \mathbf{r}\times\frac{d\mathbf{r}}{dt} = \mathbf{r}\times\frac{dr}{dt}\mathbf{n} = r\mathbf{n}\times\frac{dr}{dt}\mathbf{n}$$
$$= r\frac{dr}{dt}\mathbf{n}\times\mathbf{n} = \mathbf{0}.$$

Example 2. *If* $\mathbf{r} = \mathbf{a}\cos nt + \mathbf{b}\sin nt$ *where* **a**, **b**, n *are constants, prove*

(*i*) $\mathbf{r}\times\frac{d\mathbf{r}}{dt} = n\mathbf{a}\times\mathbf{b}$ (*ii*) $\frac{d^2\mathbf{r}}{dt^2} + n^2\mathbf{r} = \mathbf{0}$

(*iii*) $\left[\mathbf{r}, \dfrac{d\mathbf{r}}{dt}, \dfrac{d^2\mathbf{r}}{dt^2}\right] = 0;$

$$\mathbf{r} = \mathbf{a}\cos nt + \mathbf{b}\sin nt$$

$$\frac{d\mathbf{r}}{dt} = \mathbf{a}\,(-n\sin nt) + \cos nt\,\frac{d\mathbf{a}}{dt} + \mathbf{b}(n\cos nt) + \sin nt\,\frac{d\mathbf{b}}{dt}$$

$$= \mathbf{a}(-n\sin nt) + \mathbf{b}(n\cos nt) \qquad (\because \text{ other terms vanish})$$

(*i*) $\mathbf{r}\times\dfrac{d\mathbf{r}}{dt} = [\mathbf{a}\cos nt + \mathbf{b}\sin nt]\times[\mathbf{a}(-n\sin nt) + \mathbf{b}(n\cos nt)]$

$$= n\cos^2 nt\,\mathbf{a}\times\mathbf{b} - n\sin^2 nt\,\mathbf{b}\times\mathbf{a} = n\mathbf{a}\times\mathbf{b}.$$

(*ii*) $\dfrac{d^2\mathbf{r}}{dt^2} = \mathbf{a}(-n^2\cos nt) + (-n\sin nt)\dfrac{d\mathbf{a}}{dt}$

$$+ \mathbf{b}(-n^2\sin nt) + (n\cos nt)\,\frac{d\mathbf{b}}{dt}$$

$$= -n^2(\mathbf{a}\cos nt + \mathbf{b}\sin nt) = -n^2\mathbf{r} \qquad \because \text{ other two terms vanish}$$

$$\therefore \qquad \frac{d^2\mathbf{r}}{dt} + n^2\mathbf{r} = 0$$

(*iii*) $\left[\mathbf{r}, \dfrac{d\mathbf{r}}{dt}, \dfrac{d^2\mathbf{r}}{dt^2}\right] = \left(\mathbf{r}\times\dfrac{d\mathbf{r}}{dt}\right).\dfrac{d^2\mathbf{r}}{dt^2} = n(\mathbf{a}\times\mathbf{b}).\{-n^2(\mathbf{a}\cos nt + \mathbf{b}\sin nt)\}$

$$= -n^3\cos nt[\mathbf{a}, \mathbf{b}, \mathbf{a}] - n^3\sin nt[\mathbf{a}, \mathbf{b}, \mathbf{b}] = \mathbf{0}$$

Example 3. *Find the unit tangent vector for the curve*

$$\mathbf{r} = a\cos t\,\mathbf{i} + a\sin t\,\mathbf{j} + ct\mathbf{k}$$

Solution. $\dfrac{d\mathbf{r}}{dt} = -a\sin t\,\mathbf{i} + a\cos t\,\mathbf{j} + c\mathbf{k}$

Unit vector along the tangent is $\dfrac{d\mathbf{r}}{dt} \div \left|\dfrac{d\mathbf{r}}{dt}\right|$

$$= \frac{-a\sin t\,\mathbf{i} + a\cos t\,\mathbf{j} + c\mathbf{k}}{\sqrt{(a^2\sin^2 t + a^2\cos^2 t + c^2)}} = \frac{-a\sin t\,\mathbf{i} + a\cos t\,\mathbf{j} + c\mathbf{k}}{\sqrt{(a^2 + c^2)}}$$

Example 4. *Find the velocity vector, the speed, and the acceleration vector of the particle whose path is given by (i)* $\mathbf{r} = 3\cos 2t\,\mathbf{i} + 2\sin 3t\,\mathbf{k}$ *and*

(*ii*) $\qquad \mathbf{r} = 4t\,\mathbf{i} - 4t\,\mathbf{j} + t\,\mathbf{k}$

(*i*) $\mathbf{r} = 3\cos 2t\,\mathbf{i} + 2\sin 3t\,\mathbf{k}$; $\mathbf{v} = \dfrac{d\mathbf{r}}{dt} = -6\sin 2t\,\mathbf{i} + 6\cos 3t\,\mathbf{k}$

Speed $= |\mathbf{v}| = \sqrt{36(\cos^2 3t + \sin^2 2t)}$

$$\therefore \quad \mathbf{a} = \frac{d^2\mathbf{r}}{dt^2} = -12\cos 2t\,\mathbf{i} - 18\sin 3t\,\mathbf{k}$$

(*ii*) $\mathbf{r} = 4t\,\mathbf{i} - 4t\mathbf{j} + t\mathbf{k}$; $\mathbf{v} = \dfrac{d\mathbf{r}}{dt} = 4\mathbf{i} - 4\mathbf{j} + \mathbf{k}$

Speed $= |\mathbf{v}| = \sqrt{16+16+1} = \sqrt{33}$ $\therefore$ $\mathbf{a} = \dfrac{d\mathbf{v}}{dt} = \mathbf{O}$

Example 5. *A particle moves along a curve whose parametric equations are $x = e^{-t}$, $y = 2\cos 3t$, $z = 2\sin 3t$ where t is the time. Find velocity and acceleration at t = 0.*

Solution $\mathbf{r} = e^{-t}\,\mathbf{i} + 2\cos 3t\,\mathbf{j} + 2\sin 3t\,\mathbf{k}$

$$\mathbf{v} = \frac{d\mathbf{r}}{dt} = -e^{-t}\,\mathbf{i} - 6\sin 3t\,\mathbf{j} + 6\cos 3t\,\mathbf{k}$$

$$\mathbf{a} = \frac{d^2\mathbf{r}}{dt^2} = e^{-t}\,\mathbf{i} - 18\cos 3t\,\mathbf{j} - 18\sin 3t\,\mathbf{k}$$

$$\mathbf{v}_{\text{at } t=0} = -\mathbf{i} + 6\,\mathbf{k}$$

Speed $= \sqrt{1+36} = \sqrt{37}$; $\mathbf{a}_{\text{at } t=0} = \mathbf{i} - 18\,\mathbf{j}$

Example 6. *The position vector of a moving point is given by* $\mathbf{r} = \cos\omega t\,\mathbf{i} + \sin\omega t\,\mathbf{j}$. *Show that the velocity* **v** *of the particle is perpendicular to* **r** **(*MS. Ap. 94*)**

Solution. $\mathbf{v} = \dfrac{d\mathbf{r}}{dt} = -\omega\sin\omega t\,\mathbf{i} + \omega\cos\omega t\,\mathbf{j}$

$$\mathbf{v} \cdot \mathbf{r} = -\omega\sin\omega t\cos\omega t + \omega\sin\omega t\cos\omega t = 0$$

v is perpendicular to **r**.

Example 7. *If* $\mathbf{r} = \mathbf{a}\cosh nt + \mathbf{b}\sinh nt$ *where* **a**, **b**, *n are constants show that* $\mathbf{r} \times \dfrac{d\mathbf{r}}{dt} = n(\mathbf{a} \times \mathbf{b})$ **(*Ms. Nov. 94*)**

Solution. $\dfrac{d\mathbf{r}}{dt} = n\sinh nt\,\mathbf{a} + n\cosh nt\,\mathbf{b}$

$$\mathbf{r} \times \frac{d\mathbf{r}}{dt} = (\mathbf{a}\cosh nt + \mathbf{b}\sinh 2t) \times (n\sinh nt\,\mathbf{a} + n\cosh nt\,\mathbf{b})$$

$$= n\cosh^2 nt\,\mathbf{a}\times\mathbf{b} + n\sinh^2 nt\,\mathbf{b}\times\mathbf{a}$$

$$= n\cosh^2 nt\,\mathbf{a}\times\mathbf{b} - n\sinh^2 nt\,\mathbf{a}\times\mathbf{b}$$

$$= n(1)(\mathbf{a}\times\mathbf{b}) = n(\mathbf{a}\times\mathbf{b})$$

EXERCISES 1

1. Find the unit tangent vector to the curve $x = 3t + 2$, $y = 5t^2$, $z = 2t - 1$ at the point where $t = 1$. (**Hint** : $\mathbf{r} = x\mathbf{i} + y\mathbf{j} + z\mathbf{k}$)

2. Find the unit vector to the curve $x = t^2 + 1$, $y = 4t - 3$, $z = 2t^2 - 6t$ at the point where $t = 2$. ***(BN. Ap. 94)***

3. Find the derivatives of (*i*) $\dfrac{\mathbf{r} + \mathbf{a}}{\mathbf{r}^2 + \mathbf{a}^2}$ (*ii*) $\dfrac{\mathbf{r} \times \mathbf{a}}{\mathbf{r} . \mathbf{a}}$

4. If $\mathbf{r} = \mathbf{a}\, e^{kt} + \mathbf{b}\, e^{-kt}$, prove that $\dfrac{d^2\mathbf{r}}{dt^2} - k^2\mathbf{r} = \mathbf{0}$ where **a**, **b** are constants.

(M.M. Nov. 94)

5. If $\mathbf{r} = \mathbf{a} \cosh nt - \mathbf{b} \sinh nt$ where n, **a**, **b** are constants show that $\dfrac{d^2\mathbf{r}}{dt^2} - n^2\mathbf{r} = \mathbf{0}$.

6. If $\dfrac{d\mathbf{u}}{dt} = \mathbf{w} \times \mathbf{u}$, $\dfrac{d\mathbf{v}}{dt} = \mathbf{w} \times \mathbf{v}$, prove that $\dfrac{d}{dt}(\mathbf{u} \times \mathbf{v}) = \mathbf{w} \times (\mathbf{u} \times \mathbf{v})$.

7. Find the velocity vector, the speed, and the acceleration vector of the particle whose path is given by

   (*i*) $\mathbf{r}(t) = 3 \cos 2t\, \mathbf{i} + 2 \sin 3t\, \mathbf{k}$.

   (*ii*) $\mathbf{r}(t) = 4t\,\mathbf{i} - 4t\mathbf{j} + t\mathbf{k}$ (*iii*) $\mathbf{r}(t) = 5\, e^{-t}\,\mathbf{i} + 7\, e^{-t}\,\mathbf{j}$.

8. A particle moves along a curve whose parametric equations are $x = e^{-t}$, $y = 2 \cos 3t$, $z = 2 \sin 3t$, t being the time. Find its velocity and acceleration at $t = 0$.

9. A particle moves along the curves $x = 2t^2$, $y = t^2 - 4t$, $z = 3t - 5$ where t denotes time. Find the component of its velocity and acceleration at $t = 1$ in the direction of $\mathbf{i} - \mathbf{j} + \mathbf{k}$.

10. If $u = x y^2 z$ and $\mathbf{r} = x z\, \mathbf{i} - xy^2\mathbf{j} + y z^2\, \mathbf{k}$, find $\dfrac{\partial^2 (u\mathbf{r})}{\partial x \partial z}$ and

    $\dfrac{\partial^3}{\partial x^2 \partial z}(u\,\mathbf{r})$ *at* (2, 1, 0)

11. If **u**, **v**, **w** are functions of t, prove that

    (*i*) $\dfrac{d}{dt}[\mathbf{u}, \mathbf{v}, \mathbf{w}] = \mathbf{u} \,.\, \mathbf{v} \times \dfrac{d\mathbf{w}}{dt} + \mathbf{u} \,.\, \dfrac{d\mathbf{v}}{dt} \times \mathbf{w} + \dfrac{d\mathbf{u}}{dt} \,.\, \mathbf{u} \times \mathbf{w}$

    (*ii*) $\dfrac{d}{dt}\{\mathbf{u} \times (\mathbf{v} \times \mathbf{w})\} = \mathbf{u} \times \left(\mathbf{v} \times \dfrac{d\mathbf{w}}{dt}\right) + \mathbf{u} \times \left(\dfrac{d\mathbf{v}}{dt} \times \mathbf{w}\right) + \left\{\dfrac{d\mathbf{u}}{dt} \times (\mathbf{v} \times \mathbf{w})\right\}$

12. If $\mathbf{x} = (\cos wt)\, \mathbf{i} + (\sin wt)\, \mathbf{j}$, show that the velocity **v** of the particle is perpendicular to **r** and $\mathbf{r} \times \mathbf{v}$ is a constant vector. ***(M.S. Ap 94)***

13. If $\mathbf{r} \times \dfrac{d\mathbf{r}}{dt} = 0$, show that **r** has a constant direction. ***(B.Sc. 1982 ; MU)***

CHAPTER 2

Gradient, Divergence and Curl

The vector differential operator DEL (∇) is defined by

$$\nabla \equiv \mathbf{i}\frac{\partial}{\partial x} + \mathbf{j}\frac{\partial}{\partial y} + \mathbf{k}\frac{\partial}{\partial z}.$$

This vector operator has properties analogous to those of ordinary vectors. We define below three quantities using the del operator.

The gradient. Let $\phi(x, y, z)$ define a differentiable scalar field. That is, ϕ is differentiable at each point (x, y, z) in a certain region of space. Then the *gradient* of ϕ denoted by $\nabla\phi$ or grad ϕ is defined by

$$\nabla\phi = \left(\mathbf{i}\frac{\partial}{\partial x} + \mathbf{j}\frac{\partial}{\partial y} + \mathbf{k}\frac{\partial}{\partial z}\right)\phi = \mathbf{i}\frac{\partial\phi}{\partial x} + \mathbf{j}\frac{\partial\phi}{\partial y} + \mathbf{k}\frac{\partial\phi}{\partial z} = \sum \mathbf{i}\frac{\partial\phi}{\partial x}.$$

Evidently $\nabla\phi$ defines a *vector field.*

The divergence. If $\mathbf{f}(x, y, z)$ is defined and differentiable vector point function at each point (x, y, z) in a certain region of space, then the *divergence* of $\mathbf{f}$ denoted by $\nabla\cdot\mathbf{f}$ or div $\mathbf{f}$ is defined by

$$\text{div } \mathbf{f} = \nabla\cdot\mathbf{f} = \left(\mathbf{i}\frac{\partial}{\partial x} + \mathbf{j}\frac{\partial}{\partial y} + \mathbf{k}\frac{\partial}{\partial z}\right)\cdot\mathbf{f} = \sum \mathbf{i}\cdot\frac{\partial\mathbf{f}}{\partial x}$$

If $\mathbf{f} = f_i\,\mathbf{i} + f_2\,\mathbf{j} + f_3\,\mathbf{k}$, then

$$\text{div } \mathbf{f} = \left(\mathbf{i}\frac{\partial}{\partial x} + \mathbf{j}\frac{\partial}{\partial y} + \mathbf{k}\frac{\partial}{\partial z}\right)\cdot(f_1\,\mathbf{i} + f_2\,\mathbf{j} + f_3\,\mathbf{k}) = \frac{\partial f_1}{\partial x} + \frac{\partial f_2}{\partial y} + \frac{\partial f_2}{\partial z}$$

Evidently div $\mathbf{f}$ *is a scalar.*

Note : 1. Writing $\nabla\cdot\mathbf{f}$ as $\mathbf{f}\cdot\nabla$ is wrong. $\nabla\cdot\mathbf{f}$ is a scalar, while $\mathbf{f}\cdot\nabla$ is only an operator

2. $\nabla\cdot\mathbf{f}$ is independent of the choice of coordinate axes.

3. If $\mathbf{f}$ is a constant vector, then $\frac{\partial\mathbf{f}}{\partial x}, \frac{\partial\mathbf{f}}{\partial y}$ and $\frac{\partial\mathbf{f}}{\partial z}$ are all zeros and hence div $\mathbf{f} = 0$

Solenoidal vector. If $\mathbf{f}$ is a vector such that div $\mathbf{f} = 0$ for all points in a given region, then it is said to be a *solenoidal* vector in that region.

The curl. If $\mathbf{f}(x, y, z)$ is a differentiable vector point function in a certain region of space, then the *curl or rotation* of $\mathbf{f}$ denoted by $\nabla \times \mathbf{f}$ or curl $\mathbf{f}$ or rot $\mathbf{f}$ is defined by

$$\nabla \times \mathbf{f} = \text{curl } \mathbf{f} = \left(\mathbf{i}\frac{\partial}{\partial x} + \mathbf{j}\frac{\partial}{\partial y} + \mathbf{k}\frac{\partial}{\partial z} \right) \times \mathbf{f} = \sum \mathbf{i} \times \frac{\partial \mathbf{f}}{\partial x}.$$

If $\mathbf{f} = f_1\,\mathbf{i} + f_2\,\mathbf{j} + f_2\,\mathbf{k}$, then $\nabla \times \mathbf{f} = \left(\mathbf{i}\frac{\partial}{\partial x} + \mathbf{j}\frac{\partial}{\partial y} + \mathbf{k}\frac{\partial}{\partial z} \right) \times \left(f_1\mathbf{i} + f_2\mathbf{j} + f_2\mathbf{k} \right)$

$$= \begin{vmatrix} \mathbf{i} & \mathbf{j} & \mathbf{k} \\ \frac{\partial}{\partial x} & \frac{\partial}{\partial y} & \frac{\partial}{\partial z} \\ f_1 & f_2 & f_3 \end{vmatrix} = \left(\frac{\partial f_3}{\partial y} - \frac{\partial f_2}{\partial z} \right)\mathbf{i} + \left(\frac{\partial f_1}{\partial z} - \frac{\partial f_3}{\partial x} \right)\mathbf{j} + \left(\frac{\partial f_2}{\partial x} - \frac{\partial f_1}{\partial y} \right)\mathbf{k}$$

This curl $\mathbf{f}$ is evidently a *vector.*

Note 1. Curl $\mathbf{f}$ is independent of choice of coordinate axes.

2. If $\mathbf{f}$ is a constant vector, $\frac{\partial \mathbf{f}}{\partial x}, \frac{\partial \mathbf{f}}{\partial y}$ and $\frac{\partial \mathbf{f}}{\partial z}$ are all zeros and hence curl $\mathbf{f} = 0$.

Irrotational vector. If $\mathbf{f}$ is a vector such that curl $\mathbf{f} = \mathbf{0}$ for all points in the region, then it is called an *irrotational vector or lamellar vector* in that region.

Physical meaning. Considering a region of space filled with a moving fluid, let $\mathbf{f}$ denote a vector point function representing the instantaneous *velocity* of the fluid at the point (x, y, z). Then div $\mathbf{f}$ gives the rate at which the fluid is originating at the point per unit volume.

If $\mathbf{f}(x, y, z)$ is the linear velocity of a point (x, y, z) of a rigid body rotating about some fixed axis, then the angular velocity of the body is equal to half of curl $\mathbf{f}$.

Directional Derivative

Let f be a scalar point function in a scalar field and let $f(P), f(Q)$ be the functional values of scalar point function f at P and Q respectively of the field. Then, if

$$\underset{Q \to P}{\text{Lt}} \frac{f(Q) - f(P)}{PQ}$$ exists, it is called the directional derivative of the scalar function at P along PQ.

If $\mathbf{f}$ is a vector point function, then

$$\underset{Q \to P}{\text{Lt}} \frac{\mathbf{f}(Q) - \mathbf{f}(P)}{PQ}$$, if it exists, is called the directional derivative of a vector point function $\mathbf{f}$ at P along PQ.

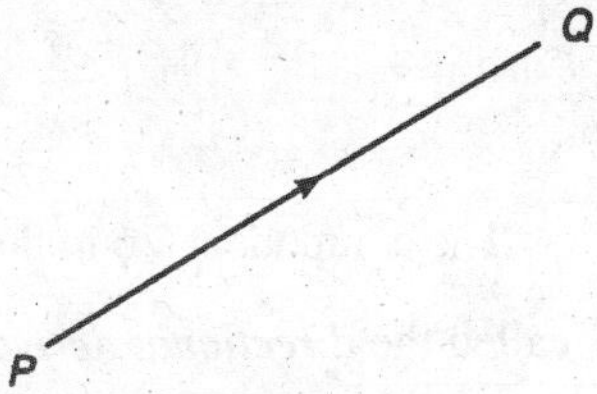

Fig. 2

Taking the direction PQ parallel to x-axis (Fig. 2.)

$$\underset{Q \to P}{\text{Lt}} \frac{f(Q) - f(P)}{PQ} = \underset{\delta x \to 0}{\text{Lt}} \frac{f(x + \delta x, y, z) - f(x, y, z)}{\delta x} = \frac{\partial f}{\partial x}$$

$\therefore$ $\frac{\partial f}{\partial x}, \frac{\partial f}{\partial y}, \frac{\partial f}{\partial z}$ **are the directional derivatives at P of the scalar function** f in the directions of coordinate axes.

Similarly $\frac{\partial \mathbf{f}}{\partial x}, \frac{\partial \mathbf{f}}{\partial y}, \frac{\partial \mathbf{f}}{\partial z}$ are directional derivatives at P of vector function $\mathbf{f}$ in the directions of coordinates axes.

Directional derivative of scalar point function along any line

Let f be the scalar point function in a scalar field.

Let P be (x, y, z) and let the direction cosines of the line PQ be l, m, n. If $PQ = r$, then Q is $(x + lr, y + mr, z + nr)$.

$$\text{Then} \quad \underset{Q \to P}{\text{Lt}} \frac{f(Q) - f(P)}{PQ} = \underset{r \to 0}{\text{Lt}} \frac{f(x + lr, y + mr, z + nr) - f(x, y, z)}{r}$$

$$= \underset{r \to 0}{\text{Lt}} \frac{1}{r} \left\{ f(x, y, z) + l\, r \frac{\partial f}{\partial x} (x_1, y_1, z_1) \right.$$

$$\left. + mr \frac{\partial f}{\partial y} (x_1, y_1, z_1) + nr \frac{\partial f}{\partial z} (x_1, y_1, z_1) - f(x, y, z) \right\}$$

where (x_1, y_1, z_1) is any point on PQ, by using Taylor's theorem

$$= l \frac{\partial f}{\partial x} (x, y, z) + m \frac{\partial f}{\partial y} (x, y, z) + n \frac{\partial f}{\partial z} (x, y, z)$$

$$= l \frac{\partial f}{\partial x} + m \frac{\partial f}{\partial y} + n \frac{\partial f}{\partial z} \qquad ...(1)$$

This is the directional derivative of scalar function f in the direction whose direction cosines are l, m, n.

If $\mathbf{f}$ is a vector point function, then the directional derivative of $\mathbf{f}$ in the direction whose direction cosines are l, m, n is

$$l \frac{\partial \mathbf{f}}{\partial x} + m \frac{\partial \mathbf{f}}{\partial y} + n \frac{\partial \mathbf{f}}{\partial z}.$$

The component $\nabla \phi$ in the direction of a vector $\mathbf{a}$ is given by $\frac{\nabla \phi \cdot \mathbf{a}}{|\mathbf{a}|}$ and is called the *directional derivative* of ϕ in the direction of $\mathbf{a}$. If ϕ is a contant $\nabla \phi = 0$ and hence the directional derivative is zero.

If the direction cosines of $\mathbf{a}$ are l, m, n then $\mathbf{a} = l\mathbf{i} + m\mathbf{j} + n\mathbf{k}$ and hence $\frac{\nabla \phi \cdot \mathbf{a}}{|\mathbf{a}|} = l \frac{\partial \phi}{\partial x} + m \frac{\partial \phi}{\partial y} + n \frac{\partial \phi}{\partial z}$ which is same as (1).

We can easily prove that $\nabla \phi$ *is a vector normal to the level surface* $\phi(x, y, z) = C$, where C is a constant.

Suppose $\mathbf{r} = x\mathbf{i} + y\mathbf{j} + z\mathbf{k}$ is the position vector of any point $A\ (x, y, z)$ on the surface $\phi(x, y, z) = c$. Then $d\mathbf{r} = \mathbf{i}dx + \mathbf{j}dy + \mathbf{k}\,dz$ lies in the tangent plane to the surface $\phi(x, y, z)$ at A.

$$\nabla\phi \cdot d\mathbf{r} = \left(\mathbf{i}\frac{\partial\phi}{\partial x} + \mathbf{j}\frac{\partial\phi}{\partial y} + \mathbf{k}\frac{\partial\phi}{\partial z}\right)\cdot(\mathbf{i}dx + \mathbf{j}dy + \mathbf{k}dz)$$

$$= \frac{\partial\phi}{\partial x}dx + \frac{\partial\phi}{\partial y}dy + \frac{\partial\phi}{\partial z}dz = d\phi = d(C) = 0.$$

$\therefore$ $\nabla\phi$ is perpendicular to $d\mathbf{r}$ and therefore to the surface $\phi(x, y, z) = c$.

Since the directional derivative of the function in any direction is the projection of the gradient of the function in that direction and $\nabla\phi$ is normal to the surface $\phi\ (x,y,z) = C$, *the maximum directional derivative of* ϕ is along the normal to $\phi\ (x,y,z) = C$ and is given in magnitude and direction by $\nabla\phi$.

Formulae :

(1) $\text{grad}\ (\phi + \psi) = \text{grad}\,\phi + \text{grad}\,\psi$

(2) $\text{div}\ (\boldsymbol{u} + \boldsymbol{v}) = \text{div}\,\boldsymbol{u} + \text{div}\boldsymbol{v}$

(3) $\text{curl}\ (\mathbf{u} + \mathbf{v}) = \text{curl}\,\mathbf{u} + \text{curl}\,\mathbf{v}$

(4) $\text{grad}\,(\phi\psi) = \phi\ \text{grad}\,\psi + \psi\ \text{grad}\,\phi$

These formulae follow immediately from the definitions.

(1) $$\nabla(\phi + \psi) = \mathbf{i}\frac{\partial}{\partial x}(\phi + \psi) + \mathbf{j}\frac{\partial}{\partial y}(\phi + \psi) + \mathbf{k}\frac{\partial}{\partial z}(\phi + \psi)$$

$$= \left(\mathbf{i}\frac{\partial\phi}{\partial x} + \mathbf{j}\frac{\partial\phi}{\partial y} + \mathbf{k}\frac{\partial\phi}{\partial z}\right) + \left(\mathbf{i}\frac{\partial\psi}{\partial x} + \mathbf{j}\frac{\partial\psi}{\partial y} + \mathbf{k}\frac{\partial\phi}{\partial z}\right) = \nabla\phi + \nabla\psi.$$

Formulae (2) and (3) can be seen on similar lines.

(4) $$\nabla(\phi\psi) = \sum \mathbf{i}\frac{\partial}{\partial x}(\phi\psi) = \sum \mathbf{i}\left(\phi\frac{\partial\psi}{\partial x} + \psi\frac{\partial\phi}{\partial x}\right)$$

$$= \phi\sum \mathbf{i}\frac{\partial\psi}{\partial x} + \psi\sum \mathbf{i}\frac{\partial\phi}{\partial x} = \phi\,\nabla\psi + \psi\,\nabla\phi$$

Level surface. Let $\phi(x, y, z)$ be a scalar point function in a certain region of space. The set of all points of the region for which ϕ becomes a constant C will be a surface called a level surface. That is, $\phi(x, y, z) = C$ represents a level surface. For different values of C, we get different level surfaces. No two level surfaces will have common points.

Note : Operator a. ∇

a. ∇ is an operator defined by $\mathbf{a}.\nabla = \mathbf{a.i}\frac{\partial}{\partial x} + \mathbf{a.j}\frac{\partial}{\partial y} + a.\mathbf{k}\frac{\partial}{\partial z}$

$$\therefore\ (\mathbf{a}.\nabla)\,\mathbf{f} = \mathbf{a.i}\frac{\partial\mathbf{f}}{\partial x} + \mathbf{a.j}\frac{\partial\mathbf{f}}{\partial y} + \mathbf{a.k}\frac{\partial\mathbf{f}}{\partial z}$$

Example 8. *Find the directional derivative of* $\phi = x^2yz + 4xz^2 + xyz$ *at* $(1, 2, 3)$ *in the direction of* $2\mathbf{i} + \mathbf{j} - \mathbf{k}$

Solution. $\nabla \phi = (2xyz + 4z^2 + yz)\,\mathbf{i} + (x^2z + xz)\,\mathbf{j} + (x^2y + 8xz + xy)\,\mathbf{k}$

$(\nabla \phi)$ at $(1, 2, 3) = 54\,\mathbf{i} + 6\,\mathbf{j} + 28\,\mathbf{k}$

$\therefore$ The directional derivative of ϕ in the direction of $2\mathbf{i} + \mathbf{j} - \mathbf{k}$ is

$$\frac{(54\mathbf{i} + 6\mathbf{j} + 28\mathbf{k}) \cdot (2\mathbf{i} + \mathbf{j} - \mathbf{k})}{\sqrt{(2)^2 + (1)^2 + (-1)^2}} = \frac{86}{\sqrt{6}}$$

Since this is positive, ϕ is increasing in this direction

Example 9. *In what direction from the point* $(1, -1, 2)$ *is the directional derivative of* $\phi = x^2y^2z^3$ *a maximum ? What is the magnitude of this maximum ?*

Solution. $\nabla \phi = \sum \mathbf{i} \dfrac{\partial \phi}{\partial x}$

$= 2xy^2z^3\,\mathbf{i} + 2x^2yz^3\,\mathbf{j} + 3x^2y^2z^2\,\mathbf{k} = 16\mathbf{i} - 16\mathbf{j} + 12\mathbf{k}$ at $(1, -1, 2)$

The directional derivative is maximum in the direction $\nabla\phi$ and the magnitude of this maximum is $|\nabla\phi| = \sqrt{256 + 256 + 144} = \sqrt{656}$

Example 10. *Find a unit normal to the surface* $x^2y + 2xz^2 = 8$ *at the point* $(1, 0, 2)$. ***(BR April 94)***

Solution. $\nabla \phi = \nabla (x^2y + 2xz^2) = (2xy + 2z^2)\,\mathbf{i} + x^2\,\mathbf{j} + 4xz\mathbf{k}$

$= 8\mathbf{i} + \mathbf{j} + 8\mathbf{k}$ at the point

$\therefore$ A unit normal to the given surface at the point is $\dfrac{\nabla \phi}{|\nabla \phi|}$

$$= \frac{8\mathbf{i} + \mathbf{j} + 8\mathbf{k}}{\sqrt{64 + 1 + 64}} = \frac{8\mathbf{i} + \mathbf{j} + 8\mathbf{k}}{\sqrt{129}}$$

Example 11. *Find the equation of the tangent plane at* $(1, 0, 2)$ *to the surface* $x^2y + 2xz^2 = 8$.

Solution. From the previous problem, $\nabla\phi = 8\,\mathbf{i} + \mathbf{j} + 8\,\mathbf{k}$

$\therefore$ The direction ratios of the normal to the tangent plane at $(1, 0, 2)$ are $(8, 1, 8)$. The tangent plane passes through $(1, 0, 2)$. Hence the equation of the tangent plane required is

$$(x - 1)\,8 + (y - 0) \cdot 1 + (z - 2)8 = 0 \quad i.e., \quad 8x + y + 8z = 24$$

Also the vector equation of the tangent plane at $(1, 0, 2)$ is

$$[(x - 1)\,\mathbf{i} + y\mathbf{j} + (z - 2)\,\mathbf{k}] \,.\, (8\mathbf{i} + j + 8\mathbf{k}) = 0$$

Example 12. *Find the angle between the surfaces* $z = x^2 + y^2 - 3$ *and* $x^2 + y^2 + z^2 = 9$ *at* $(2, -1, 2)$.

Solution. Finding the normal at the point to each surface, we get

$$\nabla \phi_1 = \nabla (x^2 + y^2 - z) = 2x\mathbf{i} + 2y\mathbf{j} - \mathbf{k} = 4\mathbf{i} - 2\mathbf{j} - \mathbf{k} \text{ at the point } (2, -1, 2)$$

$$\nabla \phi_2 = \nabla (x^2 + y^2 + z^2) = 2x\mathbf{i} + 2y\mathbf{j} + 2z\mathbf{k} = 4\mathbf{i} - 2\mathbf{j} + 4\mathbf{k}. \text{ at } (2, -1, 2)$$

Then the angle θ is given by

$$\cos\theta = \frac{\nabla\phi_1 \cdot \nabla\phi_2}{|\nabla\phi_1|\,|\nabla\phi_2|} = \frac{16 + 4 - 4}{\sqrt{21}\ \sqrt{36}} = \frac{8}{3\sqrt{21}}$$

Example 13. *If* $\nabla\phi = yz\,\mathbf{i} + zx\,\mathbf{j} + xy\,\mathbf{k}$, *find* ϕ.

Solution. $\nabla\phi = \sum \mathbf{i}\dfrac{\partial\phi}{\partial x} = yz\,\mathbf{i} + xz\,\mathbf{j} + xy\,\mathbf{k}$.

$$\therefore \quad \frac{\partial\phi}{\partial x} = yz \quad \therefore \quad \phi = xyz + \text{a function not involving } x.$$

$$\frac{\partial\phi}{\partial y} = xz \quad \therefore \quad \phi = xyz + \text{a function not involving } y.$$

$$\frac{\partial\phi}{\partial z} = xy \quad \therefore \quad \phi = xyz + \text{a function not involving } z.$$

From the last three statements, we conclude,

$$\phi = xyz + \text{a constant.}$$

Example 14. *If* $\mathbf{f} = xz^3\,\mathbf{i} - 2xyz\,\mathbf{j} + xz\mathbf{k}$, *find div* $\mathbf{f}$ *and curl* $\mathbf{f}$ *at* $(1, 2, 0)$.

Solution. $\text{div}\ \mathbf{f} = \dfrac{\partial f_1}{\partial x} + \dfrac{\partial f_2}{\partial y} + \dfrac{\partial f_3}{\partial z} = z^3 - 2xz + x$

div **f** at (1, 2, 0) is 1

$$\text{curl } \mathbf{f} = \begin{vmatrix} \mathbf{i} & \mathbf{j} & \mathbf{k} \\ \dfrac{\partial}{\partial x} & \dfrac{\partial}{\partial y} & \dfrac{\partial}{\partial z} \\ xz^3 & -2xyz & xz \end{vmatrix} = 2xy\,\mathbf{i} + (3xz^2 - z)\,\mathbf{j} + (-2yz)\,\mathbf{k} = 4\mathbf{i} \text{ at } (1, 2, 0)$$

Example 15. *Prove that* $\mathbf{A} = 3y^4z^2\,\mathbf{i} + 4x^3z^2\mathbf{j} - 3x^2y^2\,\mathbf{k}$ *is a solenoidal vector while* $\mathbf{B} = 2xy\,e^z\,\mathbf{i} + x^2e^z\,\mathbf{j} + x^2ye^z\,\mathbf{k}$ *is a irrotational vector.*

Solution. $\text{div}\,\mathbf{A} = \dfrac{\partial}{\partial x}(3y^4z^2) + \dfrac{\partial}{\partial y}(4x^3z^2) + \dfrac{\partial}{\partial z}(-3x^2y^2)$

$$= 0 + 0 + 0 = 0 \quad \therefore \ \mathbf{A} \text{ is solenoidal.}$$

$$\text{curl } \mathbf{B} = \sum\left(\frac{\partial f_3}{\partial y} - \frac{\partial f_2}{\partial z}\right)\mathbf{i}$$

$$= (x^2e^z - x^2e^z)\,\mathbf{i} + (2xye^z - 2xye^z)\,\mathbf{j} + (2xe^z - 2xe^z)\,\mathbf{k} = 0$$

$\therefore$ **B** is irrotational. **B** is a conservative force.

Example 16. *If* $\mathbf{r}$ *is the position vector of the point* $P\,(x, y, z)$,

(BR April 94)

prove (*i*) *div* $\mathbf{r} = 3$ (*ii*) *curl* $\mathbf{r} = \mathbf{0}$ (*iii*) $\nabla r^n = nr^{n-2}\,\mathbf{r}$ *where* $r = |\,\mathbf{r}\,|$

Solution. Here, $\mathbf{r} = x\mathbf{i} + y\mathbf{j} + z\mathbf{k}$

$$\text{div}\,\mathbf{r} = \frac{\partial}{\partial x}(x) + \frac{\partial}{\partial y}(y) + \frac{\partial}{\partial z}(z) = 1 + 1 + 1 = 3$$

$$\text{curl}\,\mathbf{r} = \left\{\frac{\partial}{\partial y}(z) - \frac{\partial}{\partial z}(y)\right\}\mathbf{i} + \left\{\frac{\partial}{\partial z}(x) - \frac{\partial}{\partial x}(z)\right\}\mathbf{j} + \left\{\frac{\partial}{\partial x}(y) - \frac{\partial}{\delta y}(x)\right\}\mathbf{k}$$

$$= (0)\,\mathbf{i} + (0)\,\mathbf{j} + (0)\,\mathbf{k} = \mathbf{0}$$

But $\quad r^2 = x^2 + y^2 + z^2 \therefore 2r\frac{\partial r}{\partial x} = 2x \therefore \frac{\partial r}{\partial x} = \frac{x}{r}$

Similarly, $\frac{\partial r}{\partial y} = \frac{y}{r}$ and $\frac{\partial r}{\partial z} = \frac{z}{r}$.

$$\nabla r^n = \sum \mathbf{i}\frac{\partial}{dx}(r^n) = \sum \mathbf{i}\, nr^{n-1}\frac{\partial r}{\partial x}$$

$$= \sum \mathbf{i}\, nr^{n-1}\left(\frac{x}{r}\right) = \sum nr^{n-2}\, x\,\mathbf{i}$$

$$= nr^{n-2}\,(x\,\mathbf{i} + y\,\mathbf{j} + z\,\mathbf{k}) = nr^{n-2}\,\mathbf{r}$$

Example 17. *If* $\mathbf{v} = \mathbf{w} \times \mathbf{r}$, $\mathbf{w} = \frac{1}{2}$ *curl v where w is a constant vector.*

(*B.Sc. 1983, M.U.*) (*MKU April, 94*)

Solution. Let $\mathbf{w} = \omega_1\mathbf{i} + \omega_2\,\mathbf{j} + \omega_3\mathbf{k}$; $\mathbf{r} = x\mathbf{i} + y\mathbf{j} + z\mathbf{k}$

$$\text{curl}\ \mathbf{v} = \nabla \times \mathbf{v} = \nabla \times \begin{vmatrix} \mathbf{i} & \mathbf{j} & \mathbf{k} \\ \omega_1 & \omega_2 & \omega_3 \\ x & y & z \end{vmatrix}$$

$$= \nabla \times \{ (\omega_2 z - \omega_3 y)\,\mathbf{i} + (\omega_3 x - \omega_1 z)\,\mathbf{j} + (\omega_1 y - \omega_2 x)\,\mathbf{k}\}$$

$$= \begin{vmatrix} \mathbf{i} & \mathbf{j} & \mathbf{k} \\ \frac{\partial}{\partial x} & \frac{\partial}{\partial y} & \frac{\partial}{\partial z} \\ \omega_1 z - \omega_3 y & \omega_3 x - \omega z_1 & \omega_1 y - \omega_2 x \end{vmatrix}$$

$$= 2(\omega_1\,\mathbf{i} + \omega_2\,\mathbf{j} + \omega_3\,\mathbf{k}) = 2\mathbf{w}$$

EXERCISES 2

1. Find the gradient of ϕ where ϕ is

(*i*) xyz (*ii*) $x^2 + y^2 + z^2$ (*iii*) $3x^2y - y^3z^2$ at $(1, -2, -1)$

(*iv*) $(x^2+y^2+z^2)^{\frac{3}{2}}$ (*v*) $\frac{1}{2}\log(x^2+y^2+z^2)$ (*vi*) $\frac{1}{\sqrt{x^2+y^2+z^2}}$.

2. Find a unit normal to the following surfaces at the specified points

(*i*) $x^2y+2xz=4$ at $(2,-2,3)$ (*ii*) $xy+yz+zx=7$ at $(1,1,3)$

(*iii*) $x^2+y^2-z=1$ at $(1,1,1)$ (*iv*) $x^2+y^2=z$ at $(1,2,5)$.

3. If **a** is a constant vector and **r** is the position of any point (x, y, z), prove

(*i*) $\nabla(\mathbf{a}\cdot\mathbf{r})=\mathbf{a}$ (*ii*) $\nabla\cdot(\mathbf{a}\times\mathbf{r})=0$ (*iii*) $(\mathbf{a}\cdot\nabla)\,\mathbf{r}=\mathbf{a}$

(*iv*) curl $(\mathbf{a}\times\mathbf{r})=2\mathbf{a}$ (*v*) curl $\left(\frac{a\times\mathbf{r}}{r^3}\right)=-\frac{\mathbf{a}}{r^3}+\frac{3\mathbf{r}}{r^5}(\mathbf{r}\cdot\mathbf{a})$

4. Find the directional derivative of

(*i*) x^2+2xy at $(1,-1,3)$ in the direction of $\mathbf{i}+2\mathbf{j}+2\mathbf{k}$ ***(1989 B.E., M.U.)***

(*ii*) $(4xz^2)+x^2yz$ at $(1,-2,-1)$ in the direction of $2\mathbf{i}-\mathbf{j}-2\mathbf{k}$

(*iii*) $xy+yz+zx$ at $(1,1,1)$ in the direction of $\mathbf{i}+\mathbf{j}$

(*iv*) $x^2+y^2+z^2$ at $(1,0,1)$ in the direction of $2\mathbf{i}+3\mathbf{k}$

(*v*) $4xz^3-3x^2y^2z$ at $(2,-1,2)$ in the direction of $2\mathbf{i}+3\mathbf{j}+6\mathbf{k}$

(*vi*) x^2+y^2+4xyz at $(1,-2,2)$ in the direction of $2\mathbf{i}-2\mathbf{j}+\mathbf{k}$

5. Find the maximum directional derivative of (*i*) $\phi=xyz^2$ at $(1,0,3)$ (*ii*) $\phi=x^2+y^2+z^2$ at $(1,1,1)$ (*iii*) $2xz^4-x^2y$ at $(2,-2,-1)$.

6. Find the magnitude and direction of the greatest directional derivative of (*i*) x^2yz^3 at $(2,1,-1)$ (*ii*) $3x^2y-y^3z^3$ at $(1,-2,-1)$ (*iii*) $xy+yz+zx$ at $(1,1,3)$

7. If the temperature at any point in space is given by

$T=xy+yz+zx$, find the direction in which temperature changes most rapidly with distance from the point $(1,1,1)$ and determine the maximum rate of change. ***(B.E.1963)***

8. Find the angle between the normals at points $(1,1,1)$ and $(4,1,2)$ to the surface $xy-z^2=0$. ***(Pevi. Year Ap. 2000)***

9. Find the angle between the surfaces $x^2-y^2-z^2=11$ and $xy+yz-zx-18=0$ at the point $(6,4,3)$.

10. Find *a* and *b* so that the surfaces

(*i*) $ax^2-byz=(a+2)x$ and $4x^2y+z^3=4$ cut orthogonally at $(1,-1,2)$

(*ii*) $ax^3-by^2z=(a+3)x^2$ and $4x^2y-z^3=11$ cut orthogonally at $(2,-1,-3)$

11. Find the equation of the tangent plane to the surface $xz^2+x^2y=z-1$ at $(1,-3,2)$

12. Show that the following vectors are solenoidal

(*i*) $\mathbf{a}=(x+3y)\,\mathbf{i}+(y-3z)\,\mathbf{j}+(x-2z)\,\mathbf{k}$

(*ii*) $\mathbf{a}=5y^4z^3\,\mathbf{i}+8xz^2\,\mathbf{j}-y^2x\,\mathbf{k}$ (*iii*) $\mathbf{a}=(2x-5y)\,\mathbf{i}+(x-y)\,\mathbf{j}+(3x-z)\,\mathbf{k}$

13. Show that the following vectors are irrotational

(*i*) $\mathbf{A} = (x^2 - y^2 + x)\ \mathbf{i} - (2xy + y)\ \mathbf{j}$ ***(BR. Nov. 1999)***

(*ii*) $\mathbf{B} = yz^2\ \mathbf{i} + (xz^2 - 1)\ \mathbf{j} + 2(xyz - 1)\ \mathbf{k}$ (*iii*) $\mathbf{C} = yz\ \mathbf{i} + zx\ \mathbf{j} + xy\mathbf{k}$

14. Find the constants *a, b, c* so that the following vector is irrotational.

(*i*) $\mathbf{F} = (x + 2y + az)\ \mathbf{i} + (bx - 3y - z)\ \mathbf{j} + (4x + cy + 2z)\ \mathbf{k}$

(*ii*) $\mathbf{F} = (axy + bz^3)\ \mathbf{i} + (3x^2 - cz)\ \mathbf{j} + (3xz^2 - y)\ \mathbf{k}$

(*iii*) $\mathbf{A} = (axy - z^3)\ \mathbf{i} + (a-2)x^2\ \mathbf{j} + (1-a)\ xz^2\ \mathbf{k}$

15. Find the value of *a* if $(x + 3y)\ \mathbf{i} + (y - 2z)\ \mathbf{j} + (x + az)\ \mathbf{k}$ is solenoidal.

16. If $\mathbf{F} = (3x^2y - z)\ \mathbf{i} + (xz^3 + y^4)\ \mathbf{j} - 2x^3 z^2\ \mathbf{k}$ find grad div **F** at $(2, -1, 0)$.

17. If $\mathbf{A} = x^2y\ \mathbf{i} + y^2z\ \mathbf{j} + z^2 x\mathbf{k}$, $\phi = xy + yz + zx$, find **A.** (grad ϕ) and **A** × (grad ϕ) at the point $(3, -1, 2)$. ***(B.E. 1965)***

18. If $\mathbf{F} = (x + y + 1)\ \mathbf{i} + \mathbf{j} - (x + y)\ \mathbf{k}$, prove that $\mathbf{F} \cdot \text{curl } \mathbf{F} = 0$

19. Show that $\mathbf{A} = (2x^2 + 8xy^2z)\ \mathbf{i} + (3x^3y - 3xy)\ \mathbf{j} - (4y^2z^2 + 2x^3z)\ \mathbf{k}$ is not solenoidal but $\mathbf{B} = xyz^2\ \mathbf{A}$ is solenoidal.

20. Compute curl curl **F** at $(1, 2, -3)$ where $\mathbf{F} = x^2\mathbf{i} + y^2\ \mathbf{j} + z^2\mathbf{k}$.

(B.Sc. 1983, M.U.)

Formulae involving ∇

In what follows, ϕ is a scalar point function and **u, v** are vector point functions

Prove that

1. $\text{div } (\phi\, \mathbf{u}) = \nabla \phi \,.\, \mathbf{u} + \phi \text{ div } \mathbf{u}$;

$$\text{div } (\phi\, \mathbf{u}) = \sum \mathbf{i} \cdot \frac{\partial}{\partial x} (\phi\, \mathbf{u})$$

$$= \sum \mathbf{i} \cdot \left(\phi \frac{\partial \mathbf{u}}{\partial x} + \frac{\partial \phi}{\partial x} \mathbf{u} \right)$$

$$= \phi \sum \mathbf{i} \cdot \frac{\partial \mathbf{u}}{\partial x} + \left(\sum \mathbf{i} \frac{\partial \phi}{\partial x} \right) \cdot \mathbf{u} = \phi \text{ div } \mathbf{u} + \nabla \phi \cdot \mathbf{u}$$

2. Prove that

$\text{curl } (\phi \mathbf{u}) = \nabla \phi \times \mathbf{u} + \phi \text{ curl } \mathbf{u}$;

$$\text{curl } (\phi \mathbf{u}) = \sum \mathbf{i} \times \frac{\partial}{\partial x} (\phi \mathbf{u})$$

$$= \sum \mathbf{i} \times \left(\phi \frac{\partial \mathbf{u}}{\partial x} + \frac{\partial \phi}{\partial x} \mathbf{u} \right)$$

$$= \phi \sum \mathbf{i} \times \frac{\partial \mathbf{u}}{\partial x} + \left(\sum \mathbf{i} \frac{\partial \phi}{\partial x} \right) \times \mathbf{u}$$

$$= \phi \text{ curl } \mathbf{u} + (\text{grad } \phi) \times \mathbf{u}$$

3. Prove that

div $(\mathbf{u}\times\mathbf{v}) = \mathbf{v}.$ curl $\mathbf{u} - \mathbf{u}.$ curl $\mathbf{v}$;

$$\text{div}\,(\mathbf{u}\times\mathbf{v}) = \sum \mathbf{i}.\frac{\partial}{\partial x}(\mathbf{u}\times\mathbf{v}) = \sum \mathbf{i}.\left\{\mathbf{u}\times\frac{\partial \mathbf{v}}{\partial x}+\frac{\partial \mathbf{u}}{\partial x}\times\mathbf{v}\right\}$$

$$= -\sum \mathbf{i}.\left(\frac{\partial \mathbf{v}}{\partial x}\times\mathbf{u}\right)+\sum \mathbf{i}.\left(\frac{\partial \mathbf{u}}{\partial x}\times\mathbf{v}\right)$$

$$= \left(\sum \mathbf{i}\times\frac{\partial \mathbf{u}}{\partial x}\right)\cdot\mathbf{v}-\left(\sum \mathbf{i}\times\frac{\partial \mathbf{v}}{\partial x}\right)\cdot\mathbf{u}$$

$$= (\text{curl }\mathbf{u})\cdot\mathbf{v}-(\text{curl }\mathbf{v})\cdot\mathbf{u} = \mathbf{v}\cdot\text{curl }\mathbf{u}-\mathbf{u}\cdot\text{curl }\mathbf{v}$$

4. Prove that

$$\text{curl}\,(\mathbf{u}\times\mathbf{v}) = (\mathbf{v}\cdot\nabla)\,\mathbf{u}-(\mathbf{u}\cdot\nabla)\,\mathbf{v}+\mathbf{u}\ \text{div}\,\mathbf{v}-\mathbf{v}\ \text{div}\,\mathbf{u}$$

$$\text{curl}\,(\mathbf{u}\times\mathbf{v}) = \sum \mathbf{i}\times\frac{\partial}{\partial x}(\mathbf{u}\times\mathbf{v}) = \sum \mathbf{i}\times\left\{\frac{\partial \mathbf{u}}{\partial x}\times\mathbf{v}+\mathbf{u}\times\frac{\partial \mathbf{v}}{\partial x}\right\}$$

$$= \sum \mathbf{i}\times\left(\frac{\partial \mathbf{u}}{\partial x}\times\mathbf{v}\right)+\sum \mathbf{i}\times\left(\mathbf{u}\times\frac{\partial \mathbf{v}}{\partial x}\right)$$

$$= \sum (\mathbf{i}\cdot\mathbf{v})\frac{\partial \mathbf{u}}{\partial x}-\sum\left(\mathbf{i}\cdot\frac{\partial \mathbf{u}}{\partial x}\right)\mathbf{v}$$

$$+\sum\left(i\cdot\frac{\partial v}{\partial x}\right)u-\sum (i\cdot u)\ \frac{\partial v}{\partial x}$$

$$= \left(\sum \mathbf{i}\cdot\frac{\partial \mathbf{v}}{\partial x}\right)\mathbf{u}-\left(\sum \mathbf{i}\cdot\frac{\partial \mathbf{u}}{\partial x}\right)\mathbf{v}+\mathbf{v}\cdot\left(\mathbf{i}\frac{\partial}{\partial x}+\mathbf{j}\frac{\partial}{\partial y}+\mathbf{k}\frac{\partial}{\partial z}\right)\mathbf{u}$$

$$-\mathbf{u}\cdot\left(\mathbf{i}\frac{\partial}{\partial x}+\mathbf{j}\frac{\partial}{\partial y}+\mathbf{k}\frac{\partial}{\partial \mathbf{z}}\right)\mathbf{v}$$

$$= \mathbf{u}\ \text{div}\,\mathbf{v}-\mathbf{v}\ \text{div}\,\mathbf{u}+(\mathbf{v}\cdot\nabla)\,\mathbf{u}-(\mathbf{u}\cdot\nabla)\,\mathbf{v}$$

5. Prove that

$$\text{grad}\,(\mathbf{u}\cdot\mathbf{v}) = \mathbf{v}\times\text{curl }\mathbf{u}+\mathbf{u}\times\text{curl }\mathbf{v}+(\mathbf{v}\cdot\nabla)\,\mathbf{u}+(\mathbf{u}\cdot\nabla)\mathbf{v}$$

$$\text{grad}\,(\mathbf{u}\cdot\mathbf{v}) = \sum \mathbf{i}\,\frac{\partial}{\partial x}(\mathbf{u}\cdot\mathbf{v})$$

$$= \sum \mathbf{i}\left[\frac{\partial \mathbf{u}}{\partial x}\cdot\mathbf{v}+\mathbf{u}\cdot\frac{\partial \mathbf{v}}{\partial x}\right] \qquad ...(1)$$

By formula,

$$\mathbf{v}\times\text{curl }\mathbf{u} = \mathbf{v}\times\left[\sum \mathbf{i}\times\frac{\partial \mathbf{u}}{\partial x}\right] = \sum \mathbf{v}\times\left(\mathbf{i}\times\frac{\partial \mathbf{u}}{\partial x}\right)$$

$$= \sum\left\{\left(\mathbf{v}\cdot\frac{\partial \mathbf{u}}{\partial x}\right)\mathbf{i}-(\mathbf{v}\cdot\mathbf{i})\frac{\partial \mathbf{u}}{\partial x}\right\}$$

$$= \sum\left(\mathbf{v}\cdot\frac{\partial \mathbf{u}}{\partial x}\right)\mathbf{i}-\mathbf{v}.\left(\mathbf{i}\frac{\partial}{\partial x}+\mathbf{j}\frac{\partial}{\partial y}+\mathbf{k}\frac{\partial}{\partial x}\right)\mathbf{u}$$

$$= \sum \left(\mathbf{v} \cdot \frac{\partial \mathbf{u}}{\partial x} \right) \mathbf{i} - (\mathbf{v} \cdot \nabla)\, \mathbf{u} \qquad \text{...(2)}$$

Similarly $\mathbf{u} \times \text{curl } \mathbf{v} = \sum \left(\mathbf{u} \cdot \frac{\partial \mathbf{v}}{\partial x} \right) \mathbf{i} - (\mathbf{u} \cdot \nabla)\, \mathbf{v}$ ...(3)

Adding (2) and (3)

$$\mathbf{v} \times \text{curl } \mathbf{u} + \mathbf{u} \times \text{curl } \mathbf{v} = \sum \left(\mathbf{u} \cdot \frac{\partial \mathbf{v}}{\partial x} + \mathbf{v} \cdot \frac{\partial \mathbf{u}}{\partial x} \right) \mathbf{i} - (\mathbf{v} \cdot \nabla)\, \mathbf{u} - (\mathbf{u} \cdot \nabla)\, \mathbf{v}$$

$$= \text{grad } (\mathbf{u} \cdot \mathbf{v}) - (\mathbf{v} \cdot \nabla)\, \mathbf{u} - (\mathbf{u} \cdot \nabla)\, \mathbf{v} \text{ using (1)}$$

$\therefore$ grad $(\mathbf{u} \cdot \mathbf{v}) = \mathbf{v} \times \text{curl } \mathbf{u} + \mathbf{u} \times \text{curl } \mathbf{v} + (\mathbf{v} \cdot \nabla)\, \mathbf{u} + (\mathbf{u} \cdot \nabla)\, \mathbf{v}$.

Second order Differential Operators

We shall prove a few more identities :

1. The operator $\nabla^2 = \frac{\partial^2}{\partial x^2} + \frac{\partial^2}{\partial y^2} + \frac{\partial^2}{\partial z^2}$

Div grad $\phi = \nabla \cdot (\nabla \phi) = \nabla \cdot \left(\mathbf{i} \frac{\partial \phi}{\partial x} + \mathbf{j} \frac{\partial \phi}{\partial y} + \mathbf{k} \frac{\partial \phi}{\partial z} \right)$

$$= \frac{\partial}{\partial x}\left(\frac{\partial \phi}{\partial x} \right) + \frac{\partial}{\partial y}\left(\frac{\partial \phi}{\partial y} \right) + \frac{\partial}{\partial z}\left(\frac{\partial \phi}{\partial z} \right) \quad \text{by definition}$$

$$= \left(\frac{\partial^2}{\partial x^2} + \frac{\partial^2}{\partial y^2} + \frac{\partial^2}{\partial z^2} \right) \phi = \nabla^2 \phi \quad \therefore \quad \nabla \cdot \nabla \phi = \nabla^2 \phi.$$

Note : $\nabla^2 = \frac{\partial^2}{\partial x^2} + \frac{\partial^2}{\partial y^2} + \frac{\partial^2}{\partial z^2}$ is called the Laplacian operator and read as 'del squared'.

2. Prove that curl (grad ϕ) = $\mathbf{0}$ *i.e.*, $\nabla \times (\nabla \phi) = \mathbf{0}$ ***(BR. Ap. 2000)***

$$\text{curl (grad } \phi) = \nabla \times \left(\mathbf{i} \frac{\partial \phi}{\partial x} + \mathbf{j} \frac{\partial \phi}{\partial y} + \mathbf{k} \frac{\partial \phi}{\partial z} \right)$$

$$= \begin{vmatrix} \mathbf{i} & \mathbf{j} & \mathbf{k} \\ \frac{\partial}{\partial x} & \frac{\partial}{\partial y} & \frac{\partial}{\partial z} \\ \frac{\partial \phi}{\partial x} & \frac{\partial \phi}{\partial y} & \frac{\partial \phi}{\partial z} \end{vmatrix}$$

$$= \left(\frac{\partial^2 \phi}{\partial y \partial z} - \frac{\partial^2 \phi}{\partial z \partial y} \right) \mathbf{i} + \left(\frac{\partial^2 \phi}{\partial z \partial x} - \frac{\partial^2 \phi}{\partial x \partial z} \right) \mathbf{j} + \left(\frac{\partial^2 \phi}{\partial x \partial y} - \frac{\partial^2 \phi}{\partial y \partial x} \right) \mathbf{k}$$

$$= \mathbf{0} + \mathbf{0} + \mathbf{0} = \mathbf{0}$$

(Remember as C.G. = 0)

3. Prove that div curl $\mathbf{F} = 0$ *i.e.*, $\nabla \cdot (\nabla \times \mathbf{F}) = 0$ ***(BR. Ap. 2000)***

$$\text{Curl } \mathbf{F} = \begin{vmatrix} \mathbf{i} & \mathbf{j} & \mathbf{k} \\ \frac{\partial}{\partial x} & \frac{\partial}{\partial y} & \frac{\partial}{\partial y} \\ F_1 & F_1 & F_3 \end{vmatrix} \quad \text{if } \mathbf{F} = F_1\,\mathbf{i} + F_2\,\mathbf{j} + F_3\,\mathbf{k}$$

$$= \left(\frac{\partial F_3}{\partial y} - \frac{\partial F_2}{\partial z}\right)\mathbf{i} + \left(\frac{\partial F_1}{\partial z} - \frac{\partial F_3}{\partial x}\right)\mathbf{j} + \left(\frac{\partial F_2}{\partial x} - \frac{\partial F_1}{\partial y}\right)\mathbf{k}$$

$$\therefore \text{ div curl } \mathbf{F} = \frac{\partial}{\partial x}\left(\frac{\partial F_3}{\partial y} - \frac{\partial F_2}{\partial z}\right) + \frac{\partial}{\partial y}\left(\frac{\partial F_1}{\partial z} - \frac{\partial F_3}{\partial x}\right) + \frac{\partial}{\partial z}\left(\frac{\partial F_2}{\partial x} - \frac{\partial F_1}{\partial y}\right)$$

$$= \frac{\partial^2 F_3}{\partial x \partial y} - \frac{\partial^2 F_2}{\partial x \partial z} + \frac{\partial^2 F_1}{\partial y \partial z} - \frac{\partial^2 F_3}{\partial y \partial x} + \frac{\partial^2 F_2}{\partial z \partial x} - \frac{\partial^2 F_1}{\partial z \partial y} = 0$$

This is true for any vector **F**. (Remember as d.c = 0)

4. **Prove that curl curl F = grad div F -∇^2 F and hence deduce that curl curl curl curl F = ∇^4 F if F is solenoidal.**

Let $\mathbf{F} = F_1\mathbf{i} + F_2\mathbf{j} + F_3\mathbf{k}$

$$\text{curl } \mathbf{F} = \left(\frac{\partial F_3}{\partial y} - \frac{\partial F_2}{\partial z}\right)\mathbf{i} + \left(\frac{\partial F_1}{\partial z} - \frac{\partial F_3}{\partial x}\right)\mathbf{j} + \left(\frac{\partial F_2}{\partial x} - \frac{\partial F_1}{\partial y}\right)\mathbf{k}$$

$$\text{Again, curl curl } \mathbf{F} = \begin{vmatrix} \mathbf{i} & \mathbf{j} & \mathbf{k} \\ \frac{\partial}{\partial x} & \frac{\partial}{\partial y} & \frac{\partial}{\partial z} \\ \frac{\partial F_3}{\partial y} - \frac{\partial F_2}{\partial z} & \frac{\partial F_1}{\partial z} - \frac{\partial F_3}{\partial x} & \frac{\partial F_2}{\partial x} - \frac{\partial F_1}{\partial y} \end{vmatrix}$$

$$= \left[\frac{\partial}{\partial y}\left(\frac{\partial F_2}{\partial x} - \frac{\partial F_1}{\partial y}\right) - \frac{\partial}{\partial z}\left(\frac{\partial F_1}{\partial z} - \frac{\partial F_3}{\partial x}\right)\right]\mathbf{i}$$

$$+ \left[\frac{\partial}{\partial z}\left(\frac{\partial F_3}{\partial y} - \frac{\partial F_1}{\partial z}\right) - \frac{\partial}{\partial x}\left(\frac{\partial F_2}{\partial x} - \frac{\partial F_1}{\partial y}\right)\right]\mathbf{j}$$

$$+ \left[\frac{\partial}{\partial x}\left(\frac{\partial F_1}{\partial z} - \frac{\partial F_3}{\partial x}\right) - \frac{\partial}{\partial y}\left(\frac{\partial F_3}{\partial y} - \frac{\partial F_2}{\partial z}\right)\right]\mathbf{k}$$

$$= \sum \left\{\frac{\partial}{\partial x}\left(\frac{\partial F_2}{\partial y} + \frac{\partial F_3}{\partial z}\right) - \left(\frac{\partial^2}{\partial y^2} + \frac{\partial^2}{\partial z^2}\right)F_1\right\}\mathbf{i}$$

$$= \sum \left\{\frac{\partial}{\partial x}\left(\frac{\partial F_1}{\partial x} + \frac{\partial F_2}{\partial y} + \frac{\partial F_3}{\partial z}\right) - \left(\frac{\partial^2}{\partial x^2} + \frac{\partial^2}{\partial y^2} + \frac{\partial^2}{\partial z^2}\right)F_1\right\}\mathbf{i}$$

$$= \sum \left\{\frac{\partial}{\partial x}(\text{div } \mathbf{F}) - \nabla^2 F_1\right\}\mathbf{i}$$

$$= \left(\mathbf{i}\frac{\partial}{\partial x} + \mathbf{j}\frac{\partial}{\partial x} + \mathbf{k}\frac{\partial}{\partial z} \right)(\text{div}\,\mathbf{F}) - \nabla^2(F_1\mathbf{i} + F_2\mathbf{j} + F_3\mathbf{k})$$

$$= \text{grad div}\,\mathbf{F} - \nabla^2\mathbf{F}$$

If **F** is solenoidal, div$F = 0$ In that case, curl curl $\mathbf{F} = -\nabla^2\mathbf{F}$

$\therefore$ curl curl curl curl $\mathbf{F}$ = curl curl $(-\nabla^2\mathbf{F})$

$$= \textit{grad}\ \text{div}(-\nabla^2\mathbf{F}) - \nabla^2(-\nabla^2\mathbf{F})$$

$$= \textit{grad}\ \{-\nabla^2\,\text{div}F\} + \nabla^4\mathbf{F} = \nabla^4\mathbf{F}.$$

Example 18. *If ϕ and f are differentiable scalar point functions prove that $\nabla\phi \times \nabla f$ is solenoidal.*

Sol. div $(\nabla\phi \times \nabla f) = \nabla f\,.\,\text{curl}\nabla\phi - \nabla\phi\,.\,\text{curl}\ \nabla f$ by theorem.

But curl $\nabla\phi$ = curl grad $\phi = \mathbf{0}$ and curl $\nabla f = 0$

$\therefore$ div $(\nabla\phi \times \nabla f) = 0$

Hence $\nabla\phi \times \nabla f$ is solenoidal

Example 19. *Prove* $\mathbf{w} \times \mathbf{r}$ *is solenoidal if* **w** *is a constant vector.*

Sol. div $(\mathbf{w} \times \mathbf{r}) = \mathbf{r}\,.\,\textit{curl}\ \mathbf{w} - \mathbf{w}\,.\,\textit{curl}\ \mathbf{r}$

= 0 since curl $\mathbf{w} = \mathbf{0}$ and curl $\mathbf{r} = \mathbf{0}$ for **w** is a constant vector.

$\therefore$ $\mathbf{w} \times \mathbf{r}$ is solenoidal

Example 20. *Prove that $\nabla^2 r^n = n(n+1)\,r^{n-2}$ and hence deduce $\nabla^2\left(\frac{1}{r}\right) = 0$. Also show that $\nabla \times \nabla r^n = 0$.*

Sol. $\nabla^2 r^n = \nabla\cdot(\nabla r^n) = \nabla\cdot(nr^{n-2}\,\mathbf{r}) = nr^{n-2}(\nabla\cdot\mathbf{r}) + \nabla(nr^{n-2})\cdot\mathbf{r}$

$$= nr^{n-2}(3) + n(n-2)\,r^{n-4}\,\mathbf{r}\cdot\mathbf{r} = 3nr^{n-2} + n(n-2)\,r^{n-4}\,r^2$$

$$= [3n + n(n-2)]\,r^{n-2} = n(n+1)\,r^{n-2}$$

Hence, $\nabla^2\left(\frac{1}{r}\right) = \nabla^2(r^{-1}) = (-1)(-1+1)\,r^{-3} = 0$

$$\nabla \times (\nabla r^n) = \textit{curl}\ \nabla r^n = \mathbf{0}.$$

Example 21. *Prove that div $(r^n\mathbf{r}) = (n+3)\,r^n$ and* ***(M.S. Ap. 94)***

curl $(r^n\,\mathbf{r}) = \mathbf{0}$ ***(BN. Nov. 94)***

Sol. div $(r^n\,\mathbf{r}) = \nabla r^n\cdot\mathbf{r} + r^n\,\textit{div}\ \mathbf{r}$ ***(BN. Ap. 94)***

$$= nr^{n-2}\,\mathbf{r}.\mathbf{r} + r^n 3 \quad (\because \text{div}\,\mathbf{r} = 3)$$

$$= nr^{n-2}r^2 + 3r^n = (n+3)\,r^n$$

$$\text{div}\,(r^n\mathbf{r}) = 0 \text{ if } n = -3$$

Hence, div $(r^{-3}\mathbf{r}) = (-3+3)\,r^n = 0$

$$\text{curl}\,(r^n\mathbf{r}) = \nabla r^n \times \mathbf{r} + r^n\, \text{curl}\mathbf{r} = nr^{n-2}\,\mathbf{r}\times\mathbf{r} + r^n(\mathbf{0})$$

$= \mathbf{0} + \mathbf{0} = \mathbf{0}.$ Therefore $r^n\mathbf{r}$ is irrotational.

Example 22. *Prove* $d\,\mathbf{f} = d\mathbf{r}\cdot\nabla\mathbf{f}$

Solution. We have $d\mathbf{f} = \frac{\partial \mathbf{f}}{\partial r}dx + \frac{\partial \mathbf{f}}{\partial y}dy + \frac{\partial \mathbf{f}}{\partial z}dz$ ...(1)

$$d\mathbf{r} = dx\mathbf{i} + dy\mathbf{j} + dz\mathbf{k} \quad \therefore \quad d\mathbf{r}\cdot\mathbf{i} = dx$$

$$d\mathbf{r}\cdot\mathbf{j} = dy \text{ and } d\mathbf{r}\cdot\mathbf{k} = dz$$

$\therefore$ (1) becomes $d\mathbf{f} = (d\mathbf{r}\cdot\mathbf{i})\frac{\partial \mathbf{f}}{\partial x} + (d\mathbf{r}\cdot\mathbf{j})\frac{\partial \mathbf{f}}{\partial y} + (d\mathbf{r}\cdot\mathbf{k})\frac{\partial \mathbf{f}}{\partial z}$

$$= d\mathbf{r}\cdot\left(\mathbf{i}\frac{\partial}{\partial x} + \mathbf{j}\frac{\partial}{\partial y} + \mathbf{k}\frac{\partial}{\partial z}\right)\mathbf{f} = d\mathbf{r}\cdot\nabla\mathbf{f}.$$

$$= d\mathbf{r}\cdot\nabla\,\mathbf{f}.$$

Example 23. *Determine* $f(r)$ *so that the vector* $f(r)\,\mathbf{r}$ *is both solenoidal and irrotational.* **(B.Sc. M.U. 1983) (B.N. Ap. 94)**

Solution. $f(r)\mathbf{r} = xf(r)\,\mathbf{i} + yf(r)\,\mathbf{j} + zf(r)\,\mathbf{k}$

$$\text{div}\,\{f(r)\,\mathbf{r}\} = \frac{\partial}{\partial x}\{xf(r)\} + \frac{\partial}{\partial y}\{yf(r)\} + \frac{\partial}{\partial z}\{zf(r)\}$$

$$= f(r) + xf'(r)\frac{\partial r}{\partial x} + f(r) + yf'(r)\frac{\partial r}{\partial y} + f(r) + zf'(r)\frac{\partial r}{\partial z}$$

$$= 3f(r) + f'(r)\left[\sum x\frac{\partial r}{\partial x}\right]$$

$$= 3f(r) + f'(r)\left[x\left(\frac{x}{r}\right) + y\left(\frac{y}{r}\right) + z\left(\frac{z}{r}\right)\right]$$

$$= 3f(r) + f'(r)\frac{1}{r}(x^2+y^2+z^2) = 3f(r) + r\cdot f'(r)$$

If $f(r)\mathbf{r}$ is solenoidal, then $3f(r) + r.f'(r) = 0$. $\therefore \frac{f'(r)}{f(r)} = \frac{-3}{r}$

Integrating w.r.t. r, $\log f(r) = -3\log r + \log c \quad \therefore f(r) = \frac{c}{r^3}$ ...(1)

$$\text{curl}\,\{f(r)\mathbf{r}\} = \begin{vmatrix} \mathbf{i} & \mathbf{j} & \mathbf{k} \\ \frac{\partial}{\partial x} & \frac{\partial}{\partial y} & \frac{\partial}{\partial z} \\ xf(r) & yf(r) & zf(r) \end{vmatrix} = \sum\left\{zf'(r)\frac{\partial r}{\partial y} - yf'(r)\frac{\partial r}{\partial z}\right\}\mathbf{i}$$

$$= f'(r)\sum\left(z.\frac{y}{r} - y.\frac{z}{r}\right)\mathbf{i} = \mathbf{0}$$

$\therefore$ curl $\{f(r)\mathbf{r}\} = \mathbf{0}$ for all $f(r)$ $\therefore f(r) = \frac{c}{r^3}.$

Example 24. *If* $\nabla^2\phi = 0$, *prove* $\nabla\,\phi$ *is both solenoidal and irrotational.* **(*MS. Nov. 94*)**

Solution . div $(\nabla\,\phi) = \nabla\cdot(\nabla\,\phi) = \nabla^2\,\phi = 0$ Hence $\nabla\,\phi$ is solenoidal
curl $(\nabla\,\phi)$ = curl (grad ϕ) = $\mathbf{0}$ $\therefore$ $\nabla\,\phi$ is irrotational.

Example 25. *If* **A** *and* **B** *are irrotational, prove* **(*M.K.U. Nov. 94*)**
$\mathbf{A}\times\mathbf{B}$ *is solenoidal* **(*BR. Ap. 2000*)**

Sol. Since **A** and **B** are irrotational,
curl $\mathbf{A} = \mathbf{0}$ and curl $\mathbf{B} = \mathbf{0}$
div $(\mathbf{A}\times\mathbf{B}) = \mathbf{B}.$ curl $\mathbf{A} - \mathbf{A}.$ curl $\mathbf{B} = \mathbf{B}\,.\,\mathbf{0} - \mathbf{A}\,.\,\mathbf{0} = \mathbf{0}$
Hence $\mathbf{A}\times\mathbf{B}$ is solenoidal

Example 26. *Prove curl* $(\phi\,\nabla\,\phi) = \mathbf{0}$ **(*M.K.U. Nov. 94*)**

Sol. curl $(\phi\,\nabla\,\phi) = \nabla\,\phi\times\nabla\,\phi + \phi$ curl $(\nabla\,\phi) = \mathbf{0} + \phi(\mathbf{0}) = \mathbf{0}$

Example 27. *Prove* $\mathbf{F} = (2xy+z^3)\,\mathbf{i} + x^2\,\mathbf{j} + 3xz^2\,\mathbf{k}$ *is a conservative force. Find* ϕ *so that* $\nabla\phi = \mathbf{F}$

Sol.
$$\text{curl }\mathbf{F} = \sum\left(\frac{\partial f_3}{\partial y} - \frac{\partial f_2}{\partial z}\right)\mathbf{i}$$
$$= (0-0)\,\mathbf{i} + (3z^2 - 3z^2)\,\mathbf{j} + (2x-2x)\,\mathbf{k} = \mathbf{0}$$

$\therefore$ **F** is a conservative force

$$\nabla\phi = \sum\frac{\partial\phi}{\partial x}\mathbf{i} = \mathbf{F} = (2xy+z^3)\,\mathbf{i} + x^2\,\mathbf{j} + 3xz^2\,\mathbf{k}$$

$$\therefore \qquad \frac{\partial\phi}{\partial x} = 2xy + z^3;\ \phi = \int(2xy+z^3)\,dx = x^2y + z^3x + f_1(y,z)$$

$$\frac{\partial\phi}{\partial y} = x^2 \ \therefore\ \phi = x^2y + f_2(x,z) \text{ and } \frac{\partial\phi}{\partial z} = 3xz^2 \ \therefore\ \phi = xz^3 + f_3(x,y)$$

Hence $\qquad \phi = x^2\,y + xz^3 + C$

EXERCISES 3

1. If **w** is a constant vector, prove that
   (*i*) curl $(\mathbf{w}\times\mathbf{r}) = 2\,\mathbf{w}$ (*ii*) $\nabla(\mathbf{w}\,.\,\mathbf{v}) = (\mathbf{w}\,.\,\nabla)\,\mathbf{v} + \mathbf{w}\times(\nabla\times\mathbf{v})$
   (*iii*) $\nabla\cdot(\mathbf{w}\times\mathbf{v}) = -\,\mathbf{w}\,.\,(\nabla\times\mathbf{v})$ (*iv*) $\nabla\times(\mathbf{w}\times\mathbf{v}) = \mathbf{w}\,(\nabla\,.\,\mathbf{v}) - (\mathbf{w}\cdot\nabla)\,\mathbf{v}$
2. Evaluate (*i*) $\nabla^2(\log r)$ (*ii*) div $(r^3\mathbf{r})$
   (*iii*) div $[r \text{ grad } (r^{-3})]$ (*iv*) grad div $\left(\frac{\mathbf{r}}{r}\right)$ (*v*) curl $\left(\frac{\mathbf{r}}{r^2}\right)$.
3. Prove curl (ϕ grad ϕ) = **0**.
4. Show that $\mathbf{F} = \mathbf{i} + 2z\mathbf{k}$ is a conservative field. Find its scalar potential and the work done in moving a particle from (0, 0, 0) to (0, 0, 1)
5. Prove that curl $\{f(r)\mathbf{r}\} = \mathbf{0}$ if $f(r)$ is differentiable. **(*B.N. Ap. 94*)**

6. If u, v are scalar point functions and $\mathbf{F}$ is a vector point function, then prove that $\mathbf{F}$. curl $\mathbf{F} = 0$ where $u\mathbf{F} = \nabla v$.

7. Prove div (u grad v) $= u\, \nabla^2 v + (\text{grad } u) \cdot (\text{grad } v)$

8. Show that curl (u grad v) = (grad u) $\times$ (grad v) = $-$ curl (v grad u).

9. If $\mathbf{r}_1$ and $\mathbf{r}_2$ are the vectors joining the fixed points (x_1, y_1, z_1) and (x_2, y_2, z_2) respectively to a variable point (x, y, z), then show that

   (*i*) Div $(\mathbf{r}_1 \times \mathbf{r}_2) = 0$ (*ii*) grad $(\mathbf{r}_1 \cdot \mathbf{r}_2) = \mathbf{r}_1 + \mathbf{r}_2$

   (*iii*) curl $(\mathbf{r}_1 + \mathbf{r}_2) = 2\,(\mathbf{r}_1 - \mathbf{r}_2)$.

10. Find the differentiable function $f(r)$ so that $f(r)\mathbf{r}$. is solenoidal.

11. Prove $\nabla^2 f(r) = \dfrac{d^2 f}{dr^2} + \dfrac{2}{r}\dfrac{df}{dr}$ and find $f(r)$ such that $\nabla^2 f(r) = 0$.

12. If $\mathbf{F} = (x + y + 1)\mathbf{i} + \mathbf{j} - (x + y)\mathbf{k}$, prove $\mathbf{F}$. curl $\mathbf{F} = 0$.

13. Show that grad$\left(\dfrac{\phi}{\psi}\right) = \dfrac{\psi \text{ grad } \phi - \phi \text{ grad } \psi}{\psi^2}$

14. Show that $\nabla \cdot (u\, \nabla v - v\, \nabla u) = u \nabla^2 v - v\, \nabla^2 u$.

15. If $\mathbf{F} = 3xyz^2\, \mathbf{i} + 2xy^3\, \mathbf{j} - x^2 yz\, \mathbf{k}$, and $\phi = 3x^2 - yz$ find (*i*) $\nabla \cdot \mathbf{F}$ (*ii*) $\mathbf{F} \cdot \nabla \phi$ (*iii*) $\nabla \cdot (\phi\, \mathbf{F})$ (*iv*) $\nabla \cdot (\nabla \phi)$ at the point $(1, -1, 1)$.

16. A fluid motion is given by $\mathbf{q} = (y + z)\,\mathbf{i} + (z + x)\,\mathbf{j} + (x + y)\,\mathbf{k}$. Is the motion irrotational ? If so, find the velocity potential ϕ such that $\mathbf{q} = \nabla \phi$.

17. If $\nabla \phi = 2xyz^3\, \mathbf{i} + x^2 z^3\, \mathbf{j} + 3x^2 yz^2 \mathbf{k}$, find $\phi\,(x, y, z)$ if $\phi\,(1, -2, 2) = 4$.

18. Prove $\nabla f(r) = \dfrac{f'(r)\mathbf{r}}{r}$ and find

    (*i*) f if $\nabla f = 2r^4\, \mathbf{r}$ (*ii*) $\phi(r)$ if $\nabla \phi = \dfrac{\mathbf{r}}{\mathbf{r}^5}$ and $\phi(1) = 0$.

19. If $f = x^2 yz$, $\phi = xy - 3z^2$ find (*i*) $\nabla\{\nabla f \cdot \nabla \phi\}$

    (*ii*) $\nabla \cdot [(\nabla f) \times (\nabla \phi)] = 0$ (*iii*) $\nabla \times [(\nabla f) \times (\nabla \phi)]$ at $(1, 0, 0)$.

20. If $f = x^3 + y^2 + 3xyz$, find curl grad ϕ and if

    $\mathbf{F} = x^3\, \mathbf{i} + y^3\, \mathbf{j} + z^3\, \mathbf{k}$, find div curl $\mathbf{F}$

21. Show that $\mathbf{F} = (6xy + z^3)\,\mathbf{i} + (3x^2 - z)\,j + (3xz^2 - y)\,\mathbf{k}$ is irrotational. Find ϕ such that $\mathbf{F} = \nabla\phi$ ***(BR 1995A)***

22. Show that if $\phi(x, y, z)$ is any solution of Laplace's equation, prove $\nabla\phi$ is both solution and irrotational.

23. Prove $\mathbf{F} = (y^2 \cos x + z^3)\,\mathbf{i} + (2y \sin x - 4)\,\mathbf{j} + 3xz^2\, \mathbf{k}$ is irrotational and find its scalar potential. ***(Pariyar Ap. 2000)***

24. If $\nabla\phi$ is a solenoidal vector, show that ϕ is a solution of Laplace's equation.

CHAPTER 3

Integration of Vectors

Integration as inverse of differentiation. If $\mathbf{f}(t)$ and $\mathbf{F}(t)$ be two vector functions of a scalar variable t, such that

$$\frac{d}{dt}\mathbf{F}(t) = \mathbf{f}(t), \text{ then we say}$$

$\mathbf{F}(t)$ is an integral of $\mathbf{f}(t)$ with respect to t and we write

$\int \mathbf{f}(t)dt = \mathbf{F}(t) + \mathbf{c}$ where $\mathbf{c}$, is any constant vector. $\int \mathbf{f}(t)\,dt$ is an indefinite integral.

The definite integral is $\int_a^b \mathbf{f}(t)\,dt = \mathbf{F}(b) - \mathbf{F}(a)$

Also, $\int \left[f_1(t)\,\mathbf{i} + f_2(t)\,\mathbf{j} + f_3(t)\,\mathbf{k} \right] dt = \mathbf{i}\int f_1(t)\,dt + \mathbf{j}\int f_2(t)\,dt + \mathbf{k}\int f_3(t)dt$

The following results may be useful, if remembered

1. $\int \left(\mathbf{r}\cdot\frac{d\mathbf{r}}{dt} \right) dt = \frac{1}{2}\mathbf{r}^2 + c$

2. $\int \left(\frac{d\mathbf{r}}{dt}\cdot\frac{d^2\mathbf{r}}{dt^2} \right) dt = \frac{1}{2}\left(\frac{d\mathbf{r}}{dt} \right)^2 + \mathbf{c}$

3. $\int \left[\mathbf{r}\times\frac{d^2\mathbf{r}}{dt^2} \right] dt = \mathbf{r}\times\frac{d\mathbf{r}}{dt} + \mathbf{c}$

Example 1. *The acceleration of a moving particle at time t is given by* $\mathbf{a} = 9t^2\,\mathbf{i} - 24t^2\,\mathbf{j} + 4\sin t\,\mathbf{k}$. *If* $\mathbf{r} = 2\mathbf{i} + \mathbf{j}$ *and* $\frac{d\mathbf{r}}{dt} = -\mathbf{i} - 3\mathbf{k}$ *at* $t = 0$, *find* $\mathbf{r}$

Solution. We know $\frac{d\mathbf{v}}{dt} = \mathbf{a}.$

$$\therefore\ \mathbf{v} = \int \mathbf{a}dt = \mathbf{i}\int 9t^2dt - 24\mathbf{j}\int t^2\,dt + 4\mathbf{k}\int \sin t\,dt$$

$$= 3t^3\,\mathbf{i} - 8t^3\,\mathbf{j} - 4\cos t\mathbf{k} + c$$

At $t = 0$, $\mathbf{v} = -4\mathbf{k} + \mathbf{c} = -\mathbf{i} - 3\mathbf{k}$ (given)

$$\therefore\ \mathbf{c} = -\mathbf{i} + \mathbf{k}\ \therefore\ \mathbf{v} = (3t^3 - 1)\,\mathbf{i} - 8t^3\,\mathbf{j} - (4\cos t - 1)\,\mathbf{k}$$

$$\mathbf{r} = \int \mathbf{v}dt = \left(\frac{3t^4}{4} - t\right)\mathbf{i} - 2t^4\,\mathbf{j} - (4\sin t - t)\,\mathbf{k} + \mathbf{c}'$$

At $t = 0$, $\mathbf{r} = \mathbf{c}' = 2\,\mathbf{i} + \mathbf{j}$

Hence, $$\mathbf{r} = \left(\frac{3t^4}{4} - t + 2\right)\mathbf{i} + (1 - 2t^4)\,\mathbf{j} + (t - 4\sin t)\,\mathbf{k}.$$

The line integral

Let $\mathbf{F}(x, y, z)$ be a vector point function defined throughout some region of space and let C be any curve in that region.

Let P and Q be two neighbouring points on C and let $\overrightarrow{OP} = \mathbf{r}$ and $\overrightarrow{OQ} = \mathbf{r} + \delta\mathbf{r}$, so that $\overrightarrow{PQ} = \delta\mathbf{r}$. (Fig. 3.)

If $\mathbf{F}(x, y, z)$ acts at P in a direction making an angle θ with PQ direction, then

$$F \cdot d\,\mathbf{r} = \mathbf{F}\cos\theta\; dr.$$

The integral $\int_C \mathbf{F} \cdot d\mathbf{r}$ is defined as the line integral of $\mathbf{F}$ along C. It is also equal to $\int_C \mathbf{F}\cos\theta\; dr$ which is the integral of the tangential component of $\mathbf{F}$ along the curve C. $\int_A^B \mathbf{F} \cdot d\mathbf{r}$ will mean that it is the tangential line integral over C from the point A to the point B.

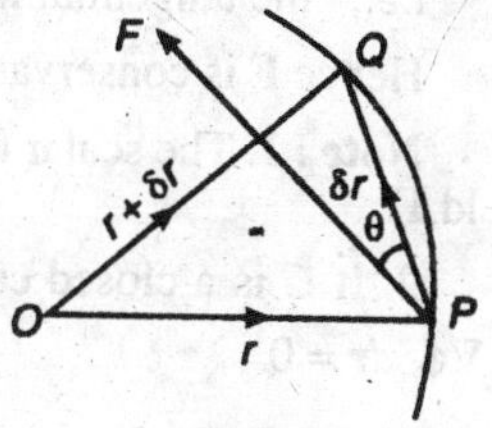

Fig. 3

If $\mathbf{F} = F_1\,\mathbf{i} + F_2\,\mathbf{j} + F_3\,\mathbf{k}$

then, $\mathbf{F} \cdot d\mathbf{r} = (F_1\,\mathbf{i} + F_2\,\mathbf{j} + F_3\,\mathbf{k}) \cdot (\mathbf{i}dx + \mathbf{j}dy + \mathbf{k}dz)$

$$(\because\; d\mathbf{r} = dx\;\mathbf{i} + dy\,\mathbf{j} + dz\,\mathbf{k})$$

$$= F_1\,dx + F_2\,dy + F_3\,dz.$$

Hence $$\int_C \mathbf{F} \cdot d\mathbf{r} = \int_C (F_1\,dx + F_2\,dy + F_3\,dz) \qquad \text{...(1)}$$

Form (1) is the tangential line integral in rectangular cartesian system. The integrals $\int_C \mathbf{F} \times d\mathbf{r}$ and $\int_C \phi\; d\mathbf{r}$ where ϕ is a scalar point function are also line integrals.

In general, the line integral $\int_A^B \mathbf{F} \cdot d\mathbf{r}$ depends not only on the path C but also on the terminal points A and B. If the integral depends only on the end points but not on the path C, $\mathbf{F}$ is called a *conservative* vector field.

Theorem : *If a vector point function* $\mathbf{F}$ *is expressible as the gradient of* *a scalar point function* ϕ, *then* **F** *is conservative.*

Let $$\mathbf{F} = \nabla \phi = \mathbf{i}\frac{\partial \phi}{\partial x} + \mathbf{j}\frac{\partial \phi}{\partial y} + \mathbf{k}\frac{\partial \phi}{\partial z}$$

$$\mathbf{F}\cdot d\mathbf{r} = \left(\mathbf{i}\frac{\partial \phi}{\partial x} + \mathbf{j}\frac{\partial \phi}{\partial y} + \mathbf{k}\frac{\partial \phi}{\partial z}\right)\cdot(\mathbf{i}\,dx + \mathbf{j}\,dy + \mathbf{k}\,dz)$$

$$= \frac{\partial \phi}{\partial x}dx + \frac{\partial \phi}{\partial y}dy + \frac{\partial \phi}{\partial z}\cdot dz = d\phi$$

$$\therefore \quad \int_A^B \mathbf{F}\cdot d\mathbf{r} = \int_A^B d\phi = \phi(B) - \phi(A)$$

where $\phi(B)$, $\phi(A)$ are the values of ϕ at B and A respectively.

i.e., the tangential line integral of **F** depends on the end points only.

Hence **F** is conservative.

Note : 1. The scalar function ϕ is called the scalar potential of the vector field **F**

2. If C is a closed curve $\phi(A) = \phi(B)$ since A, B coincide. In this case $\int_C \nabla\phi\cdot d\mathbf{r} = 0$

3. The converse of the theorem is also true viz., if a vector point function **F** is conservative, then there exists a scalar function ϕ such that $\mathbf{F} = \nabla\phi$.

4. If **F** is conservative, then curl **F**=curl grad $\phi = 0$ and hence **F** is irrotational.

Circulation : If **F** represents the velocity of a fluid particle then the line integral $\int_C$ **F**.**dr** is called the *circulation* of **F** around the curve C.When the circulation of **F** arround every closed curve in a regtion R vanishes, **F** is said to be irrotational in R. The total work done by **F** during a displacement from A to B is given by the line integral $\int_A^B$ **F**. **dr**.

Example 2. *If* $\mathbf{F} = (3x^2 + 6y)\,\mathbf{i} - 14yz\,\mathbf{j} + 20xz^2\,\mathbf{k}$, *evaluate* $\int_C \mathbf{F}\cdot d\mathbf{r}$ *from* $(0, 0, 0)$ *to* $(1, 1, 1)$ *along the curve* $x = t,\ y = t^2,\ z = t^3$.

Solution. The end points are (0, 0, 0) and (1, 1, 1)

These points correspond to $t = 0$ and $t = 1$

$$\therefore \quad dx = dt,\ dy = 2t\,dt\ ;\ dz = 3t^2\,dt$$

$$\therefore \quad \int \mathbf{F}\cdot d\mathbf{r} = \int_{t=0}^{t=1} (3x^2 + 6y)\,dx - 14yz\ dy + 20\,xz^2\,dz$$

$$= \int_{t=0}^{t=1} (3t^2 + 6t^2)\,dt - 14t^5\,(2t\,dt) + 20t^7\,(3t^2\,dt)$$

$$= \int_0^1 (9t^2 - 28t^6 + 60t^9)\, dt = \left(3t^3 - 4t^7 + 6t^{10} \right)_0^1 = 5$$

Example 3. *If* $\mathbf{A} = (5xy - 6x^2)\ \mathbf{i} + (2y - 4x)\ \mathbf{j}$, *evaluate* $\int_C \mathbf{A} \cdot d\mathbf{r}$ *where C is the curve* $y = x^3$ *in the xy plane from the point (1, 1) to (2, 8).*

Solution. Since $y = x^3$; $dy = 3x^2\, dx$

$$\int_C \mathbf{A} \cdot d\mathbf{r} = \int_{x=1}^{2} (5xy - 6x^2)\, dx + (2y - 4x)\, dy$$

$$= \int_1^2 (5x^4 - 6x^2)\, dx + (2x^3 - 4x)\ (3x^2\, dx)$$

$$= \int_1^2 (5x^4 - 6x^2 + 6x^5 - 12x^3)\, dx$$

$$= \left(x^5 - 2x^3 + x^6 - 3x^4 \right)_1^2 = 35.$$

Example 4. *Find the work done when a force* $\mathbf{F} = (x^2 - y^2 + x)\ \mathbf{i} - (2xy + y)\ \mathbf{j}$ *moves a particle in the xy plane from (0, 0) to (1, 1) along the curve* $y^2 = x$. *Is the work done different when the path is the straight line* $y = x$?

Solution. Work done $= \int \mathbf{F} \cdot d\mathbf{r}$

$$= \int_{(0,0)}^{(1,1)} (x^2 - y^2 + x)\, dx - (2xy + y) dy$$

$$= \int_0^1 \left[\left(y^4 - y^2 + y^2 \right) 2y dy - (2y^3 + y)\, dy \right] \text{since } x = y^2$$

$$= \left[\frac{2y^6}{6} - \frac{2y^4}{4} - \frac{y^2}{2} \right]_0^1 = \frac{1}{3} - \frac{1}{2} - \frac{1}{2} = -\frac{2}{3}$$

Solution. Work done on $y = x$ is (Put $y = x$) $= \int_0^1 x\, dx - \int_0^1 (2x^2 + x)\, dx$

$$= \left(\frac{x^2}{2} - \frac{2x^3}{3} - \frac{x^2}{2} \right)_0^1 = -\frac{2}{3}$$

The work done is same

Note : Infact, curl $\mathbf{F} = \mathbf{0}$ and hence the force is conservative and hence the work done by the force is independent of the path and depends only on the end points.

So, the work done by the force on any other path joining the two points is also the same.

Exercises 4

1. The acceleration of a particle at time $t \geq 0$ is given by

(*i*) $\mathbf{a} = 12\cos 2t\,\mathbf{i} - 8\sin 2t\,\mathbf{j} + 16t\,\mathbf{k}$

(*ii*) $\mathbf{a} = \cos 2t\,\mathbf{i} + \sin 2t\,\mathbf{j} + 2t\,\mathbf{k}$ (*iii*) $\mathbf{a} = 18\cos 3t\,\mathbf{i} - 8\sin 2t\,\mathbf{j} + 6t\,\mathbf{k}$

If the velocity $\mathbf{v}$ and the displacement $\mathbf{r}$ be zero at $t = 0$, find $\mathbf{v}$ and $\mathbf{r}$ at any time.

2. If $\dfrac{d^2\mathbf{r}}{dt^2} = -n^2\mathbf{r}$, prove that $\left(\dfrac{d\mathbf{r}}{dt}\right)^2 = C - n^2\mathbf{r}^2$ where C is a scalar constant.

3. If $\dfrac{d^2\mathbf{r}}{dt^2} = \lambda\mathbf{r}$ where λ is a scalar, prove $\mathbf{r} \times \dfrac{d\mathbf{r}}{dt}$ is constant.

4. If $\mathbf{F} = 2y\,\mathbf{i} - z\,\mathbf{j} + x\,\mathbf{k}$, evaluate $\int_C \mathbf{F} \times d\mathbf{r}$ along the curve $x = \cos t$,

$y = \sin t$, $z = 2\cos t$ from $t = 0$ to $t = \dfrac{\pi}{2}$.

5. Find the work done when a force $\mathbf{F} = (x^2 - 2y^3 + x)\,\mathbf{i} - (6y^2x + y)\,\mathbf{j}$ moves a particle in the xy plane from (0, 0) to (1, 1) along the curve $y^2 = x$. Is the work done different when the path is the straight line $y = x$?

$\left(\textbf{Hint}: \text{Work done} = \int_C \mathbf{F}\cdot d\mathbf{r}\right)$

6. Find the work done in moving a particle in the force field $\mathbf{F} = 3x^2\,\mathbf{i} + (2xz - y)\,\mathbf{j} - z\mathbf{k}$ from t=0 to t=1 along the curve $x = 2t^2$, $y = t$, $z = 4t^3$.

7. Evaluate $\int \mathbf{A}\cdot d\mathbf{r}$ in the following cases

(*i*) $\mathbf{A} = x^2\,\mathbf{i} + y^3\,\mathbf{j}$ along $y = x^2$ in the xy plane from (0,0) to (1, 1)

(*ii*) $\mathbf{A} = (x^2 + y^3)\,\mathbf{i} + (x^3 - y^2)\,\mathbf{j}$ along the st. line joining (1, 0) and (0, 1)

(*iii*) $\mathbf{A} = 3xy\,\mathbf{i} - y^2\,\mathbf{j}$ along $y = 2x^2$ in the xy plance from (0, 0) to (1, 2).

(*iv*) $\mathbf{A} = 3xy\,\mathbf{i} - 5z\,j + 10x\,\mathbf{k}$ along the curve

$x = 1 + t^2$, $y = 2t^2$, $z = t^3$ from $t = 1$ to $t = 2$.

(*v*) $\mathbf{A} = (2y + 3)\,\mathbf{i} + xz\,\mathbf{j} + (yz - x)\,\mathbf{k}$ along the line joining origin and the point (2, 1, 1).

8. Evaluate $\int_C \phi\, d\mathbf{r}$ where C is the curve $x = t^2$, $y = 2t$, $z = t^3$ from $t = 0$ to $t = 1$;

given $\phi = 2xyz^2$.

9. Evaluate $\int_C \mathbf{F} \times d\mathbf{r}$ along the curve $x = \cos t$, $y = 2\sin t$, $z = \cos t$ from $t=0$ to $t = \pi/2$; given $\mathbf{F} = 2x\,\mathbf{i} + y\,\mathbf{j} + z\,\mathbf{k}$.

10. Show that the vector field **F** given by
$\mathbf{F} = (y + \sin z)\ \mathbf{i} + x\,\mathbf{j} + x\cos z\,\mathbf{k}$ is conservative. Find its scalar potential.

11. Show that $\mathbf{F} = (2xy + z^3)\,\mathbf{i} + x^2\,\mathbf{j} + 3xz^2\,\mathbf{k}$ is a conservative force field. Find the scalar potential and the work done by **F** in moving an object in this field from $(1, -2, 1)$ to $(3, 1, 4)$.

[**Hint** : A necessary and sufficient condition for a force to be conservative is that curl $\mathbf{F} = \mathbf{0}$]

12. If $\phi = 2xyz^2 \cdot \mathbf{A} = xy\,\mathbf{i} - z\,\mathbf{j} + x^2\,\mathbf{k}$, evaluate

(*i*) $\int_C \phi d\mathbf{r}$ (*ii*) $\int_C \mathbf{A} \times d\mathbf{r}$ where C is the curve given by

$x = t^2$, $y = 2t$, $z = t^3$ from $(0, 0, 0)$ to $(4, 4, 8)$

13. Show that the following fields are conservative

(*i*) $\mathbf{F} = r^2\,\mathbf{r}$ (*ii*) $\mathbf{E} = r^n\,\mathbf{r}$

14. Find $\int_C \mathbf{F} \cdot d\mathbf{r}$ where C is the circle $x^2 + y^2 = 9$ in the xy-plane if

$\mathbf{F} = (2x - y + z)\,\mathbf{i} + (x + y - z^2)\,\mathbf{j} + (3x - 2y + 4z)\,\mathbf{k}$.

Surface integral

Consider a two sided surface S. If S is a closed surface, we will choose the outside surface as the positive one. Consider ds, an element of area, in the surface and let **n** be the *unit vector* normal to ds drawn on the positive direction, the outward drawn normal. Let **F** be the vector function at P making an angle θ with **n**. (Fig. 4)

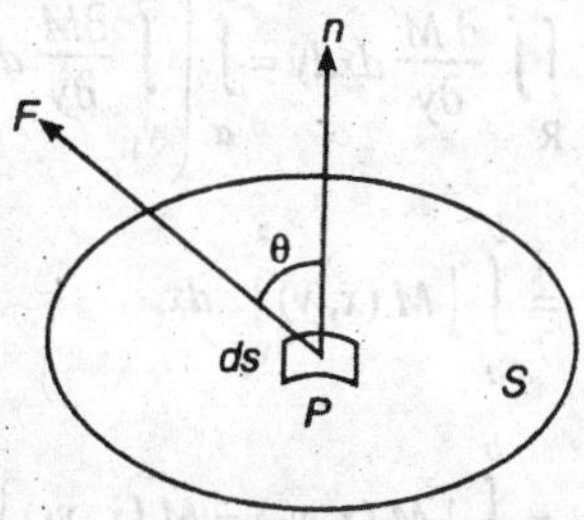

Fig. 4

$$\mathbf{F}.\mathbf{n} = F\cos\theta$$

The integral of this normal component, namely $\mathbf{F}\cdot\mathbf{n}$, through an element of area ds over the surface S is called the surface integral of **F** over S.

Hence, surface integral $= \iint_S (\mathbf{F}\cdot\mathbf{n})\, ds$

If **ds** is a vector whose magnitude is ds and whose direction is that of **n**, then $\mathbf{ds} = \mathbf{n}ds$

Hence, the surface integral $\iint_S \mathbf{F}\cdot\mathbf{n}\, ds$ can also be written as $\iint_S \mathbf{F}\cdot d\mathbf{s}$.

Note : $\iint_S \phi\, d\mathbf{s}$, $\iint_S \mathbf{F} \times d\mathbf{s}$, $\iint_S \mathbf{u}\cdot\mathbf{v}\, ds$, $\iint_S \mathbf{u} \times \mathbf{v}\ ds$

are also surface integrals.

Physical meaning. Let **F** denote the velocity vector of a moving fluid in which a fixed surface S is imagined. Now, $\mathbf{F}\cdot\mathbf{n}\, ds$ denotes the amount of fluid passing normally through the element ds in unit time. The surface integral $\iint_S \mathbf{F}\cdot d\mathbf{s}$ will represent the *total flux* of the vector **F** through the whole surface S.

GREEN'S THEOREM IN THE PLANE. If R is a closed region of the xy plane bounded by a simple closed curve C and if M and N are continuous functions of x and y having continuous derivatives in R, then

$$\int_C M\,dx + N\,dy = \iint_R \left(\frac{\partial N}{\partial x} - \frac{\partial M}{\partial y} \right) dx\,dy$$

where C is traversed in the anticlockwise direction (positive direction).

Let C be the closed curve such that any straight line parallel to the coordinate axes cuts C in at most two points.

Let the lines $x = a$, $x = b$ parallel to y axes touch the curve C at P and Q. (Fig. 5)

Let the equations of curves PBQ and PAQ be $y = y_1(x)$ and $y = y_2(x)$ respectively. Let R denote the region bounded by C. Then,

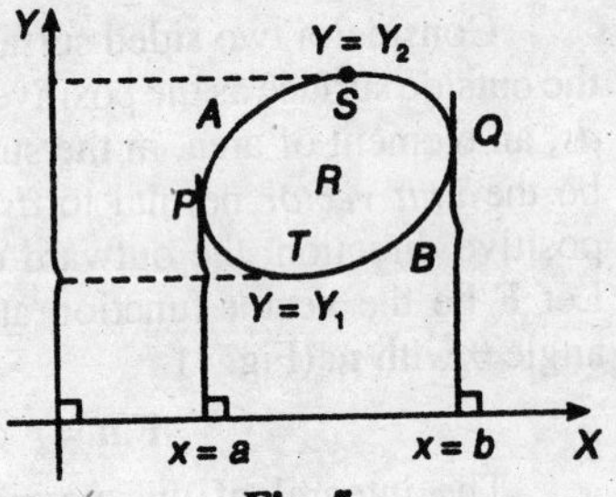

Fig. 5

$$\iint_R \frac{\partial M}{\partial y}\,dxdy = \int_a^b \left(\int_{y_1}^{y_2} \frac{\partial M}{\partial y}\,dy \right) dx$$

$$= \int_a^b \Big\{ M(x, y) \Big\}_{y_1}^{y_2} dx$$

$$= \int_a^b \Big\{ M(x, y_2) - M(x, y_1) \Big\} dx$$

$$= -\int_a^b M(x, y_1)\,dx - \int_b^a M(x, y_2)\,dx = -\int_C M\,dx \qquad \ldots(1)$$

Let $y = e$, $y = d$ touch the curve C, at T and S respectively.

Let the equations of TPS and SQT be $x = x_1(y)$ and $x = x_2(y)$ respectively.

Then, $$\iint_R \frac{\partial N}{\partial x}\,dxdy = \int_e^d \left(\int_{x_1}^{x_2} \frac{\partial N}{\partial x}\,dx \right) dy$$

$$= \int_e^d \left\{ N(x, y) \Big|_{x_1}^{x_2} \right\} dy = \int_e^d \left\{ N(x_2, y) - N(x_1, y) \right\} dy$$

$$= \int_e^d N(x_2, y)\, dy + \int_d^e N(x_1, y)\, dy = \int_C N(x, y)\, dy \qquad \text{...(2)}$$

From (1) and (2), $\int_C M\, dx + N\, dy = \iint_R \left(\frac{\partial N}{\partial x} - \frac{\partial M}{\partial y} \right) dxdy.$

Note. 1. The theorem can be extended to the curves C for which the lines parallel to the coordinate axes may cut C in more than two points.

Note. 2. If $\mathbf{A} = M\mathbf{i} + N\mathbf{j}$, and $d\mathbf{r} = \mathbf{i}dx + \mathbf{j}dy$, then $\mathbf{A} \cdot d\mathbf{r} = M\, dx + Ndy$

and $(\text{curl } \mathbf{A}) \cdot \mathbf{k} = \frac{\partial N}{\partial x} - \frac{\partial M}{\partial y}$. ∴ The Green's theorem becomes,

$$\int_C \mathbf{A} \cdot d\mathbf{r} = \iint_R (\text{curl } \mathbf{A}) \cdot \mathbf{k}\ dxdy.$$

This is a special case of Stoke's theorem which follows.

Hence Green's theorem can be deduced from Stoke's theorem.

Note. 3. If $\mathbf{A} = N\mathbf{i} - M\mathbf{j}$ and $\mathbf{n} =$ unit outward vector normal to C then

$$\mathbf{n} = \cos\alpha\, \mathbf{i} + \cos\left(\frac{\pi}{2} + \alpha\right) \mathbf{j} = \cos\alpha\, \mathbf{i} - \sin\alpha\mathbf{j} = \frac{dy}{ds}\, \mathbf{i} - \frac{dx}{ds}\, \mathbf{j}$$

Hence, Green's theorem becomes

$\int_C \mathbf{A} \cdot \mathbf{n}\, dx = \iint_R \text{div } \mathbf{A}\, dxdy$; which is a two dimensional form of the divergence theorem to be studied later.

Example 5. *Prove that the area bounded by a simple closed curve C is given by* $\frac{1}{2} \int_C (xdy - ydx)$. *Hence, find the area of an ellipse.* **(*Periyar Ap. 2000*)**

Solution. Taking $M = -y$ and $N = x$ and applying Green's theorem

$$\int_C (xdy - ydx) = \iint_R \left\{ \frac{\partial(x)}{\partial x} - \frac{\partial}{\partial y}(-y) \right\} dx\, dy = 2 \iint_R dxdy = 2A.$$

$$\therefore \qquad A = \frac{1}{2} \int_C (xdy - ydx).$$

To find area of the ellipse, put $x = a\cos\theta$, $y = b\sin\theta$.

Then, $\quad \text{Area} = \frac{1}{2} \int_C (xdy - ydx)$

$$= \frac{1}{2}\int_0^{2\pi} ab\ (\cos^2\theta + \sin^2\theta)\, d\theta = \frac{1}{2} ab\ (\theta)_0^{2\pi} = \pi\, ab.$$

Example 6. *Find the common area between $y^2 = 4x$ and $x^2 = 4y$ by using Green's theorem.*

Solution. The common area is enclosed by the curves given and their points of intersection are (0, 0) and (4, 4). The boundary of the area consists of two curves.

$$\text{Common area} = \frac{1}{2}\int_{C_1} (xdy - ydx) + \frac{1}{2}\int_{C_2} (xdy - ydx)$$

where C_1 is the curve $x^2 = 4y$ from (0, 0) to (4, 4) and C_2 is the curve $y^2 = 4x$ from (4, 4) to (0, 0)

$$A = \frac{1}{2}\int_0^4 \left(\frac{1}{2}x^2\, dx - \frac{x^2}{4}\, dx\right) + \frac{1}{2}\int_4^0 \left(\frac{y^2}{4} dy - \frac{y^2}{2} dy\right)$$

$$= \frac{1}{8}\left(\frac{x^3}{3}\right)_0^4 - \frac{1}{8}\left(\frac{y^3}{3}\right)_4^0 = \frac{8}{3} + \frac{8}{3} = \frac{16}{3} \text{ sq. units}$$

Gauss's Divergence Theorem

The surface integral of the normal component of a vector function **F** *taken over a closed surface S enclosing a volume V is equal to the volume integral of the divergence of* **F** *taken throughout the volume V.*

In other words, if the vector function **F** has continuous first partial derivatives in *V*, then $\iint_S \mathbf{F}\cdot d\mathbf{s} = \iiint_V \text{div } \mathbf{F}\, dv$

where dv is the volume element.

If $\mathbf{n} = \cos\alpha\ \mathbf{i} + \cos\beta\ \mathbf{j} + \cos\gamma\ \mathbf{k}$ denotes the unit vector along the outward normal to ds, then theorem can be written as

$$\iint_S \mathbf{F}\cdot\mathbf{n}\ ds = \iiint_V \text{div } \mathbf{F}\ dv \text{ since } d\mathbf{s} = \mathbf{n}\, ds$$

Taking $\mathbf{F} = F_1\,\mathbf{i} + F_2\,\mathbf{j} + F_3\,\mathbf{k}$, the above relation becomes,

$$\iint_S (F_1\,\mathbf{i} + F_2\,\mathbf{j} + F_3\,\mathbf{k}) \cdot \mathbf{n} ds$$

$$= \iiint_V \left(\frac{\partial F_1}{\partial x} + \frac{\partial F_2}{\partial y} + \frac{\partial F_3}{\partial z}\right) dxdydz \qquad \text{...(1)}$$

Let us prove the theorem given in the form (1).

Let S be such a surface that a line drawn parallel to Z axis meets it atmost in two points only. Denote the lower and upper portions of S by S_1 and S_2 and let their equations be $Z = f_1(x, y)$ and $Z = f_2(x, y)$ respectively. Let the projections of S on the xy plane be R. (Fig. 6.)

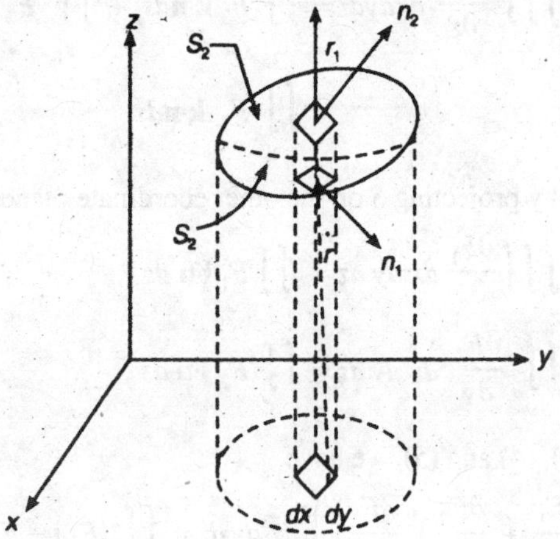

Fig. 6

Then,

$$\iiint_V \frac{\partial F_3}{\partial z} dxdydz = \iint_R \left(\int_{f_1(x, y)}^{f_2(x, y)} \frac{\partial F_3}{\partial z} dz \right) dx\, dy$$

$$= \iint_R \Big\{ F_3(x, y, z) \Big\}_{f_1}^{f_2} dx\, dy$$

$$= \iint_R \Big\{ F_3(x, y, f_2) - F_3(x, y, f_1) \Big\} dx\, dy \qquad ...(2)$$

Let γ_1, γ_2 be the angles made by the normal vector $\mathbf{n}$ with Z-axis.

Then, for upper portion S_2, $dx\, dy = ds_2 \cdot \cos\gamma_2 = \mathbf{k} \cdot \mathbf{n}\, ds_2$

and for lower portion S_1, $dx\, dy = ds_1 \cos\gamma_1 = -\mathbf{k} \cdot \mathbf{n}\, ds_1$.

The negative sign is taken in the last equation, since the normal to ds_1 makes an obtuse angle $(\pi - \gamma_1)$ with $\mathbf{k}$.

Hence, $\quad \iint_R F_3(x, y, f_2)\, dxdy = \iint_{S_2} F_3\, \mathbf{k} \cdot \mathbf{n} ds_2$

and $$\iint_R F_3(x, y, f_1)\, dx\, dy = -\iint_{S_1} F_3\, \mathbf{k}\cdot\mathbf{n}\, ds_1$$

Substituting these values in (2), we get,

$$\iiint_V \frac{\partial F_3}{\partial z}\, dxdydz = \iint_{S_2} F_3\, \mathbf{k}\cdot\mathbf{n}ds_2 + \iint_{S_1} F_3\, \mathbf{k}\cdot\mathbf{n}ds_1$$

$$= \iint_S F_3\, \mathbf{k}\cdot\mathbf{n}ds \qquad ...(3)$$

Similarly, by projecting S on the other coordinate planes, we get.

$$\iiint \frac{\partial F_1}{\partial x}\, dx\, dy\, dz = \iint F_1\, \mathbf{i}\cdot\mathbf{n}\, ds \qquad ...(4)$$

$$\iiint_V \frac{\partial F_2}{\partial y}\, dx\, dy\, dz = \iint_S F_2\, \mathbf{j}\cdot\mathbf{n}\, ds \qquad ....(5)$$

Adding (3), (4) and (5), we have,

$$\iiint_V \left(\frac{\partial F_1}{\partial x} + \frac{\partial F_2}{\partial y} + \frac{\partial F_3}{\partial z} \right) dx\, dy\, dz = \iint_S (F_1\mathbf{i} + F_2\mathbf{j} + F_3\mathbf{k})\cdot\mathbf{n}\, ds$$

$$\text{i.e.,} \quad \iiint_V \text{div } \mathbf{F}\ dv = \iint_S \mathbf{F}\cdot\mathbf{n}\, ds$$

Note. 1. This theorem relates a surface integral and a volume integral.

Note. 2. Even if the line parallel to z-axis cuts S in more than two points, still the theorem is valid.

Cor : Let $\mathbf{F} = P\mathbf{i} + Q\mathbf{j} + R\mathbf{k}$

Then $$\text{div } \mathbf{F} = \frac{\partial P}{\partial x} + \frac{\partial Q}{\partial y} + \frac{\partial R}{\partial z}$$

$$\mathbf{F}\cdot\mathbf{n}ds = P\mathbf{i}\cdot\mathbf{n}ds + Q\mathbf{j}\cdot\mathbf{n}ds + R\mathbf{k}\cdot\mathbf{n}ds$$

$$= Pdydz + Qdxdz + Rdxdy \qquad (\because\ \mathbf{k}\cdot\mathbf{n}ds = dxdy)$$

$\therefore$ Gauss's divergence Theorem can be written as

$$\iiint_V \left(\frac{\partial P}{\partial x} + \frac{\partial Q}{\partial y} + \frac{\partial R}{\partial z} \right) dxdydz = \iint_S (Pdydz + Qdxdz + Rdxdy)$$

Example 7. *If S is any closed surface enclosing a volume V and if* $\mathbf{A} = ax\,\mathbf{i} + by\,\mathbf{j} + cz\,\mathbf{k}$, *prove that* $\iint_S \mathbf{A}\cdot\mathbf{n}ds = (a + b + c)\, V$. ***(BR. Ap. 1999)***

Solution. $\text{div } \mathbf{A} = \frac{\partial}{\partial x}(ax) + \frac{\partial}{\partial y}(by) + \frac{\partial}{\partial z}(cz) = a + b + c$

Hence by Gauss's divergence theorem,

$$\iint_S \mathbf{A}\cdot\mathbf{n}\, ds = \iiint_V \text{div } \mathbf{A}\, dv$$

$$= (a+b+c)\iiint_V dv = (a+b+c)\ V.$$

Example 8. *Prove that* $\iint_S \frac{\mathbf{r}\cdot\mathbf{n}}{r^2}\, ds = \iiint_V \frac{1}{r^2}\, dv$

Solution. By Gauss's divergence theorem,

$$\iint_S \left(\frac{\mathbf{r}}{r^2}\right)\cdot\mathbf{n}\ ds = \iiint_V \text{div } \frac{\mathbf{r}}{r^2}\, dv$$

$$= \iiint_V \frac{1}{r^2}\, dv \quad (\because\ \text{div}\,(r^n\ \mathbf{r}) = (n+3)\ r^n)$$

and $\text{div}\left(\frac{\mathbf{r}}{r^2}\right) = \frac{1}{r^2}$

Example 9. *If* $\mathbf{A} = curl\ \mathbf{F}$, *prove that* $\iint_S \mathbf{A}\cdot\mathbf{n}\, ds = 0$ *for any closed surface S.*

By Gauss's theorem,

$$\iint_S \mathbf{A}\cdot\mathbf{n}ds = \iiint_V \text{div } \mathbf{A}\, dv = \iiint_V \text{div (curl } \mathbf{F})\, dv = \iiint_V 0\cdot dv = 0$$

Example 10. *Prove that* $\iint_S \mathbf{r}\cdot d\mathbf{s} = 3V$ *where S is any closed surface enclosing a volume V.*

Solution. We know $\text{div } \mathbf{r} = \frac{\partial}{\partial x}(x) + \frac{\partial}{\partial y}(y) + \frac{\partial}{\partial z}(z) = 3$

Hence, $\iint_S \mathbf{r}\cdot d\mathbf{s} = \iiint_V \text{div } \mathbf{r}\ dv = \iiint_V 3\ dv = 3V$

Example 11. *Evaluate.* $\iint_S \mathbf{A}\cdot\mathbf{n}\, ds$ *where* $\mathbf{A} = 4xz\,\mathbf{i} - y^2\,\mathbf{j} + yz\,\mathbf{k}$

where S is the surface of the cube bounded by $x=0,\ x=1,\ y=0,\ y=1,\ z=0$ *and* $z=1$

Solution. $\text{div } \mathbf{A} = \frac{\partial}{\partial x}(4xz) + \frac{\partial}{\partial y}(-y^2) + \frac{\partial}{\partial z}(yz) = (4z - y)$

Hence by Gauss's divergence theoem,

$$\iint_S \mathbf{A}\cdot\mathbf{n}ds = \iiint_V \text{div } \mathbf{A}\, dv = \iiint_V (4z-y)\, dv$$

$$= \int_0^1\int_0^1\int_0^1 (4z - y)\, dx\, dy\, dz$$

$$= \int_0^1\int_0^1 \left(2z^2 - yz\right)_0^1 dy\, dx = \int_0^1\int_0^1 (2 - y)\, dy\, dx = \frac{3}{2}.$$

Example 12. *Prove that* $\iint_S (\phi \nabla \psi - \psi \nabla \phi) \cdot \mathbf{n}\, ds$

$$= \iiint_V (\phi \nabla^2 \psi - \psi \nabla^2 \phi)\, dv$$

In the theorem, $\iint_S \mathbf{F}\cdot\mathbf{n} ds = \iiint_V \text{div}\, \mathbf{F}\, dv,$ ...(1)

set $\mathbf{F} = \phi \nabla \psi.$

Since, div $\mathbf{F} = \text{div}(\phi \nabla \psi) = \nabla\phi \cdot \nabla\psi + \phi\, \text{div}\, \nabla \psi = \nabla \phi \cdot \nabla \psi + \phi \nabla^2 \psi$

The equation (1) becomes,

$$\iint_S \phi\nabla\psi\cdot \mathbf{n}\, ds = \iiint_V (\phi\nabla^2 \psi + \nabla \phi \cdot \nabla \psi)\, dv \qquad ...(2)$$

Interchanging ϕ and ψ in (2), we get

$$\iint_S \psi\nabla \phi\cdot\mathbf{n} ds = \iiint_V (\psi\nabla^2 \phi + \nabla\psi\cdot\nabla\phi)\, dv \qquad ...(3)$$

(2)—(3) gives, $\iint_S (\phi\nabla\psi - \psi\nabla\phi)\cdot\mathbf{n} ds = \iiint_V (\phi\nabla^2 \psi - \psi \nabla^2 \phi)\, dv$

Example 13. *Show that* $\iiint_V \nabla \phi\, dv = \iint_S \phi\mathbf{n} ds$

Sol. Assume $\mathbf{F} = \phi\mathbf{a}$ where $\mathbf{a}$ is an arbitrary constant vector in Gauss's divergence theorem.

Then, $\iiint_V \text{div}\, (\phi\, \mathbf{a})\, dv = \iint_S (\phi\mathbf{a})\cdot\mathbf{n}\, ds$ ...(1)

$\text{div}\,(\phi\, \mathbf{a}) = \nabla \phi \cdot\mathbf{a} + \phi\, \text{div}\mathbf{a} = \nabla \phi\cdot\mathbf{a} = \mathbf{a}\cdot\nabla\phi$ (since div $\mathbf{a} = 0$)

Hence (1) becomes, $\iiint_V \mathbf{a}\cdot (\nabla \phi)\, dv = \iint_S \mathbf{a}\cdot(\phi\, \mathbf{n})\, ds$

i.e., $\mathbf{a}\cdot\iiint_V \nabla \phi\, dv = \mathbf{a}\cdot\iint_S \phi\mathbf{n}\, ds$

Since $\mathbf{a}$ is arbitrary, we get $\iiint_V \nabla \phi\, dv = \iint_S \phi\mathbf{n} ds$

Example 14. *Verify Gauss's divergence theorem for*

$$\mathbf{F} = x^2\,\mathbf{i} + y^2\,\mathbf{j} + z^2\,\mathbf{k}$$

taken over the cube bounded by planes $x = 0, x = 1, y = 0, y = 1, z = 0$ and $z = 1$.

In the cube, Equations of the planes $OBFG$ and $ACEH$ are $x = 0$ and $x = 1$ respectively.

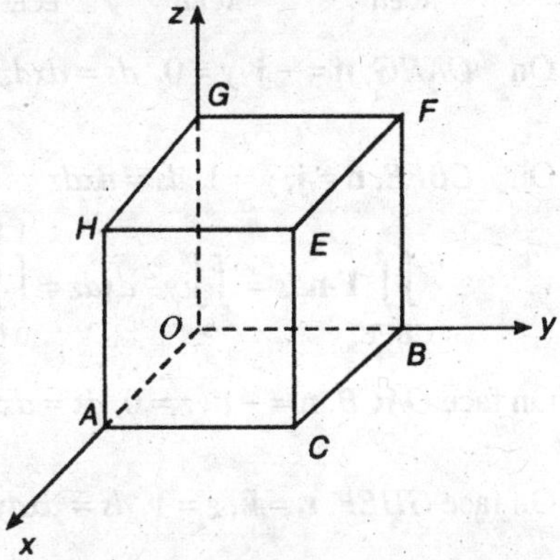

Fig. 7

(Fig. 7) $\text{div } \mathbf{F} = \frac{\partial}{\partial x}(x^2) + \frac{\partial}{\partial y}(y^2) + \frac{\partial}{\partial z}(z^2) = 2x + 2y + 2z$

The divergence theorem is

$$\iiint_V \text{div } \mathbf{F}\, dv = \iint_S \mathbf{F}\cdot\mathbf{n}\, ds.$$

In the problem,

$$\iiint_V \text{div } \mathbf{F}\, dv = 2\iiint_V (x + y + z)\, dx\, dy\, dz$$

$$= 2\int_0^1\int_0^1 \left(\int_0^1 (x + y + z)\, dz\right) dx\, dy$$

$$= 2\int_0^1\int_0^1 \left(x + y + \frac{1}{2}\right) dx\, dy$$

$$= 2\int_0^1 \left(\frac{1}{2} + y + \frac{1}{2}\right) dy = 2\left(\frac{1}{2} + \frac{1}{2} + \frac{1}{2}\right) = 3 \qquad \text{...(1)}$$

Now, the surface integral is to be evaluated over the six faces of the cube.

$$\iint_S \mathbf{F}\cdot\mathbf{n}\, ds = \iint_{OAGH} \mathbf{F}\cdot\mathbf{n}\, ds + \iint_{BCEF} \mathbf{F}\cdot\mathbf{n}\, ds + \iint_{ACEH} \mathbf{F}\cdot\mathbf{n}\, ds + \iint_{OBFG} \mathbf{F}\cdot\mathbf{n} ds + \iint_{OACB} \mathbf{F}\cdot\mathbf{n}\, ds + \iint_{GHEF} \mathbf{F}\cdot\mathbf{n} ds \qquad \text{...(2)}$$

On the face $OBFG$, $\mathbf{n} = -\,\mathbf{i}$, $x = 0$, $ds = dy\, dz$

$$\therefore \quad \iint_{OBFG} \mathbf{F}\cdot\mathbf{n}\, ds = \iint_{OBFG} -x^2\, ds = \iint_{OBFG} O ds = 0$$

On the face $OCEH$, $\mathbf{n} = \mathbf{i}$, $x = 1$, $ds = dy\,dz$

$$\therefore \quad \iint_{ACEH} \mathbf{F}\cdot\mathbf{n}\,ds = \iint_{ACEH} x^2\,ds = \iint_{ECEH} ds = \int_0^1\int_0^1 dy\,dz = 1$$

On $OAHG$, $\mathbf{n} = -\mathbf{j}\cdot y = 0$, $ds = dxdz$ $\displaystyle\iint_{OAHG} \mathbf{F}\cdot\mathbf{n}ds = \iint_S -y^2\,ds = 0$

On $CBFE$, $\mathbf{n} = \mathbf{j}$, $y = 1$, $ds = dxdz$

$$\therefore \quad \iint_{CBFE} \mathbf{F}\cdot\mathbf{n}ds = \iint_S y^2\,dxdz = \int_0^1\int_0^1 dxdz = 1.$$

On face $OACB$, $\mathbf{n} = -\mathbf{k}$, $z = 0$, $ds = dxdy$ $\displaystyle\iint_{OACB} \mathbf{F}\cdot\mathbf{n}ds = \iint_{OACB} -z^2\,dx\,dy = 0$

On face $GHEF$, $\mathbf{n} = \mathbf{k}$, $z = 1$, $ds = dxdy$

$$\iint_{GHEF} \mathbf{F}\cdot\mathbf{n}ds = \iint_{GHEF} z^2\,dxdy = \int_0^1\int_0^1 dx\,dy = 1$$

Therefore, using (2), $\displaystyle\iint_S \mathbf{F}\cdot\mathbf{n}ds = 0 + 1 + 0 + 1 + 0 + 1 = 3$ ...(3)

From (1) and (3), $\displaystyle\iiint_V \text{dv}\ \mathbf{F}\,\text{dv} = \iint_S \mathbf{F}\cdot\mathbf{n}\,ds = 3.$

Example 15. *Evaluate* $\displaystyle\iint_S (x\ dy\,dz + y\,dz\,dx + z\,dx\,dy)$ *over the surface of a sphere of radius a.*

Sol. Referring to the corollary under divergence theorem,

$P = x,\ Q = y,\ R = z.$ $\therefore$ The given surface integral is equal to

$$\iiint_V \left[\frac{\partial}{\partial x}(x) + \frac{\partial}{\partial y}(y) + \frac{\partial}{\partial z}(z) \right] dxdydz$$

$$= \iiint_V 3dv = 3\left(\frac{4}{3}\pi a^3 \right) = 4\pi a^3.$$

Stoke's theorem. The surface integral of the normal component of the curl of a vector point function **F** over an open surface S is equal to the line integral of the tangential component of **F** around the closed curve C bounding the open surface S. In other words, if **F** is a vector function with continuous first partial derivatives and C is a simple closed curve bounding the open surface S, then

$$\int_C \mathbf{F}\cdot d\mathbf{r} = \iint_S \text{curl}\,\mathbf{F}\cdot d\mathbf{s} = \iint_S \text{curl}\ \mathbf{F}\cdot\mathbf{n}ds$$

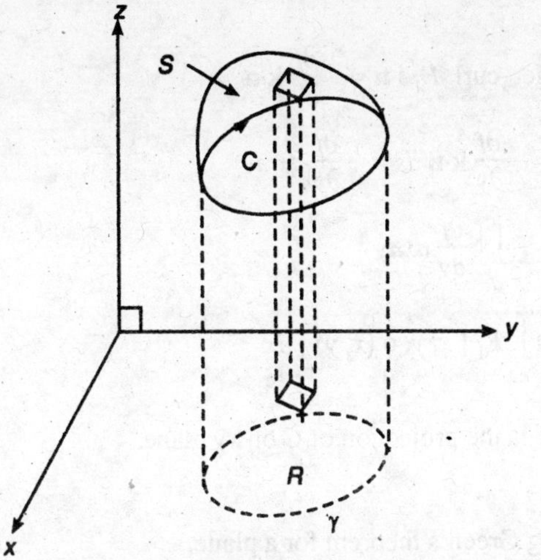

Fig. 8

Let S be a surface such that its projection on the xy, yz, zx planes be regions bounded by simple closed curve. (Fig. 8.) Let the equation of S be $Z = \phi(x, y)$. Let the projection of S on the xy plane be R.

Let $\mathbf{F} = F_1\,\mathbf{i} + F_2\,\mathbf{j} + F_3\,\mathbf{k}$;

$$\text{Curl } F_1\,\mathbf{i} = \begin{vmatrix} \mathbf{i} & \mathbf{j} & \mathbf{k} \\ \dfrac{\partial}{\partial x} & \dfrac{\partial}{\partial y} & \dfrac{\partial}{\partial z} \\ F_1 & 0 & 0 \end{vmatrix} = \frac{\partial F_1}{\partial z}\mathbf{j} - \frac{\partial F_1}{\partial y}\mathbf{k}$$

$$\therefore\ \text{Curl } F_1\,\mathbf{i}\cdot\mathbf{n} = \frac{\partial F_1}{\partial z}\mathbf{j}\cdot\mathbf{n} - \frac{\partial F_1}{\partial y}\mathbf{k}\cdot\mathbf{n} \qquad \text{...(1)}$$

Any point of S is given by $\mathbf{r} = x\,\mathbf{i} + y\,\mathbf{j} + z\,\mathbf{k} = x\mathbf{i} + y\,\mathbf{j} + \phi(x, y)\,\mathbf{k}$

$$\therefore \qquad \frac{\partial \mathbf{r}}{\partial y} = \mathbf{j} + \frac{\partial \phi}{\partial y}\mathbf{k} = \mathbf{j} + \frac{\partial z}{\partial y}\mathbf{k}$$

$$\frac{\partial \mathbf{r}}{\partial y}\cdot\mathbf{n} = \mathbf{j}\cdot\mathbf{n} + \frac{\partial z}{\partial y}\mathbf{k}\cdot\mathbf{n}$$

But $\dfrac{\partial \mathbf{r}}{\partial y}\cdot\mathbf{n} = 0$, since $\dfrac{\partial \mathbf{r}}{\partial y}$ is a tangent vector and $\mathbf{n}$ is a normal vector to S.

$$\therefore \qquad \mathbf{j}\cdot\mathbf{n} + \frac{\partial z}{\partial y}\mathbf{k}\cdot\mathbf{n} = 0;\ \mathbf{j}\cdot\mathbf{n} = -\frac{\partial z}{\partial y}\mathbf{k}\cdot\mathbf{n}$$

Substituting this value in (1), we have,

$$\text{curl}\quad F_1\,\mathbf{i}\cdot\mathbf{n} = -\left(\frac{\partial F_1}{\partial z}\frac{\partial z}{\partial y} + \frac{\partial F_1}{\partial y}\right)\mathbf{k}\cdot\mathbf{n} \qquad \text{...(2)}$$

Since F_1 is a function of x, y, z and $z = \phi(x, y)$ on S,

$$F_1(x, y, z) = F_1[x, y, \phi(x, y)] = f(x, y), \text{ say.}$$

$$\therefore \qquad \frac{\partial f}{\partial y} = \frac{\partial \cdot F_1(x, y, z)}{\partial y} = \frac{\partial F_1}{\partial y} + \frac{\partial F_1}{\partial z}\cdot\frac{\partial z}{\partial y} \qquad \text{...(3)}$$

Using this in (2), we get curl $F_1\ \mathbf{i}\cdot\mathbf{n} = -\dfrac{\partial f}{\partial y}\mathbf{k}\cdot\mathbf{n}$

$$\text{curl } F_1\ \mathbf{i}\cdot\mathbf{n}\ ds = -\frac{\partial f}{\partial y}\mathbf{k}\cdot\mathbf{n}\ ds = -\frac{\partial f}{\partial y}\,dx\,dy$$

$$\therefore \iint_s \text{curl } F_1\ \mathbf{i}\cdot\mathbf{n}\ ds = -\iint_R \frac{\partial f}{\partial y}\,dxdy \qquad ...(4)$$

$$\int_c F_1(x, y, z)\,dx = \int_c F_1[x, y, \phi(x, y)]\,dx$$

$=\int_\Upsilon f(x, y)\,dx$ where Υ is the projection of C on xy plane.

$$= -\iint_R \frac{\partial f}{\partial y}\,dxdy. \text{ (using Green's theorem for a plane.)} \qquad ...(5)$$

Here R *is* the region of the xy plane whose boundary is Υ.

From (4) and (5) $\iint \text{curl } F_1\ \mathbf{i}\cdot\mathbf{n}\ ds = \int_c F_1\,dx$ ...(6)

Similarly, projecting s on the other coordinate planes, we get

$$\iint \text{curl } F_2\ \mathbf{j}\cdot\mathbf{n}\ ds = \int_c F_2\,dy \qquad ...(7)$$

and
$$\iint_s \text{curl } F_3\ \mathbf{k}\cdot\mathbf{n}\ ds = \int_c F_3\,dz \qquad ...(8)$$

Adding (6), (7), (8), we get,

$$\iint_s \text{curl } (F_1\,\mathbf{i} + F_2\,\mathbf{j} + F_3\,\mathbf{k})\cdot\mathbf{n}ds = \int_c (F_1dx + F_2dy + F_3dz)$$

i.e.,
$$\iint_s \text{curl } \mathbf{F}\cdot\mathbf{n}\ ds = \int_c \mathbf{F}\cdot d\mathbf{r}$$

Note 1. Stoke's theorem relates a line integral in space to the surface integral.

Note 2. If **F** is irrotational, curl **F** = 0 and hence by Stoke's theorem $\int_c \mathbf{F}\cdot d\mathbf{r} = \iint_s \text{curl } \mathbf{F}\cdot\mathbf{n}\ ds = 0$ and hence **F** is conservative.

Note 3. If s is a closed surface, divide s into two portions s_1 and s_2 by a closed curve c. Now c is common boundary to s_1 and s_2.

$$\therefore \qquad \iint_{s_1} \text{curl } \mathbf{F}\cdot\mathbf{n}\ ds = \int_c \mathbf{F}\cdot d\mathbf{r} \qquad ...(1)$$

and $$\iint_{s_2} \text{curl } \mathbf{F}\cdot\mathbf{n}\, ds = -\int_c \mathbf{F}\cdot d\mathbf{r} \qquad ...(2)$$

Since the positive direction of C on s_1 is opposite to positive direction of C on s_2, adding (1) and (2), $\iint_s \text{curl } \mathbf{F}\cdot\mathbf{n}\, ds = 0$

Note 4. Setting $\mathbf{F} = \phi\mathbf{a}$ where $\mathbf{a}$ is an arbitrary constant vector in Stoke's theorem, we get $\int_c \phi d\mathbf{r} = \iint \mathbf{n}\times\nabla\phi\, ds$

Note 5. Replacing $\mathbf{F}$ by $\mathbf{a}\times\mathbf{F}$ where $\mathbf{a}$ is a constant vector in Stoke's theorem, we get $\int_c d\mathbf{r}\times\mathbf{F} = \iint_s (\mathbf{n}\times\nabla)\times\mathbf{F}\, ds$

Note 6. Let $\mathbf{F} = P\mathbf{i} + Q\mathbf{j} + R\mathbf{k}$

Then $\mathbf{F}\cdot d\mathbf{r} = P\,dx + Q\,dy + R\,dz$

$$\text{curl } \mathbf{F} = \left(\frac{\partial R}{\partial y} - \frac{\partial Q}{\partial z}\right)\mathbf{i} + \left(\frac{\partial P}{\partial z} - \frac{\partial R}{\partial x}\right)\mathbf{j} + \left(\frac{\partial Q}{\partial x} - \frac{\partial P}{\partial y}\right)\mathbf{k}$$

$$\text{curl } \mathbf{F}\cdot\mathbf{n}\, ds = \left(\frac{\partial R}{\partial y} - \frac{\partial Q}{\partial z}\right)dydz + \left(\frac{\partial P}{\partial z} - \frac{\partial R}{\partial x}\right)dxdz + \left(\frac{\partial Q}{\partial x} - \frac{\partial P}{\partial y}\right)dxdy.$$

Hence Stoke's theorem reduces to

$$\int_c (Pdx + Qdy + Rdz) = \iint_s \left\{\left(\frac{\partial R}{\partial y} - \frac{\partial Q}{\partial z}\right)dy\,dz + \left(\frac{\partial P}{\partial z} - \frac{\partial R}{\partial x}\right)dzdx + \left(\frac{\partial Q}{\partial x} - \frac{\partial P}{\partial y}\right)dxdy\right\}$$

Example 16. *Using Stoke's theorm, prove that*

(*i*) *div curl* $\mathbf{F} = 0$ *and* (*ii*) *curl grad* $\phi = \mathbf{0}$. **(*BR. Ap. 94*)**

Solution. (*i*) Let s be a closed surface enclosing a volume v

Then by divergence theorem,

$$\iiint_v \text{div (curl } \mathbf{F})\, dv = \iint_s (\text{curl } \mathbf{F})\cdot\mathbf{n}ds \qquad ...(1)$$

But by Note 3 of Stoke's theorem, $\iint_s (\text{curl } \mathbf{F})\cdot\mathbf{n}\, ds = 0$ ...(2)

From (1) and (2), $\iiint_v \text{div (curl } \mathbf{F})\, dv = 0$

Since v is a rbitrary, div (curl $\mathbf{F}$) = 0.

(*ii*) Suppose s to be an open surface bounded by a simple closed curve c.

Then by Stoke's theorem,

$$\iint_s \text{curl (grad } \phi) \cdot \mathbf{n}\, ds = \int_c \text{grad } \phi \cdot d\mathbf{r} = \int_c \nabla \phi \cdot d\mathbf{r} = \int d\phi = 0.$$

Since s is arbitrary, curl grad $\phi = \mathbf{0}$

Example 17. *Verify Stoke's theorem for* $\mathbf{F} = (2x - y)\,\mathbf{i} - yz^2\,\mathbf{j} - y^2 z\,\mathbf{k}$ *where s is the upper half surface of the sphere* $x^2 + y^2 + z^2 = 1$ *and c is the circular boundary on* $z = 0$ *plane.*

Solution. $\mathbf{F} \cdot d\mathbf{r} = (2x - y)\, dx - yz^2\, dy - y^2 z dz$

On c, $z = 0$ and let $x = \cos\theta$, $y = \sin\theta$

$$\therefore \quad \int_c \mathbf{F} \cdot d\mathbf{r} = \int_0^{2\pi} (2\cos\theta - \sin\theta)(-\sin\theta)\, d\theta$$

$$= \int_0^{2\pi} \left(\frac{1 - \cos 2\theta}{2} - \sin 2\theta \right) d\theta = \pi \qquad \text{...(1)}$$

$$\text{curl } \mathbf{F} = \begin{vmatrix} \mathbf{i} & \mathbf{j} & \mathbf{k} \\ \dfrac{\partial}{\partial x} & \dfrac{\partial}{\partial y} & \dfrac{\partial}{\partial z} \\ 2x - y & -yz^2 & -y^2 z \end{vmatrix} = \mathbf{k}$$

$$\therefore \iint_S \text{curl } \mathbf{F} \cdot \mathbf{n}\, ds = \iint_S \mathbf{k} \cdot \mathbf{n}\, ds = \iint_R dxdy \text{ since } dxdy = \mathbf{k} \cdot \mathbf{n}\, ds$$

(where R is the region enclosed by the circle $x^2 + y^2 = 1$ on $z = 0$)

$$= \text{area of the circle} = \pi \qquad \text{...(2)}$$

From (1) and (2), we get, $\int_c \mathbf{F} \cdot d\mathbf{r} = \iint_c \text{curl } \mathbf{F} \cdot \mathbf{n}\, ds.$

Example 18. *Verify Stoke's theorem for a vector field* $\mathbf{A} = (x^2 - y^2)\,\mathbf{i} + 2xy\,\mathbf{j}$ *in the rectangular region of the* $z = 0$ *plane bounded by the lines* $x = 0, x = a, y = 0$ *and* $y = b$. ***(B.E. 63)***

Solution.

$$\text{curl } \mathbf{A} = \begin{vmatrix} \mathbf{i} & \mathbf{j} & \mathbf{k} \\ \dfrac{\partial}{\partial x} & \dfrac{\partial}{\partial y} & \dfrac{\partial}{\partial z} \\ x^2 - y^2 & 2xy & 0 \end{vmatrix} = 4y\,\mathbf{k}$$

$$\text{curl } \mathbf{A} \cdot \mathbf{n}\, ds = 4y\, \mathbf{k} \cdot \mathbf{k}\, dxdy \quad \because \mathbf{n} = \mathbf{k} \text{ on } z = 0 \text{ plane.}$$

$$= 4y\, dxdy$$

$$\therefore \iint_S \text{curl } \mathbf{A}\cdot\mathbf{n}\, ds = 4\int_0^b\int_0^a y\, dxdy = 4\int_0^a dx\int_0^b ydy$$

$$= 4\,(a)\left(\frac{b^2}{2}\right) = 2ab^2 \qquad ...(1)$$

$\mathbf{A}\cdot d\mathbf{r} = (x^2 - y^2)\, dx + 2\, xydy.$

on OA, $y = 0$ and $dy = 0$

$$\therefore \int_{OA} \mathbf{A}\cdot d\mathbf{r} = \int_0^a x^2 dx = \frac{a^3}{3}$$

On AB, $x = a$, $dx = 0$

$$\therefore \int_{AB} \mathbf{A}\cdot d\mathbf{r} = \int_0^b 2aydy = ab^2$$

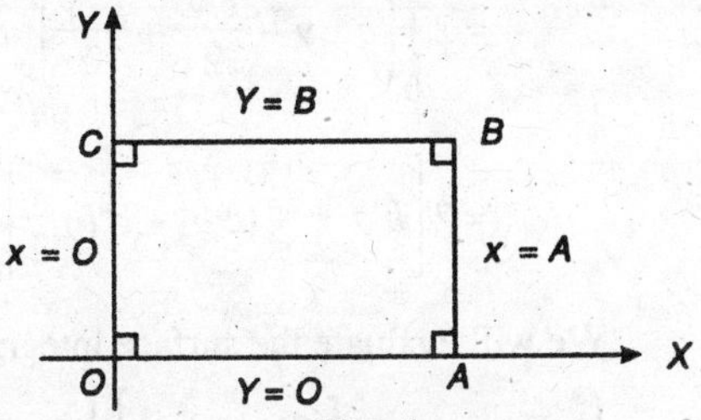

Fig. 9

On BC, $y = b$; $dy = 0$

$$\therefore \int_{BC} \mathbf{A}\cdot d\mathbf{r} = \int_a^0 (x^2 - b^2)\, dx = \left(\frac{x^3}{3} - b^2 x\right)_a^0 = a\,b^2 - \frac{a^3}{3}$$

On CO, $x = 0$, $dx = 0$ $\therefore \int_{CO} \mathbf{A}\cdot d\mathbf{r} = \int_b^0 0\cdot dy = 0$

Adding, $$\int_{OABCO} \mathbf{A}\cdot d\mathbf{r} = \frac{a^3}{3} + ab^2 + ab^2 - \frac{a^3}{3} = 2ab^2 \qquad ...(2)$$

From (1) and (2), $\iint_S \text{curl } \mathbf{A}\cdot\mathbf{n}\, ds = \int_C \mathbf{A}\cdot d\,\mathbf{r} = 2ab^2$

Example 19. *Evaluate* $\int_c (yzdx + zxdy + xydz)$ *where* c *is the circle* $x^2 + y^2 + z^2 = 1$ *and* $z = 0$.

Using the Note 6 under Stoke's Theorem, $P = yz$, $Q = zx$, $R = xy$, we have

$$\int_C (yzdx + zxdy + xydz) = \iint_S (x - x)\, dydz + (y - y)\, dzdx + (z - z)\, dxdy = 0$$

where s is any surface with c as boundary.

Example 20. *Verify Gauss divergence theorem for* $\mathbf{F} = (x^2 - yz)\,\mathbf{i} + (y^2 - zx)\,\mathbf{j} + (z^2 - xy)\,\mathbf{k}$ *taken over the rectangular parallele piped enclosed by* $x = 0$, $x = a$, $y = 0$, $y = b$, $z = 0$ *and* $z = c$ ***(BN Ap. 94) (BR. Nov. 1999)***

Refer to Figure 7 of this chapter

$$\text{div } \mathbf{F} = \frac{\partial F_1}{\partial x} + \frac{\partial F_2}{\partial y} + \frac{\partial F_3}{\partial z} = 2\,(x + y + z).$$

$$\iiint_V \text{div}\, \mathbf{F}\, dv = 2\iiint_V (x+y+z)\, dx\, dy\, dz = 2\int_0^a\int_0^b\int_0^c (x+y+z)\, dz\, dy\, dx$$

$$= 2\int_0^a\int_0^b \left(xz + yz + \frac{z^2}{2}\right)_0^c dy\, dx = 2\int_0^a\int_0^b \left(cx + cy + \frac{c^2}{2}\right) dy\, dx$$

$$= 2\int_0^a \left(cxy + \frac{cy^2}{2} + \frac{c^2 y}{2}\right)_0^b dx = 2\int_0^a \left(bcx + \frac{cb^2}{2} + \frac{c^2 b}{2}\right) dx$$

$$= 2\left[bc\frac{x^2}{2} + (cb^2 + c^2 b)\frac{x}{2}\right]_0^a = abc\,(a+b+c) \qquad ...(1)$$

We will evaluate the surface integrals over the six faces.

$$\iint_S \mathbf{F}\,.\,\mathbf{n}\, ds = \iint_{OAHG} \mathbf{F}\,.\,\mathbf{n}\, ds + \iint_{BCEF} \mathbf{F}\,.\,\mathbf{n}\, ds + \iint_{ACEH} \mathbf{F}\,.\,\mathbf{n}\, ds + \iint_{OBFG} \mathbf{F}\,.\,\mathbf{n}\, ds$$

$$+ \iint_{OACB} \mathbf{F}\,.\,\mathbf{n}\, ds + \iint_{GHEF} \mathbf{F}\,.\,\mathbf{n}\, ds \qquad ...(2)$$

$$= I_1 + I_2 + I_3 + I_4 + I_5 + I_6$$

On the face OAHG, $\mathbf{n} = -\mathbf{j}$, $y = 0$, $ds = dx\, dz$

$$\therefore\ I_1 = \iint (-y^2 + xz)\, dx\, dz = \int_0^a\int_0^c xz\, dz = \left(\frac{x^2}{2}\right)_0^a \left(\frac{z^2}{2}\right)_0^c = \frac{a^2c^2}{4}$$

On face BCEF, $n = \mathbf{j}$, $y = b$, $ds = dx\, dz$

$$I_2 = \iint (y^2 - xz)\, dx\, dz = \int_0^a\int_0^c (b^2 - xz)\, dx\, dz = b^2 ac - \frac{a^2c^2}{4}$$

On face ACEH, $n = \mathbf{i}$, $x = a$, $ds = dy\, dz$

$$I_3 = \iint (a^2 - yz)\, dy\, dz = a^2\int_0^b\int_0^c dy\, dz - \int_0^b\int_0^c yz\, dy\, dz = a^2 bc - \frac{b^2c^2}{4}$$

On face OBFG, $n = -\mathbf{i}$, $x = 0$, $ds = dy\, dz$; $I_4 = \int_0^b\int_0^c (yz)\, dy\, dz = \frac{b^2c^2}{4}$

On face, OACB, $n = -\mathbf{k}$, $z = 0$, $ds = dx\, dy$; $I_5 = \int_0^b\int_0^a (xy)\, dx\, dy = \frac{a^2b^2}{4}$

On face, GHEF, $\mathbf{n} = \mathbf{k}$, $z = c$, $ds = dx\, dy$; $I_6 = \int_0^b \int_0^a (c^2 - xy)\, dy\, dy = c^2ab - \frac{a^2b^2}{4}$

Adding all surface integrals,

$$\iint_S \mathbf{F} \,.\, \mathbf{n}\, ds = \frac{a^2c^2}{4} + b^2ac - \frac{a^2c^2}{4} + a^2bc - \frac{b^2c^2}{4}$$

$$+ \frac{b^2c^2}{4} + \frac{a^2b^2}{4} + c^2ab - \frac{a^2b^2}{4}$$

$$= abc\,(a + b + c) \qquad ...(3)$$

From (1) and (3) $\iiint_V \text{div}\, \mathbf{F}\, dv = \iint_S \mathbf{F} \,.\, \mathbf{n}\, ds$ is true.

Hence verified.

Example 21. *Verify Stoke's theorem for* $\mathbf{F} = (y - z)\,\mathbf{i} + yz\,\mathbf{j} - xz\,\mathbf{k}$ *where S is the surface bounded by the planes* $x = 0, x = 1, y = 0, y = 1, z = 0$ *and* $z = 1$ *above the xoy plane.* *(BN. Nov. 1994)*

Solution. Again refer to the fig. 7 of this chapter.

$$\mathbf{F} \cdot d\mathbf{r} = (y - z)\, dx + yz\, dy - xz\, dz$$

$$\int_C \mathbf{F} \cdot d\mathbf{r} = \int_{OA} \mathbf{F} \cdot d\mathbf{r} + \int_{AC} \mathbf{F} \cdot d\mathbf{r} + \int_{CB} \mathbf{F} \cdot d\mathbf{r} + \int_{BO} \mathbf{F} \cdot d\mathbf{r} = I_1 + I_2 + I_3 + I_4$$

On OA, $y = 0, z = 0, dy = 0, dz = 0$; $\therefore\ I_1 = \int_0^1 0\, dx = 0$

On AC, $x = 1, z = 0, dx = 0, dz = 0$; $I_2 = \int_0^1 0\, dy = 0$

On CB, $y = 1, dy = 0, z = 0, dz = 0$; $I_3 = \int_1^0 1\, dx = (x)_1^0 = -1$

On BO, $x = 0, z = 0, dx = 0, dz = 0.$ $I_4 = \int_1^0 0 = 0$

$$\int_C \mathbf{F} \,.\, d\mathbf{r} = 0 + 0 - 1 + 0 = -1 \qquad ...(1)$$

$$\text{curl } \mathbf{F} = \begin{vmatrix} \mathbf{i} & \mathbf{j} & \mathbf{k} \\ \frac{\partial}{\partial x} & \frac{\partial}{\partial y} & \frac{\partial}{\partial z} \\ y-z & yz & -xz \end{vmatrix} = (-y)\,\mathbf{i} + (z-1)\,\mathbf{j} + (-1)\,\mathbf{k}$$

$$\iint_S \text{curl } \mathbf{F} \cdot \mathbf{n}\, ds = \iint_{OAHG} + \iint_{BCEF} + \iint_{ACEH} + \iint_{OBFG} + \iint_{GHEF}$$

$$= J_1 + J_2 + J_3 + J_4 + J_5 \qquad ...(2)$$

On OAGH, $n = -\mathbf{j}$, $y = 0$, $ds = dx\,dz$

$$\therefore \qquad J_1 = \int_0^1\int_0^1 (1-z)\,dx\,dz = \int_0^1 \left(z - \frac{z^2}{2}\right)_0^1 dx = \frac{1}{2}$$

On BCEF, $n = \mathbf{j}$, $y = 1$, $ds = dx\,dz$

$$J_2 = \int_0^1\int_0^1 (z-1)\,dx\,dz = \int_0^1 \left(\frac{z^2}{2} - z\right)_0^1 dx = -\frac{1}{2}$$

On ACEH, $n = \mathbf{i}$, $x = 1$, $ds = dy\,dz$

$$J_3 = \int_0^1\int_0^1 (-y)\,dydz = \left(\frac{-y^2}{2}\right)_0^1 (z)_0^1 = -\frac{1}{2}$$

On OBFG, $n = -\mathbf{i}$, $x = 0$, $ds = dy\,dz$; $J_4 = \int_0^1\int_0^1 y\,dydz = \left(\frac{y^2}{2}\right)_0^1 (z)_0^1 = \frac{1}{2}$

On GHEF, $n = \mathbf{k}$, $z = 1$, $ds = dx\,dy$; $J_5 = \int_0^1\int_0 - dxdy = (-1)(x)_0^1\,(y)_0^1 = -1$

Adding all J's, $\iint_S \text{curl } \mathbf{F} \cdot \mathbf{n}\, ds = \frac{1}{2} - \frac{1}{2} - \frac{1}{2} + \frac{1}{2} - 1 = -1 \qquad ...(3)$

From (1), and (3), $\iint_S \text{curl } \mathbf{F} \cdot \mathbf{n}\, ds = \int_c \mathbf{F} \cdot d\mathbf{r}$ and the theorem is verified.

Example 22. *Evaluate* $\int_c (xdx + ydy + zdz)$ *where C is the the circle* $x^2 + y^2 + z^2 = a^2, z = 0.$ ***(BR. Ap. 2000)***

Solution. By stoke's theorem,

$$\int_c xdx + ydy + zdz = \iint_s \left[\left(\frac{\partial}{\partial y}(z) - \frac{\partial}{\partial z}(y)\right) dy\,dz + \left(\frac{\partial}{\partial z}(x) - \frac{\partial}{\partial x}(z)\right) dz\,dx\right.$$

$$+\left(\frac{\partial}{\partial x}(y)-\frac{\partial}{\partial y}(x)\right)dx\,dy\Bigg]=0.$$

Example 23. *Evaluate* $\iint_S (x^3\,dy\,dz + x^2 y\,dz dx + x^2 z\,dx dy)$

where S is the surface of the cube $x = 0, y = 0, z = 0, x = 1, y = 1, z = 1$.

Sol. By Gauss divergence theorem,

$$\iint_S (x^2\,dydz + x^2 y\,dzdx + x^2 z\,dxdy)$$

$$= \iiint_V \left[\frac{\partial}{\partial x}(x^3) + \frac{\partial}{\partial y}(x^2 y) + \frac{\partial}{\partial z}(x^2 z)\right] dx\,dy\,dz$$

$$= 5\iiint_V x^2\,dx\,dy\,dz = 5\left(\int_0^1 x^2\,dx\right)\left(\int_0^1 dy\right)\left(\int_0^1 dz\right) = 5/3$$

Exercises 5

1. **Using Green's theorem for a plane, find the area of a circle of radius *a*.**
2. **Evaluate $\int_c (\sin x - y)\,dx - \cos x\,dy$ where *c* is the triangle whose vertices are $(0, 0), \left(\frac{\pi}{2}, 0\right)$ and $\left(\frac{\pi}{2}, 1\right)$. (BR. Ap. 1997)**
3. **Prove $\iiint_V \text{curl}\,\mathbf{F}\,dv = -\iint_S \mathbf{F} \times \mathbf{n}\,ds$**

   (Hint: In divergence theorem, replace F by a × F where a is a constant vector).
4. **Evaluate $\iint_S \mathbf{F}\cdot\mathbf{n}\,ds$ where $\mathbf{F} = yz\,\mathbf{i} + zx\,\mathbf{j} + xy\,\mathbf{k}$ and *s* is the surface of the sphere in the first octant.**
5. **Verify Green's theorem in the plane for $\int_c (3x^2 - 8y^2)\,dx + (4y - 6xy)\,dy$ where *c* is the boundary of the region defined by** ***(BR. Ap. 2000)***

   (*i*) $x = 0, y = 0, x + y = 1$ (*ii*) $x = y^2, y = x^2$
6. **Evaluate $\int_c (3x^2 + 2y)\,dx + (x + 3\cos y)\,dy$ where *c* is the perimeter of the parallelogram whose vertices are (0, 0), (2, 0), (3, 1), (1, 1).**
7. **Prove the following : (*s* is any closed surface)**

   (*i*) $\iint_S (\mathbf{f} \times \nabla\phi)\cdot\mathbf{n}\,ds = \iiint_V \nabla\phi\cdot\text{curl}\,\mathbf{f}\,dv$. (*ii*) $\iint_S \nabla\left(\frac{1}{r}\right)\cdot ds = 0$

(iii) $\iint_s (\nabla f \times \nabla g) \cdot \mathbf{n}\, ds = 0$ (iv) $\iint_s \mathbf{r} \times \mathbf{n}\, ds = 0$ (v) $\iint_s \nabla r^2 \cdot \mathbf{n}\, ds = 6V$

(vi) $\iint_s \mathbf{n}\, ds = 0$ (vii) $\iint_s r^5\, \mathbf{n}\, ds = \iiint_v 5r^4\, \mathbf{r}\, dv$

8. **If s be any closed surface enclosing a volume v, and $\mathbf{A} = x\,\mathbf{i} + 2y\,\mathbf{j} + 3z\,\mathbf{k}$, show that $\iint_s \mathbf{A} \cdot \mathbf{n}\, ds = 6v$.**

9. **Verify the divergence theorem for the function $\mathbf{A} = x^2\,\mathbf{i} + z\,\mathbf{j} + yz\,\mathbf{k}$ over the cube $x = \pm 1, y = \pm 1, z = \pm 1$.**

10. **Verify the divergence theorem for**

(i) $\mathbf{F} = 4xz\,\mathbf{i} - y^2\,\mathbf{j} + yz\,\mathbf{k}$ **(Periyar Ap. 2000)**

(ii) $\mathbf{F} = (2x - z)\,\mathbf{i} + x^2y\,\mathbf{j} - xz^2\,\mathbf{k}$ **over the cube bounded by** $x = 0, x = 1, y = 0, y = 1, z = 0, z = 1$ **(B.Sc., M.U., 1982)**

11. **Verify the divergence theorem for the vector field $\mathbf{F} = (2xy + z)\,\mathbf{i} + x^2y\,\mathbf{j} - (x + 3y)\,\mathbf{k}$ taken over the region bounded by $2x + 2y + z = 6,\ x = 0, y = 0, z = 0$.** **(B.E. 65)**

12. **Verify the divergence theorem for $\mathbf{F} = xy^2\,\mathbf{i} + yz^2\,\mathbf{j} + zx^2\,\mathbf{k}$ over the region bounded by $x = 0, x = 1, y = 0, y = 2, z = 0, z = 3$.**

13. **Verify the divergence theorem for the function $\mathbf{F} = 2xz\,\mathbf{i} + yz\,\mathbf{j} + z^2\,\mathbf{k}$ over the upper half of the sphere $x^2 + y^2 + z^2 = a^2$**

14. **Evaluate $\iint_s (\nabla \times \mathbf{F}) \cdot \mathbf{n}\, ds$ where $\mathbf{F} = (x + 2y)\,\mathbf{i} - 3z\,\mathbf{j} + x\,\mathbf{k}$ where s is the surface bounded by $x = 0, y = 0, z = 0, x + y + z = 1$.**

15. **Verify the divergence theorem for the function $\mathbf{A} = y\,\mathbf{i} + x\,\mathbf{j} + \mathbf{k}$ over the region bounded by $x^2 + y^2 = a^2, z = 0$ and $z = h$.**

17. **Evaluate $\iint_s \mathbf{A} \cdot \mathbf{n}\, ds$ over the region bounded by $x = 0, x = a, y = 0, y = b\ z = 0, z = c$ where $\mathbf{A} = (x^2 - yz)\,\mathbf{i} + (y^2 - zx)\,\mathbf{j} + (z^2 - xy)\,\mathbf{k}$.**

18. **Given the vector $\mathbf{F} = (x^2 - y^2)\,\mathbf{i} + 2xy\,\mathbf{j} + (y^2 - xy)\,\mathbf{k}$** **(BR. 1999 Nov.)** **verify the divergence theorem over the cube with centre at the origin and of side length a.**

19. **Prove that**

(i) $\int_c \mathbf{r} \cdot d\mathbf{r} = 0$ (ii) $\int_c u\nabla v \cdot d\mathbf{r} = -\int_a v\, \nabla u \cdot d\mathbf{r}$

(iii) $\int_0 \mathbf{r} \times d\mathbf{r} = 2\iint_s \mathbf{n}\, ds$ **(Hint : Take $\mathbf{F} = \mathbf{a} \times \mathbf{r}$ in Stoke's theorem)** (*iv*) $\int_c \phi \nabla \phi \cdot dr = 0$

20. **Verify Stoke's theorem for $\mathbf{F} = (y - z + 2)\,\mathbf{i} + (yz + 4)\,\mathbf{j} - xz\,\mathbf{k}$ where s is the open surface of the cube $x = 0, y = 0, z = 0, x = 2, y = 2, z = 2$. above the xy plane.** ***(BR 1995 A)***

21. **Verify Stoke's theorem for $\mathbf{A} = x^2\,\mathbf{i} + xy\,\mathbf{j}$ in the square region in xy plane bounded by lines $x = 0, y = 0, x = a, y = a$.**

22. **Verify Stoke's theorem for $\mathbf{F} = xz\,\mathbf{i} - y\,\mathbf{j} + x^2y\,\mathbf{k}$ where s is the surface of the region bounded by $x = 0, y = 0, z = 0, 2x + y + 2z = 8$ which is not included in xz plane.**

23. **Verify Stoke's theorem for $\mathbf{f} = y^2\,\mathbf{i} + xy\,\mathbf{j} + xz\,\mathbf{k}$ over the upper half of the sphere $x^2 + y^2 + z^2 = a^2$.**

24. **Verify Stoke's theorem for $\mathbf{F} = x^2\,\mathbf{i} - xy\,\mathbf{j}$ integrated around the square in the xy plane bounded by the lines $x = 0, y = 0, x = a, y = a$.**

25. **Using Stoke's theorem evaluate $\int_c \mathbf{F} \cdot dr$, where $\mathbf{F} = (2x + y - 2z)\,\mathbf{i} + (2x - 4y + z^2)\,\mathbf{j} + (x - 2y - z^2)\,\mathbf{k}$ where c is the circle with centre (0, 0, 3) and radius 5 in the plane $z = 3$.**

26. **Verify Stoke's theorem for (*i*) $\mathbf{A} = z\,\mathbf{i} + x\,\mathbf{j} + y\,\mathbf{k}$**

(*ii*) $\mathbf{A} = (2x - y)\,\mathbf{i} - yz^2\,\mathbf{j} - y^2z\,\mathbf{k}$

where s is the upper half of the sphere $x^2 + y^2 + z^2 = 1$ and c is its boundary in the xy plane. ***(B.Sc. M.U. 1983)***

27. **If $\mathbf{F} = \text{grad}\,\phi$ and $\nabla^2\phi = -4\pi\rho$, prove that $\iint_S \mathbf{F} \cdot \mathbf{n}\,ds = -4\pi \iiint_V \rho\,dv$.**

28. **If $\mathbf{F} = x^3\,\mathbf{i} + y^3\,\mathbf{j} + z^3\,\mathbf{k}$ prove $\iint_S \mathbf{F} \cdot \mathbf{n}\,ds = \frac{12}{5}\pi a^5$ where S is the surface of sphere $x^2 + y^2 + z^2 = a^2$.**

29. **Evaluate $\iint_s (x^3\,dydz + x^2y\,dzdx + x^2z\,dxdy)$**

where s is the surface of the closed cylinder $z = 0, z = h, x^2 + y^2 = a^2$

30. **Evaluate $\iint_s \{(x^3 - yz)\,dy\,dz - 2x^2y\,dzdx + z\,dx\,dy\}$**

where s is the surface of the cube $x = 0, y = 0, z = 0, x = 1, y = 1, z = 1$.

31. **Evaluate $\int_c (x\,dx + y\,dy + z\,dz)$ where c is the circle $x^2 + y^2 + z^2 = a^2, z = 0$.**

32. **Show that the following vector functions $\mathbf{f}$ are solenoidal and find the function $\mathbf{F}$ such that $\mathbf{f} = \text{curl}\,\mathbf{F}$**

$$\mathbf{f} = (y - z)\,\mathbf{i} + (z - x)\,\mathbf{j} + (x - y)\,\mathbf{k}$$

33. **Prove** $\iint_s$ curl $\mathbf{F} \cdot \mathbf{n}\, ds = 0$, **$s$ being any closed surface (Do by two methods).**

34. **Evaluate** $\iint_s \frac{\mathbf{r}}{r^3} \cdot ds$ **if origin is a point outside s.**

Using Stoke's theórem:

35. **Evaluate** $\int_C (yz\, dx + zx\, dy + xy\, dz)$ **where c is the closed curve** $x^2 + y^2 = 1,\ z = y^2$

36. **Evaluate** $\int_c (xy\, dx + xy^2\, dy)$ **where c is the square whose vertices are (–1, 0), (1, 0) (0, 1) and (0, –1).**

37. **Evaluate** $\int_c \mathbf{A} \,.\, d\mathbf{r}$ **where** $\mathbf{A} = y^2\,\mathbf{i} + x^2\,\mathbf{j} - (x + z)\,\mathbf{k}$ **and C is the boundry of the triangle whose vertices are (0, 0, 0) (1, 0, 0) and (1, 1, 0).**

38. **Calculate** $\int_c (\sin y\, dz - \cos x\, dy + \sin z\, dx)$ **where c is the rectangle** $0 \le x \le \pi$, $0 \le y \le 1,\ z = 3$.

39. **Verify Stoke's theorem if** $\mathbf{A} = y\,\mathbf{i} + z\,\mathbf{j} + x\,\mathbf{k}$ **where S is the upper half of the sphere** $x^2 + y^2 + z^2 = 1$ **and c is the boundary on** $z = 0$.

40. **Verify Stoke's theorem if** $\mathbf{A} = (x^2 + y^2)\,\mathbf{i} - 2xy\,\mathbf{j}$ **taking the boundary of the rectangle** $x = \pm a,\ y = 0,\ y = b$ **as c.**

41. **Verify Stoke's theorem if** $\mathbf{A} = 2y\,\mathbf{i} + 3x\,\mathbf{j} - z^2\,\mathbf{k}$ **where S is the upper half of sphere** $x^2 + y^2 + z^2 = 9$ **and c is the circle on** $z = 0$.

42. **Prove** $\int_c (r^n\,\mathbf{r}) \,.\, d\mathbf{r} = 0$ **where c is any closed curve and** $r = |\mathbf{r}|$ **and** $\mathbf{r} = x\,\mathbf{i} + y\,\mathbf{j} + z\,\mathbf{k}$

43. **If** $\mathbf{F} = (x^2 + y - 4)\,\mathbf{i} + 3xy\,\mathbf{j} + (2xz + z^2)\,\mathbf{k}$ **evaluate** $\iint_S$ **curl F . n** ds **where S is the upper half of sphere** $x^2 + y^2 + z^2 = 16$ **above xy plane.**

44. **Show that** $\iint_S (ax^2 + by^2 + cz^2)\, dx$ **over the sphere** $x^2 + y^2 + z^2 = r^2$ **is** $\frac{4}{3}\pi\,(a + b + c)\, r^4$.

[Hint: $\mathbf{F} = ax\,\mathbf{i} + by\,\mathbf{j} + cz\,\mathbf{k};\ \mathbf{n} = \frac{1}{r}(x\,\mathbf{i} + y\,\mathbf{j} + z\,\mathbf{k})$**]**

FURTHER WORKED EXAMPLES

In what follows, we use the following notations :

$$\mathbf{r} = x\mathbf{i} + y\mathbf{j} + z\mathbf{k} \text{ and } |\mathbf{r}| = r = \sqrt{x^2 + y^2 + z^2}$$

Example 23. *Prove (i)* $\mathbf{r} \cdot \frac{d\mathbf{r}}{dt} = \frac{r\,dr}{dt}$ (*i*) $\mathbf{r} \cdot \frac{d\mathbf{r}}{dt} = 0$ if $|\mathbf{r}|$ = constant

Solution. We know $\mathbf{r} \cdot \mathbf{r} = r^2$; Differentiate w.r.t. t,

$$\mathbf{r} \cdot \frac{d\mathbf{r}}{dt} + \mathbf{r} \cdot \frac{d\mathbf{r}}{dt} = 2r\frac{dr}{dt}$$

$$\therefore \qquad \mathbf{r} \cdot \frac{d\mathbf{r}}{dt} = \frac{r\,dr}{dt}$$

Hence, problem (*i*) becomes $\mathbf{r} \cdot \frac{d\mathbf{r}}{dt} = 0$ if $\mathbf{r}$ = constant

Or

$$\mathbf{r} \cdot \mathbf{r} = |\mathbf{r}|^2 = \text{constant.}$$

Differentiating, $\mathbf{r} \cdot \frac{d\mathbf{r}}{dt} + \frac{d\mathbf{r}}{dt} \cdot \mathbf{r} = 0$

$$\therefore \qquad \mathbf{r} \cdot \frac{d\mathbf{r}}{dt} = 0$$

Example 24. *Find* $\nabla\phi$ *where* $\phi = \frac{1}{2}\log(x^2 + y^2 + z^2)$ *at (1, 1, 1).*

(*BR. Nov., 1999*)

Solution. $\frac{\partial\phi}{\partial x} = \frac{1}{2} \cdot \frac{2x}{x^2 + y^2 + z^2} = \frac{x}{x^2 + y^2 + z^2}$

$$\nabla\phi = \Sigma\,\mathbf{i}\frac{\partial\phi}{\partial x} = \frac{1}{x^2 + y^2 + z^2}(x\mathbf{i} + y\mathbf{j} + z\mathbf{k})$$

$\nabla\phi$ at $(1, 1, 1) = \frac{1}{3}(\mathbf{i} + \mathbf{j} + \mathbf{k})$

Example 25. *If* $\mathbf{r} = \mathbf{a}\cosh nt + \mathbf{b}\sinh nt$ *where* n, $\mathbf{a}, \mathbf{b}$ *are constants prove* $\frac{d^2\mathbf{r}}{dt^2} - n^2\mathbf{r} = 0.$ (*BR. Nov., 1999*)

Solution. $\mathbf{r} = \mathbf{a}\cosh nt + \mathbf{b}\sinh nt$

$$\frac{d\mathbf{r}}{dt} = \mathbf{a}n\sinh nt + \mathbf{b}n\cosh nt$$

$$\frac{d^2\mathbf{r}}{dt^2} = \mathbf{a}n^2\cosh nt + \mathbf{b}n^2\sinh nt = n^2\mathbf{r}$$

$$\frac{d^2\mathbf{r}}{dt^2} - n^2\mathbf{r} = 0.$$

Example 26. *If* $\mathbf{r} = \mathbf{a}e^{nt} + \mathbf{b}e^{-nt}$, *prove* $\frac{d^2\mathbf{r}}{dt^2} - n^2\mathbf{r} = 0$ *where* **a**, **b** *are constants.*

Solution. $\frac{d\mathbf{r}}{dt} = \mathbf{a}ne^{nt} - \mathbf{b}ne^{-nt}$

$$\frac{d^2\mathbf{r}}{dt^2} = \mathbf{a}n^2e^{nt} + \mathbf{b}n^2e^{-nt} = n^2\mathbf{r}$$

Example 27. *Find the unit tangent vector to the curve.*

$$\mathbf{r} = (t^2+1)\,\mathbf{i} + (4t-3)\,\mathbf{j} + (2t^2-6t)\,\mathbf{k} \text{ at } t = 1$$

Solution. $\frac{d\mathbf{r}}{dt} = 2t\mathbf{i} + 4\mathbf{j} + (4t-6)\,\mathbf{k}$

At $t = 1$, $\frac{d\mathbf{r}}{dt} = 2\mathbf{i} + 4\mathbf{j} - 2\mathbf{k}$

$$\text{Unit vector along the tangent} = \frac{\frac{d\mathbf{r}}{dt}}{\left|\frac{d\mathbf{r}}{dt}\right|} = \frac{2\mathbf{i}+4\mathbf{j}-2\mathbf{k}}{\sqrt{24}}$$

$$= \frac{\mathbf{i}+2\mathbf{j}-\mathbf{k}}{\sqrt{6}}$$

Example 28. *Define gradient of a scalar function and divergence, curl of a vector function.*

Solution. If $\phi\,(x, y, z)$ is a differentiable scalar point function then the gradient of ϕ is

$$\nabla\phi = \frac{\partial\phi}{\partial x}\mathbf{i} + \frac{\partial\phi}{\partial y}\mathbf{j} + \frac{\partial\phi}{\partial z}\mathbf{k}$$

If $\mathbf{f}\,(x, y, z)$ is a differentiable vector point function.
the divergence of **f** is

$$\text{div}\,\mathbf{f} = \nabla\cdot\mathbf{f} = \mathbf{i}\cdot\frac{\partial\mathbf{f}}{\partial x} + \mathbf{j}\cdot\frac{\partial\mathbf{f}}{\partial y} + \mathbf{k}\cdot\frac{\partial\mathbf{f}}{\partial z}$$

and $\text{curl}\,\mathbf{f} = \nabla\times\mathbf{f} = i\times\frac{\partial\mathbf{f}}{\partial x} + \mathbf{j}\times\frac{\partial\mathbf{f}}{\partial y} + \mathbf{k}\times\frac{\partial\mathbf{f}}{\partial z}$

Example 29. *Write down the formulae for curl* **f** *and div* **f** *in terms of components.*

Solution. If $\mathbf{f} = f_1\mathbf{i} + f_2\mathbf{j} + f_3\mathbf{k}$ then

$$\text{div }\mathbf{f} = \frac{\partial f_1}{\partial x} + \frac{\partial f_2}{\partial y} + \frac{\partial f_3}{\partial z}$$

$$\text{curl}\,\mathbf{f} = \left(\frac{\partial f_3}{\partial y} - \frac{\partial f_2}{\partial z}\right)\mathbf{i} + \left(\frac{\partial f_1}{\partial z} - \frac{\partial f_3}{\partial x}\right)\mathbf{j} + \left(\frac{\partial f_2}{\partial x} - \frac{\partial f_1}{\partial y}\right)\mathbf{k}$$

Example 30. *Define solenoidal vector and irrotational vector.*

Solution. If **f** (x, y, z) is a vector point function, then

(*i*) **f** is solenoidal if div $\mathbf{f} = 0$ at all points.

(*ii*) **f** is irrotational if curl $\mathbf{f} = \mathbf{0}$ at all points.

Example 31. *Define directional derivative of* $\phi(x, y, z)$. *In what direction is it maximum ?*

Solution. Directional derivative of $\phi(x, y, z)$ in any direction is the projection of grad ϕ in that direction. It is maximum along the normal to $\phi(x, y, z) = c$.

Example 32. *Find the directional derivative of* $\phi = 4xz^2 + x^2yz$ *at* $(1, -2, 1)$ *in direction of* $2\mathbf{i} + 3\mathbf{j} + 4\mathbf{k}$.

Solution. $\nabla\phi = (4z^2 + 2xyz)\,\mathbf{i} + (x^2z)\,\mathbf{j} + (8xz + x^2y)\,\mathbf{k}$

$\nabla\phi$ at $(1, -2, 1) = 0\mathbf{i} + \mathbf{j} + 6\mathbf{k}$

Directional derivative in the direction $2\mathbf{i} + 3\mathbf{j} + 4\mathbf{k}$ is

$$= \frac{\nabla\phi \cdot \mathbf{a}}{|\mathbf{a}|} = \frac{(\mathbf{j} + 6\mathbf{k}) \cdot (2\mathbf{i} + 3\mathbf{j} + 4\mathbf{k})}{\sqrt{4 + 9 + 16}}$$

$$= \frac{27}{\sqrt{29}}$$

Example 33. *Find the maximum directional derivative of* $\phi = xyz^2$ *at* $(1, 0, 3)$.

Solution. $\nabla\phi = (yz^2)\,\mathbf{i} + (xz^2)\,\mathbf{j} + 2xyz\,\mathbf{k}$

$(\nabla\phi)_{(1,\,0,\,3)} = 9\hat{\mathbf{j}}$

It is maximum in the direction of y-axis and maximum value is 9.

Example 34. *Find the unit normal at* $(6, 4, 3)$ *to* $xy + yz + zx = 54$.

Solution. $\nabla\phi = (y + z)\,\mathbf{i} + (x + z)\,\mathbf{j} + (x + y)\,\mathbf{k}$

$= 7\vec{\mathbf{i}} + 9\mathbf{j} + 10\mathbf{k}$ at $(6, 4, 3)$

$$\text{Unit normal} = \frac{\nabla\phi}{|\nabla\phi|} = \frac{7\mathbf{i} + 9\mathbf{j} + 10\mathbf{k}}{\sqrt{49 + 81 + 100}} = \frac{7\mathbf{i} + 9\mathbf{j} + 10\mathbf{k}}{\sqrt{230}}$$

Example 35. *Find the tangent plane at* $(6, 4, 3)$ *to* $xy + yz + zx = 54$.

Solution. $(\nabla\phi)_{(6,\,4,\,3)} = 7\mathbf{i} + 9\mathbf{j} + 10\mathbf{k}$

Direction ratios of the normal to the tangent plane at $(6, 4, 3)$ are $(7, 9, 10)$.

Tangent plane at $(6, 4, 3)$ is $7(x - 6) + 9(y - 4) + 10(z - 3) = 0$

i.e., $7x + 9y + 10z = 108$

Example 36. *Prove* $\mathbf{F} = 3y^4z^2\,\mathbf{i} + 4x^3z^2\,\mathbf{j} - 3x^2y^2\,\mathbf{k}$ *is solenoidal.*

(BR. U. 94N)

Solution. $$\text{div } \mathbf{F} = \frac{\partial}{\partial x}(3y^4z^2) + \frac{\partial}{\partial y}(4x^3z^2) + \frac{\partial}{\partial z}(-3x^2y^2)$$

$$= 0 + 0 + 0 = 0$$

Therefore, **F** is solenoidal.

Example 37. *Prove* $yz\mathbf{i} + zx\mathbf{j} + xy\mathbf{k} = \mathbf{F}$ *is irrotational.*

Solution. $\text{curl } \mathbf{F} = \begin{vmatrix} \mathbf{i} & \mathbf{j} & \mathbf{h} \\ \frac{\partial}{\partial x} & \frac{\partial}{\partial y} & \frac{\partial}{\partial z} \\ yz & zx & xy \end{vmatrix} = (x - x)\,\mathbf{i} + (y - y)\,\mathbf{j} + (z - z)\,\mathbf{k}$

$= 0$. Therefore, **F** is irrotational.

Example 38. *Write down the identities.*

Solution. $\text{div}(\phi\mathbf{u}) = \nabla\phi \cdot \mathbf{u} + \phi \text{ div } \mathbf{u}$

$\text{curl}(\phi\mathbf{u}) = \nabla\phi \times \mathbf{u} + \phi \text{ curl } \mathbf{u}$

$\text{div}(\mathbf{u} \times \mathbf{v}) = \mathbf{v} \cdot \text{curl } \mathbf{u} - \mathbf{u} \cdot \text{curl } \mathbf{v}$

Example 39. *Find* ∇r. ***(BR. Nov., 1999)***

Solution. Here, $r^2 = x^2 + y^2 + z^2$

$2r\frac{\partial r}{\partial x} = 2x$ therefore, $\frac{\partial r}{\partial x} = \frac{x}{r}$ and similar results.

$$\nabla r = \sum \mathbf{i}\frac{\partial r}{\partial x} = \sum \frac{x}{r}\mathbf{i} = \frac{1}{r}(x\mathbf{i} + y\mathbf{j} + z\mathbf{k}) = \frac{\mathbf{r}}{r}$$

Example 40. *Find curl* ∇r. ***(BR. U., 94N)***

Solution. curl ∇r = curl grad $r = \mathbf{0}$ since curl grad $\phi = \mathbf{0}$

Example 41. *If* $u = x^2 + y^2 + z^2$ *prove curl* $\nabla u = 0$.

Solution. It is true since curl grad $\phi = 0$

(or calculate grad u and proceed)

Example 42. *Prove* $\nabla(\phi\psi) = \psi\nabla\phi + \phi\nabla\psi$

Solution. $\nabla(\phi\psi) = \sum \mathbf{i}\frac{\partial}{\partial x}(\phi\psi)$

$$= \sum \mathbf{i}\left[\phi\frac{\partial\psi}{\partial x} + \psi\frac{\partial\phi}{\partial x}\right]$$

$$= \phi\sum \mathbf{i}\frac{\partial\psi}{\partial x} + \psi\sum \mathbf{i}\frac{\partial\phi}{\partial x}$$

$$= \phi\nabla\psi + \psi\nabla\phi$$

Example 43. *Prove curl* $(\phi \text{ grad } \phi) = \mathbf{0}$.

Solution. $\text{curl}(\phi \text{ grad. } \phi) = \phi \text{ curl } \nabla\phi + (\text{grad}\,\phi) \times \nabla\phi$

$$= \phi(\mathbf{0}) + \mathbf{0} = \mathbf{0}$$

Since curl grad $\phi = \mathbf{0}$ and $\nabla\phi \times \nabla\phi = \mathbf{0}$

Example 44. *If* $\nabla\phi$ *is solenoidal, prove* $\nabla^2\phi = 0$.

Solution. $\text{div}(\nabla\phi) = 0$

$\nabla \cdot (\nabla\phi) = 0$. Therefore, $\nabla^2\phi = 0$

Example 45. *Find curl curl* **F** *at (1, 2, – 3) where*

$$\mathbf{F} = x^2\,\mathbf{i} + y^2\,\mathbf{j} + z^2\,\mathbf{k}$$

Solution. $\text{curl }\mathbf{F} = \begin{vmatrix} \mathbf{i} & \mathbf{j} & \mathbf{k} \\ \frac{\partial}{\partial x} & \frac{\partial}{\partial y} & \frac{\partial}{\partial z} \\ x^2 & y^2 & z^2 \end{vmatrix} = \Sigma\,(0-0)\,\mathbf{i} = \mathbf{0}$

$\text{curl curl }\mathbf{F} = \mathbf{0}$

Example 46. *If* $\nabla\phi = 2xyz^3\,\mathbf{i} + x^2z^3\,\mathbf{j} + 3x^2yz^2\,\mathbf{k}$*, find* ϕ.

Solution. $\nabla\phi = \sum \mathbf{i}\,\frac{\partial\phi}{\partial x} = 2xyz^3\,\mathbf{i} + x^2z^3\,\mathbf{j} + 3x^2yz^2\,\mathbf{k}$

Therefore, $\frac{\partial\phi}{\partial x} = 2xyz^3$ therefore $\phi = \int 2xyz^3\,dx = x^2yz^3 + f(y, z)$

$$\frac{\partial\phi}{\partial y} = x^2z^3 \text{ therefore } \phi = x^2yz^3 + F(x, z)$$

$$\frac{\partial\phi}{\partial z} = 3x^2yz^2 \text{ therefore } \phi = x^2yz^3 + \psi(x, y)$$

Therefore, $\phi = x^2yz^3 + c$

Example 47. *Evaluate* $\int_C \mathbf{F}\cdot d\mathbf{r}$ if

$$\mathbf{F} = 3xy\,\mathbf{i} - 5z\,\mathbf{j} + 10x\,\mathbf{k}$$

along the curve $x = t^2,\ y = 2t^2,\ z = t^3$ *from* $t = 0$ *to* $t = 1$

Solution. $\int_C \mathbf{F}\cdot d\mathbf{r} = \int_C (3xy\,dx - 5z\,dy + 10x\,dz)$

$$= \int_0^1 (12t^5 - 20t^4 + 30t^4)\,dt$$

$$= \left[2t^6 + 2t^5\right]_0^1 = 4$$

Example 48. *State Green's theorem in a plane.*

Solution. If R is a closed region of the xy plane, bounded by a simple closed curve c and if M and N are continuous functions of x and y having continuous derivatives, then

$$\int_C (M\,dx + N\,dy) = \int\int_R \left(\frac{\partial N}{\partial x} - \frac{\partial M}{\partial y}\right) dxdy$$

Example 49. *State Gauss' divergence theorem.*

Solution. The surface integral of the normal component of a vector function **F** taken over a closed surface S enclosing a volume V is equal to the volume integral of divergence **F** taken throughout the volume V.

$$\int\int_S \mathbf{F}\cdot\mathbf{n}\,ds = \int\int\int_V \text{div }\mathbf{F}\,dv$$

Example 50. *State Stoke's theorem.*

Solution. The surface integral of the normal component of the curl of a vector point function **F** over an open surface S is equal to the line integral of the tangential component of **F** around the closed curve C bounding the open surface S.

$$\int_C \mathbf{F}\cdot d\mathbf{r} = \int\int_S \text{curl}\,\mathbf{F}\cdot d\mathbf{S} = \int\int_S \text{curl}\,\mathbf{F}\cdot \mathbf{n}\, dS$$

Example 51. *Prove* $\int\int_s \mathbf{r}\cdot\mathbf{n}\, ds = 3v$ **(BR. U. 94 N)**

Solution. $\int\int_s \mathbf{r}\cdot\mathbf{n}\, ds = \int\int\int_v \text{div}\,\mathbf{r}\, dv = 3v$ since $\text{div}\,\mathbf{r} = 3$

Example 52. *If s is any closed surface, evaluate*

$$\int\int_s (ax\,\mathbf{i} + by\,\mathbf{j} + cz\,\mathbf{k})\cdot\mathbf{n}\, ds$$ **(BR. U. 94 A)**

Solution. $I = \int\int\int_v \text{div}\,\mathbf{F}\, dv$

$= \int\int\int_v (a+b+c)\, dv$ since $\text{div}\,\mathbf{F} = a+b+c$

$= (a+b+c)\, v$

If s is surface of a sphere,

$$I = (a+b+c)\frac{4}{3}\pi r^3$$

Fill up the blanks :

Example 53. $\nabla^2\left(\frac{1}{r}\right) =$ .... **[Ans.** zero]

Example 54. *If* $\nabla\phi = yz\,\mathbf{i} + zx\,\mathbf{j} + xy\,\mathbf{k}$, *then* ϕ *is* .... **[Ans.** $xyz + c$]

Example 55. $\int\int_S (x\, dydz + y\, dzdx + z\, dxdy)$ *where s is the surface of a sphere of radius a is* .... **[Ans.** $4\pi a^3$]

Example 56. *If a vector point function* **F** *is expressible as the gradient of a scalar point function* ϕ, *then* **F** *is* .... **[Ans.** conservative]

State True or False :

Example 57. *The area bounded by a simple closed curve C is* $\frac{1}{2}\int_c (x\, dy - y\, dx)$ **[Ans.** True]

Example 58. *Curl* $(r^n\,\mathbf{r}) = 0$ **[Ans.** True]

Example 59. *If c is any closed curve, then* $\int_c \nabla\phi\cdot d\mathbf{r} = 0$ **[Ans.** True]

Pick the correct answer :

Example 60. *div* $(r^n \mathbf{r})$ *is*

(*a*) $(n+3)\,\mathbf{r}$ (*b*) $(n+3)\,r^n$ (*c*) *0* (*d*) 0 **[Ans.** (*b*)**]**

Example 61. *Using Stoke's theorem, prove curl grad* $\phi = 0$

Solution. $\int\int_s \text{curl}\,(\text{grad}\,\phi)\cdot\mathbf{n}\,ds = \int_c \text{grad}\,\phi\cdot d\mathbf{r} = \int_c \nabla\phi\cdot d\mathbf{r} = \int_c d\phi = 0$

Since s is arbitrary, curl (grad ϕ) = 0

Example 62. *Find 'a' so that the vector*

$\mathbf{A} = (ax^2 - y^2 + x)\,\mathbf{i} - (2xy + y)\,\mathbf{j}$ *is irrotational.* ***(BR. Nov., 1999)***

Solution. curl $A = \begin{vmatrix} \mathbf{i} & \mathbf{j} & \mathbf{k} \\ \dfrac{\partial}{\partial x} & \dfrac{\partial}{\partial y} & \dfrac{\partial}{\partial z} \\ ax^2 - y^2 + x & -(2xy+y) & 0 \end{vmatrix}$

$$= (0)\,\mathbf{i} - (0)\,\mathbf{j} + (-2y + 2y)\,\mathbf{k}$$

$= \mathbf{0}$ for all values of 'a'

$\therefore$ **A** is irrotational for all 'a'.

Example 63. *Find 'a' so that*

$\mathbf{F} = (x+z)\,\mathbf{i} + (3x + ay)\,\mathbf{j} + (x - 5z)\,\mathbf{k}$ *is solenoidal.*

Solution. Since solenoidal, div $\mathbf{F} = 0$

$$\text{"}\frac{\partial F_1}{\partial x} + \frac{\partial F_2}{\partial y} + \frac{\partial F_3}{\partial z} = 0\text{"}$$

$\therefore$ $1 + a - 5 = 0$ *i.e.*, $a = 4$

Example 64. *If* $\mathbf{F} = F_1\,\mathbf{i} + F_2\,\mathbf{j} + F_3\,\mathbf{k}$ *and* **F** *is irrotational, find the conditions on* F_1, F_2, F_3.

Solution. Since **F** is irrotational, curl $\mathbf{F} = \mathbf{0}$

$$\therefore \quad \text{curl } \mathbf{F} = \left(\frac{\partial F_3}{\partial y} - \frac{\partial F_2}{\partial z}\right)\mathbf{i} + \left(\frac{\partial F_1}{\partial z} - \frac{\partial F_3}{\partial x}\right)\mathbf{j} + \left(\frac{\partial F_2}{\partial x} - \frac{\partial F_1}{\partial y}\right)\mathbf{k} = \mathbf{0}$$

implies $\dfrac{\partial F_3}{\partial y} = \dfrac{\partial F_2}{\partial z}$, $\dfrac{\partial F_1}{\partial z} = \dfrac{\partial F_3}{\partial x}$, $\dfrac{\partial F_2}{\partial x} = \dfrac{\partial F_1}{\partial y}$

These are the required conditions.

Example 65. *Find a, b, c if*

$(x + y + az)\,\mathbf{i} + (bx + 2y - z)\,\mathbf{j} + (-z + cy + 2z)\,\mathbf{k}$ *is irrotational.*

(BN. 1994)

Solution. Since irrotational, $\dfrac{\partial F_3}{\partial y} = \dfrac{\partial F_2}{\partial z}$, $\dfrac{\partial F_1}{\partial z} = \dfrac{\partial F_3}{\partial x}$, $\dfrac{\partial F_2}{\partial x} = \dfrac{\partial F_1}{\partial y}$

Hence, $c = -1$, $a = -1$, $b = 1$.

Example 66. *Find the unit normal to* $xy = z^2$ *at* $(1, 9, -3)$

(BN. 1994)

Solution. Let $\phi = xy - z^2 = 0$

$$\nabla\phi = \sum \mathbf{i}\frac{\partial\phi}{\partial x} = (y)\,\mathbf{i} + (x)\,\mathbf{j} + (-2z)\,\mathbf{k}$$

$$\nabla\phi_{(1,\,9,\,-3)} = 9\mathbf{i} + \mathbf{j} + 6\mathbf{k}$$

Unit normal at the point $= \dfrac{\nabla\phi}{|\nabla\phi|} = \dfrac{9\mathbf{i}+\mathbf{j}+6\mathbf{k}}{\sqrt{81+1+36}} = \dfrac{9\mathbf{i}+\mathbf{j}+6\mathbf{k}}{\sqrt{118}}$

Example 67. *Find div* **F** *and curl* **F** *at (1, – 1, 0) given*

$$\mathbf{F} = xz^3\,\mathbf{i} - 2xy\,\mathbf{j} + xz\,\mathbf{k}$$

Solution. div $\mathbf{F} = z^3 - 2x + x$; $(\text{div. }\mathbf{F})_{(1,\,1,\,1)} = 1 - 2 + 1 = 0$

$$\text{curl }\mathbf{F} = \begin{vmatrix} \mathbf{i} & \mathbf{j} & \mathbf{k} \\ \dfrac{\partial}{\partial x} & \dfrac{\partial}{\partial y} & \dfrac{\partial}{\partial z} \\ xz^3 & -2xy & xz \end{vmatrix}$$

$$= (0)\,\mathbf{i} - (z - 3xz^2)\,\mathbf{j} + (-2y)\,\mathbf{k}$$

$$(\text{curl }\mathbf{F})_{\text{at }(1,\,1,\,1)} = 2\mathbf{j} - 2\mathbf{k}$$

Example 68. *Find* λ *if* $(2x + y)\,\mathbf{i} + (z - \lambda y)\,\mathbf{j} + (2\lambda z - x)\,\mathbf{k}$ *is solenoidal.*

Solution. div $\mathbf{F} = \dfrac{\partial}{\partial x}(2x + y) + \dfrac{\partial}{\partial y}(z - \lambda y) + \dfrac{\partial}{\partial z}(2\lambda z - x) = 0$

$\therefore\quad 2 - \lambda + 2\lambda = 0 \quad \therefore\ \lambda = -2$

Example 69. *A vector field is given by* $\mathbf{F} = (x^2 - y^2 + x)\,\mathbf{i} - (2xy + y)\,\mathbf{j}$. *Show that the field is irrotational and find its scalar potential. Hence, evaluate the line integral from (1, 2) to (2, 1).*

(BR. Nov., 1999)

Solution. curl $\mathbf{F} = \sum\left(\dfrac{\partial F_3}{\partial y} - \dfrac{\partial F_2}{\partial z}\right)\mathbf{i}$

Here, $\quad F_1 = (x^2 - y^2 + x),\ F_2 = -(2xy + y),\ F_3 = 0$

$$\text{curl }\mathbf{F} = (0 - 0)\,\mathbf{i} + (0 - 0)\,\mathbf{j} + (-2y + 2y)\,\mathbf{k}$$
$$= 0$$

$\therefore\quad$ **F** is irrotational or conservative.

Let $\quad \mathbf{F} = \nabla\phi = \Sigma\,\mathbf{i}\dfrac{\partial\phi}{\partial x} = (x^2 - y^2 + x)\,\mathbf{i} - (2xy + y)\,\mathbf{j}$

$\therefore\quad \dfrac{\partial\phi}{\partial x} = x^2 - y^2 + x;\ \phi = \dfrac{x^3}{3} - xy^2 + \dfrac{x^2}{2} + f_1(y,\ z)$

$\dfrac{\partial\phi}{\partial y} = -2xy - y;\ \phi = -xy^2 - \dfrac{y^2}{2} + f_2(x,\ z)$

$\dfrac{\partial\phi}{\partial z} = 0;\ \phi = f_3(x,\ y)$

$$\therefore \qquad \phi = \frac{x^3}{3} - xy^2 + \frac{x^2}{2} - \frac{y^2}{2} + C$$

$$\int_A^B \mathbf{F} \cdot d\mathbf{r} = \phi(B) - \phi(A)$$

$$= \phi(2, 1) - \phi(1, 2)$$

$$= \left(\frac{8}{3} - 2 + 2 - \frac{1}{2} + C\right) - \left(\frac{1}{3} - 4 + \frac{1}{2} - 2 + C\right)$$

$$= 22/3$$

Example 70. *Find a and b so that the surfaces $ax^2 - byz = (a+2)x$ is orthogonal to $4x^2y + z^3 = 4$ at (1, – 1, 2).* ***(BR. April, 1999)***

Solution. Let $\phi_1 = ax^2 - byz - (a+2)x = 0$

$\phi_2 = 4x^2y + z^3 - 4 = 0$ be the two surfaces.

$$\nabla\phi_1 = \nabla(ax^2 - byz - ax - 2x)$$

$$= (2ax - a - 2)\,\mathbf{i} + (-bz)\,\mathbf{j} + (-by)\,\mathbf{k}$$

$$\nabla\phi_1(1, -1, 2) = (a-2)\,\mathbf{i} + (-2b)\,\mathbf{j} + b\mathbf{k} \qquad \text{...(1)}$$

$$\nabla\phi_2 = \nabla(4x^2y + z^3 - 4) = (8xy)\,\mathbf{i} + (4x^2)\,\mathbf{j} + (3z^2)\,\mathbf{k}$$

$$\nabla\phi_2(1, -1, 2) = -8\mathbf{i} + 4\mathbf{j} + 12\mathbf{k} \qquad \text{...(2)}$$

Since surfaces are orthogonal at the point $\nabla\phi_1 \cdot \nabla\phi_2 = 0$

$-8(a-2) - 8b + 12b = 0$, *i.e.*, $b - 2a + 4 = 0$ is the condition. But the point (1, – 1, 2) lies on both surfaces.

Substituting in the first surface equation,

$$a + 2b - a - 2 = 0 \quad \therefore \quad b = 1; \text{ using } b - 2a + 4 = 0$$

we get $a = 5/2$.

Example 71. *Evaluate the line integral* $\int_C (x^2 + xy)\,dx + (x^2 + y^2)\,dy$ *where C is the square formed by the lines $y = \pm 1$ and $x = \pm 1$.*

(BR. April, 1999)

Solution.

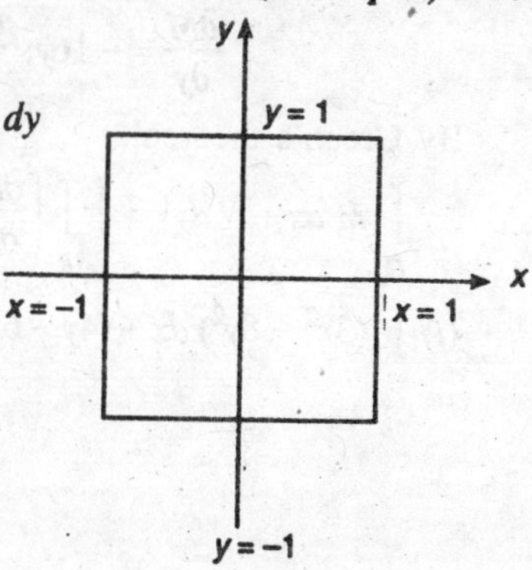

$$\int_C [M\,dx + N\,dy] = \iint_R \left(\frac{\partial N}{\partial x} - \frac{\partial M}{\partial y}\right) dx\,dy$$

$$\therefore \quad \int_C [(x^2 + xy)\,dx + (x^2 + y^2)\,dy]$$

$$= \iint_R (2x - x)\,dx\,dy$$

$$= \iint_R x\,dx\,dy$$

$$= \int_{-1}^{1} \int_{-1}^{1} x\, dx\, dy$$

$$= \left(\int_{-1}^{1} x\, dx \right) \left(\int_{-1}^{1} dy \right) = 0$$

Example 72. *Prove* $\mathbf{F} = (y^2 \cos x + z^3)\,\mathbf{i} + (2y \sin x - 4)\,\mathbf{j} + 3xz^2\,\mathbf{k}$ *is irrotational. Find* ϕ *so that* $\nabla\phi = \mathbf{F}$. **(BR. April, 1999)**

Solution. curl $\mathbf{F} = \sum \left(\dfrac{\partial f_3}{\partial y} - \dfrac{\partial f_2}{\partial z} \right) \mathbf{i}$

$$= (0 - 0)\,\mathbf{i} + (3z^2 - 3z^2)\,\mathbf{j} + (2y \cos x - 2y \cos x)\,\mathbf{k}$$

$$= \mathbf{0}$$

Hence, $\mathbf{F}$ is irrotational.

$$\mathbf{F} = \nabla\phi = \Sigma\, \mathbf{i} \frac{\partial \phi}{\partial x} \text{ implies}$$

$$\frac{\partial \phi}{\partial x} = y^2 \cos x + z^3;\ \phi = \int (y^2 \cos x + z^3)\, \partial x$$

$$= y^2 \sin x + xz^3 + f_1(y, z)$$

$$\frac{\partial \phi}{\partial y} = 2y \sin x - 4;\ \phi = y^2 \sin x - 4y + f_2(x, z)$$

$$\frac{\partial \phi}{\partial z} = 3xz^2;\ \phi = xz^3 + f_3(x, y)$$

Hence, $\phi = y^2 \sin x + xz^3 - 4y + C$

Example 73. *Evaluate* $\int_C (3x^2 - 8y^2)\, dx + (4y - 6xy)\, dy$ *where C is the*

(*i*) *boundary of the square bounded by lines* $x = 0,\ x = 1,\ y = 0,\ y = 1$.

(*ii*) *boundary of the region bounded by* $y = x^2$ *and* $x = y^2$.

Solution. Here, $M = 3x^2 - 8y^2,\ N = 4y - 6xy$

$$\frac{\partial M}{\partial y} = -16y;\ \frac{\partial N}{\partial x} = -6y$$

By Green's theorem,

$$\int_C [M\, dx + N\, dy] = \int\int_R \left(\frac{\partial N}{\partial x} - \frac{\partial M}{\partial y} \right) dx\, dy$$

(*i*) $\displaystyle\int_C [(3x^2 - 8y^2)\, dx + (4y - 6xy)\, dy] = \int\int_R (-6y + 16y)\, dx\, dy$

$$= 10 \int_0^1 \int_0^1 y\, dx\, dy$$

$$= 10\left(\int_0^1 y\,dy\right)\left(\int_0^1 dx\right)$$

$$= 10\left(\frac{1}{2}\right)(1) = 5.$$

(*ii*) Point of intersection of two parabolas is (1, 1).

$$\int_C [(3x^2 - 8y^2)\,dx + (4y - 6xy)\,dy]$$

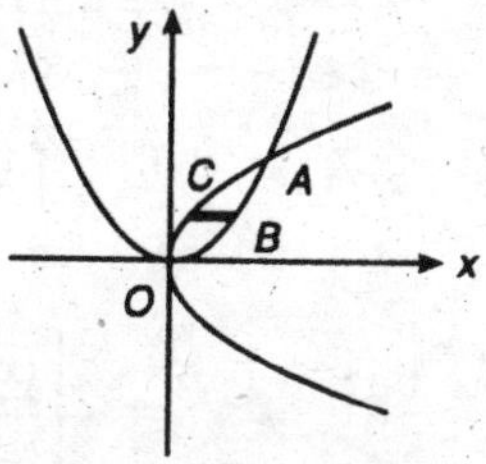

$$= \int\int_R 10y\,dx\,dy$$

$$= 10\int_0^1 \int_{y^2}^{\sqrt{y}} y\,dx\,dy$$

$$= 10\int_0^1 (x)_{y^2}^{\sqrt{y}}\, y\,dy$$

$$= 10\int_0^1 (\sqrt{y} - y^2)\,y\,dy$$

$$= 10\left[\frac{2}{5}y^{5/2} - \frac{y^4}{4}\right]_0^1$$

$$= 10\left(\frac{2}{5} - \frac{1}{4}\right) = \frac{3}{2}$$

Example 74. *Evaluate* $\int\int_S \mathbf{F}\cdot\mathbf{n}\,ds$ *where*

$\mathbf{F} = (2x^2 - 3z)\,\mathbf{i} + 2y\,\mathbf{j} - 4xz\,\mathbf{k}$ *where S is the surface of the solid bounded by planes* $x = 0,\ y = 0,\ z = 0$ *and* $2x + 2y + z = 4$.

Solution. div. $\mathbf{F} = 4x + 2 - 4x = 2$

$$\int\int_S \mathbf{F}\cdot\mathbf{n}\,dx = \int\int\int_V \text{div. }\mathbf{F}\,dv$$

$$= \int_0^2 \int_0^{2-x} \int_0^{4-2x-2y} 2\,dx\,dy\,dz$$

$$= 2\int_0^2 \int_0^{2-x} (z)_0^{4-2x-2y}\,dy\,dx$$

$$= 2\int_0^2\int_0^{2-x}(4-2x-2y)\,dy\,dx$$

$$= 4\int_0^2\left(2y - xy - \frac{y^2}{2}\right)_0^{2-x} dx$$

$$= 4\int_0^2\left[2(2-x) - x(2-x) - \frac{1}{2}(2-x)^2\right] dx$$

$$= 4\int_0^2 \frac{(2-x)^2}{2}\,dx$$

$$= 2\left[\frac{(x-2)^3}{3}\right]_0^2 = \frac{16}{3}$$

Example 75. *Evaluate* $\iint_S \mathbf{F}\cdot\mathbf{n}\,ds$ *for* $\mathbf{F} = 4x\,\mathbf{i} - 2y^2\,\mathbf{j} + z^2\,\mathbf{k}$ *taken over the surface region bounded by cylinder* $x^2+y^2=4$ *and* $z=0,\ z=3.$ **(MKU., 1996)**

Solution. The surface region of integration is shown in the figure.

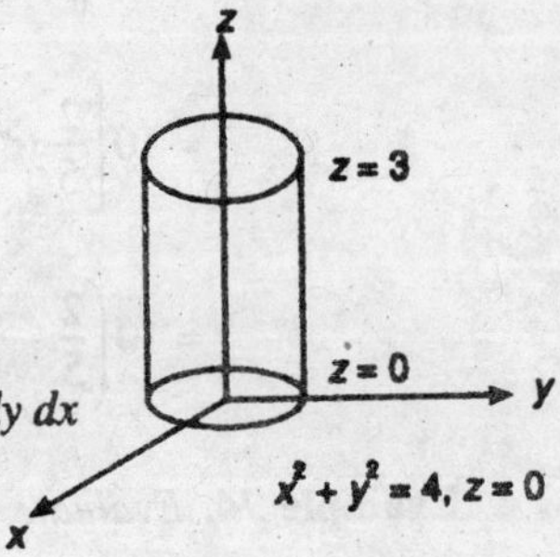

$$\text{div}\,\mathbf{F} = (4 - 4y + 2z)$$

$$\iint_S \mathbf{F}\cdot\mathbf{n}\,ds = \iiint_V \text{div.}\,\mathbf{F}\,dv$$

$$= \int_{-2}^{2}\int_{-\sqrt{4-x^2}}^{\sqrt{4-x^2}}\int_0^3 (4-4y+2z)\,dz\,dy\,dx$$

$$= \int_{-2}^{2}\int_{-\sqrt{4-x^2}}^{\sqrt{4-x^2}} (4z - 4yz + z^2)_0^3\,dy\,dx$$

$$= \int_{-2}^{2}\int_{-\sqrt{4-x^2}}^{\sqrt{4-x^2}} (12 - 12y + 9)\,dy\,dx$$

$$= \int_{-2}^{2}\int_{-\sqrt{4-x^2}}^{\sqrt{4-x^2}} (21 - 12y)\,dy\,dx$$

$$= \int_{-2}^{2}\int_{-\sqrt{4-x^2}}^{\sqrt{4-x^2}} 21\,dy\,dx \qquad \text{since } \int_{-a}^{a} y\,dy = 0$$

$$= 21 \times 2 \int_{-2}^{2} \int_{0}^{\sqrt{4-x^2}} dy\, dx$$

$$= 42 \int_{-2}^{2} (y)_0^{\sqrt{4-x^2}}\, dx$$

$$= 42 \int_{-2}^{2} \sqrt{4-x^2}\, dx$$

$$= 84 \int_{0}^{2} \sqrt{4-x^2}\, dx$$

$$= 84 \left[\frac{4}{2} \sin^{-1} \frac{x}{2} + \frac{x}{2} \sqrt{4-x^2}\right]_0^2$$

$$= 84 \left[2 \cdot \frac{\pi}{2}\right] = 84\, \pi$$

Example 76. *Prove* $\mathbf{F} = (e^x z - 2xy)\,\mathbf{i} - (x^2 - 1)\,\mathbf{j} + (e^x + z)\,\mathbf{k}$ *is conservative. Find* ϕ *so that* $\nabla\phi = \mathbf{F}$.

Hence, find $\int_C \mathbf{F} \cdot d\mathbf{r}$ *where ends of the curve C are A (0, 1, − 1) and B (2, 3, 0). Also find the work done by* **F** *in moving a particle from A to B.* **(MS. April, 1994)**

Solution. $\text{curl } \mathbf{F} = \sum \left(\frac{\partial F_3}{\partial y} - \frac{\partial F_2}{\partial z}\right) \mathbf{i}$

$$= (0 - 0)\,\mathbf{i} + (e^x - e^x)\,\mathbf{j} + (-2x + 2x)\,\mathbf{k}$$

$$= \mathbf{0}$$

$\therefore$ **F** is conservative.

$$\mathbf{F} = \nabla\phi = \Sigma\, \mathbf{i} \frac{\partial \phi}{\partial x}$$

$\therefore$ $\frac{\partial \phi}{\partial x} = e^x z - 2xy \quad \therefore \quad \phi = e^x z - x^2 y + f_1(y, z)$

$$\frac{\partial \phi}{\partial y} = -x^2 + 1 \quad \therefore \quad \phi = -x^2 y + y + f_2(x, z)$$

$$\frac{\partial \phi}{\partial z} = e^x + z \quad \therefore \quad \phi = z e^x + \frac{z^2}{2} + f_3(y, x)$$

$$\therefore \quad \phi = z e^x - x^2 y + y + \frac{z^2}{2} + C$$

$$\int_A^B \mathbf{F}\cdot d\mathbf{r} = \phi(B) - \phi(A)$$

$$= \left(ze^x - x^2y + y + \frac{z^2}{2} + C\right)_{(0,1,-1)}^{(2,3,0)}$$

$$= (-12 + 3 + C) - \left(-1 + 1 + \frac{1}{2} + C\right)$$

$$= -\frac{19}{2}$$

Work done by $\mathbf{F} = \int_A^B \mathbf{F}\cdot d\mathbf{r} = -\frac{19}{2}$

Example 77. *Evaluate* $\int\int_S \mathbf{F}\cdot\mathbf{n}\, ds$ *where* $\mathbf{F} = (x + 2y^2)\,\mathbf{i} - 2x\,\mathbf{j} + 2yz\,\mathbf{k}$ *where S is the surface of the plane* $2x + y + 2z = 6$ *in the first octant.*

Solution. Let $\phi = 2x + y + 2z - 6 = 0$

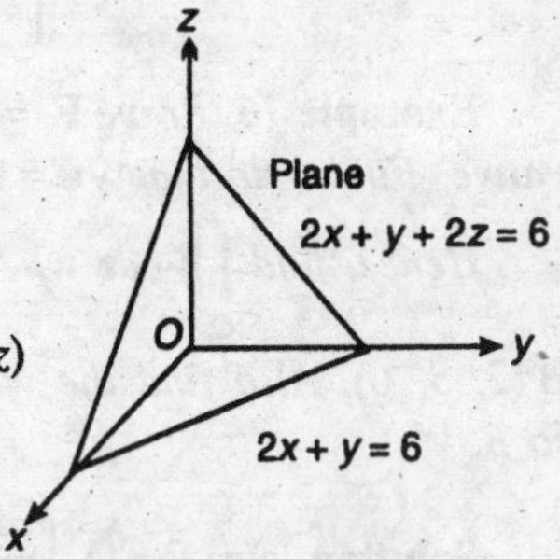

$$\nabla\phi = 2\mathbf{i} + \mathbf{j} + 2\mathbf{k}$$

$$\hat{\mathbf{n}} = \frac{\nabla\phi}{|\nabla\phi|} = \frac{1}{3}(2\mathbf{i} + \mathbf{j} + 2\mathbf{k})$$

$$\mathbf{F}\cdot\mathbf{n} = \frac{1}{3}[2(x + 2y^2) - 2x + 4yz)$$

$$= \frac{2}{3}(2y^2 + 2yz)$$

$$= \frac{2}{3}\left[2y^2 + 2y\left(\frac{6 - 2x - y}{2}\right)\right] \quad \text{since } 2x + y + 2z = 6$$

$$= \frac{2}{3}(2y^2 + 6y - 2xy - y^2)$$

$$= \frac{2}{3}(y^2 + 6y - 2xy)$$

Projecting dS on $z = 0$ plane,

$$dS = \frac{dx\,dy}{|\mathbf{n}\cdot\mathbf{k}|} = \frac{dx\,dy}{2/3}$$

$$\therefore \int\int_S \mathbf{F}\cdot\mathbf{n}\,ds = \frac{2}{3}\int\int_{z=0}(y^2 + 6y - 2xy)\left(\frac{3}{2}\,dx\,dy\right)$$

$$= \int\int_{z=0}(y^2 + 6y - 2xy)\,dx\,dy$$

$$= \int_0^3 \int_0^{6-2x} (y^2 + 6y - 2xy)\, dx\, dy$$

$$= \int_0^3 \left(\frac{y^3}{3} + 3y^2 - xy^2 \right)_0^{6-2x} dx$$

$$= \int_0^3 \left\{ y^2 \left[\frac{y}{3} + 3 - x \right] \right\}_0^{6-2x} dx$$

$$= \int_0^3 \frac{1}{3} (6 - 2x)^2 (15 - 5x)\, dx$$

$$= \frac{20}{3} \int_0^3 (3 - x)^3\, dx$$

$$= \frac{20}{3} \left[\frac{(3-x)^4}{-4} \right]_0^3 = 135$$

Example 78. *Evaluate* $\iint_S \mathbf{F} = \mathbf{n}\, ds$ *where* $\mathbf{F} = z\mathbf{i} + x\mathbf{j} - y^2 z\mathbf{k}$ *and S is the curved surface of the cylinder* $x^2 + y^2 + 1 = 0$ *included in the first octant between the planes* $z = 0$ *and* $z = 2$.

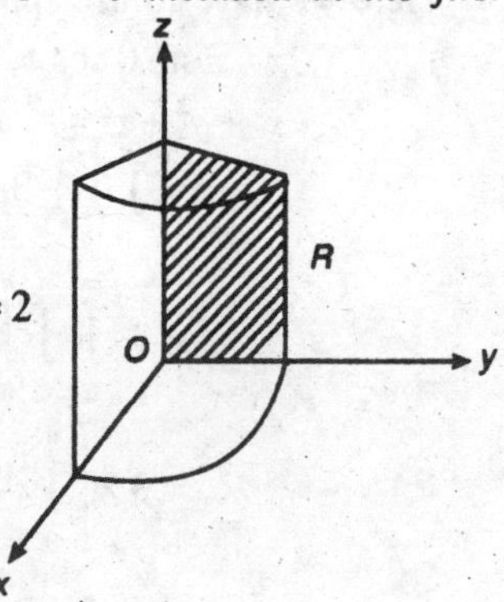

Solution. Curved surface equation is

$$\phi = x^2 + y^2 - 1 = 0$$

$$\nabla\phi = 2x\mathbf{i} + 2y\mathbf{j}$$

$$|\nabla\phi| = 2\sqrt{x^2 + y^2} = 2 \times 1 = 2$$

Hence on the curved surface

$$\mathbf{n} = \frac{\nabla\phi}{|\nabla\phi|} = x\mathbf{i} + y\mathbf{j}$$

$$\mathbf{F} \cdot \mathbf{n} = xz + xy$$

$$I = \iint_S \mathbf{F} \cdot \mathbf{n}\, ds = \iint_R x(y + z) \frac{dy\, dz}{|\mathbf{n} \cdot \mathbf{i}|}$$

$$= \iint_R x(y + z) \frac{dy\, dz}{x}$$

$$= \iint_R (y + z)\, dy\, dz$$

where R is the projected region of curved surface on yz plane which is a rectangle.

$$I = \int_0^2 \int_0^1 (y + z)\, dy\, dz$$

$$= \int_0^2 \left(\frac{y^2}{2} + zy\right)_0^1 dz$$

$$= \int_0^2 \left(z + \frac{1}{2}\right) dz = \left(\frac{z^2}{2} + \frac{1}{2}z\right)_0^2 = 3$$

Example 79. *Use Gauss divergence theorem to find* $\iint_S \mathbf{F} \cdot \mathbf{n}\, ds$ *where* $\mathbf{F} = x^3\,\mathbf{i} + y^3\,\mathbf{j} + z^3\,\mathbf{k}$ *where* S *is the surface of the sphere* $x^2 + y^2 + z^2 = a^2$.

Solution. div. $\mathbf{F} = \frac{\partial}{\partial x}(x^3) + \frac{\partial}{\partial y}(y^3) + \frac{\partial}{\partial z}(z^3) = 3\,(x^2 + y^2 + z^2)$

$$I = \iint_S \mathbf{F} \cdot \mathbf{n}\, ds = \iiint_V \text{div.}\mathbf{F}\, dv$$

$$= \iiint_V 3\,(x^2 + y^2 + z^2)\, dx\, dy\, dz \qquad ...(1)$$

Here we have to use spherical polar coordinates

$$x = r \sin\theta \cos\phi,\ y = r \sin\theta \sin\phi,\ z = r\cos\theta$$

$$x^2 + y^2 + z^2 = r^2 \text{ and } dx\, dy\, dz = r^2 \sin\theta\, dr\, d\theta\, d\phi$$

$$\therefore \qquad I = \iiint_V 3r^2 \cdot r^2 \sin\theta\, dr\, d\theta\, d\phi$$

$$= 3 \int_0^{2\pi} \int_0^{\pi} \int_0^{a} r^4 \sin\theta\, dr\, d\theta\, d\phi$$

$$= 3 \times \left(\int_0^a r^4\, dr\right)\left(\int_0^{\pi} \sin\theta\, d\theta\right)\left(\int_0^{2\pi} d\phi\right)$$

$$= 3\frac{a^5}{5} \cdot (-\cos\theta)_0^{\pi}\, (\phi)_0^{2\pi}$$

$$= \frac{3}{5} a^5\, (2)\, (2\pi) = \frac{12}{5}\pi a^5$$

Example 80. *Verify Stoke's theorem for* $\mathbf{F} = xy\mathbf{i} + xy^2\mathbf{j}$ *where the boundary* C *is the square whose vertices are* $(1, 0)$, $(0, 1)$, $(-1, 0)$ *and* $(0, -1)$.

Solution. $\mathbf{F}\cdot d\mathbf{r} = F_1 dx + F_2 dy + F_3 dz$

$$= xy\,dx + xy^2 dy$$

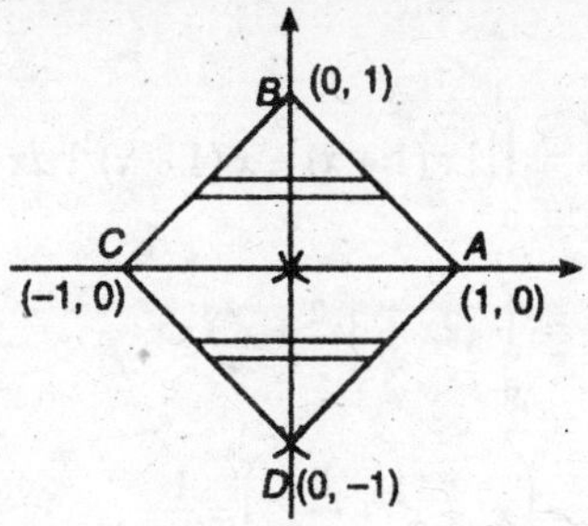

$$\text{curl } \mathbf{F} = \begin{vmatrix} \mathbf{i} & \mathbf{j} & \mathbf{k} \\ \frac{\partial}{\partial x} & \frac{\partial}{\partial y} & \frac{\partial}{\partial z} \\ xy & xy^2 & 0 \end{vmatrix} = (y^2 - x)\,\mathbf{k}$$

For this figure, $\mathbf{n} = \mathbf{k}$

$\therefore$ $\quad$ $\text{curl } \mathbf{F}\cdot\mathbf{n}\,ds = (y^2 - x)\,\mathbf{k}\cdot\mathbf{n}\,ds = (y^2 - x)\,dx\,dy$

since $\quad$ $\mathbf{k}\cdot\mathbf{n}\,ds = dx\,dy$

we have to prove $\int_C \mathbf{F}\cdot d\mathbf{r} = \int\int_S \text{curl } \mathbf{F}\cdot\mathbf{n}\,ds$ ...(1)

First, we will evaluate the left hand side of (1).

$$\int_C \mathbf{F}\cdot d\mathbf{r} = \int_C (xy\,dx + xy^2\,dy)$$

$$= \int_{AB} + \int_{BC} + \int_{CD} + \int_{DA}$$

$$= I_1 + I_2 + I_3 + I_4 \qquad ...(2)$$

Equation of AB is $\frac{x}{1} + \frac{y}{1} = 1$. That is, $x + y = 1$ $\therefore$ $dy = -dx$; $x = 1 - y$

Equation of BC is $-x + y = 1$ $\therefore$ $dy = dx$; $y = 1 + x$; $x = y - 1$

Equation of CD is $-x - y = 1$ $\therefore$ $dy = -dx$; $y = -1 - x$; $x = -1 - y$

Equation of DA is $x - y = 1$ $\therefore$ $dy = dx$; $y = x - 1$; $x = y + 1$

$$I_1 = \int_{AB} (xy\,dx + xy^2\,dy)$$

$$= \int_1^0 [x(1-x)\,dx - x(1-x)^2\,dx]$$

$$= \int_1^0 (x^2 - x^3)\, dx = \left(\frac{x^3}{3} - \frac{x^4}{4}\right)_1^0 = -\frac{1}{12}$$

$$I_2 = \int_{BC} = \int_0^{-1} [x(1+x) + x(1+x)^2]\, dx$$

$$= \int_0^{-1} (2x + 3x^2 + x^3)\, dx$$

$$= \left(x^2 + x^3 + \frac{1}{4}x^4\right)_0^{-1} = \frac{1}{4}$$

$$I_3 = \int_{CD} = \int_{-1}^0 (xy\, dx + xy^2\, dy)$$

$$= \int_{-1}^0 [x(-1-x)\, dx - x(1+x)^2\, dx]$$

$$= \int_{-1}^0 (-2x - 3x^2 - x^3)\, dx$$

$$= \left[x^2 + x^3 + \frac{x^4}{4}\right]_0^{-1} = \frac{1}{4}$$

$$I_4 = \int_{DA} = \int_0^1 (xy\, dx + xy^2\, dy)$$

$$= \int_0^1 [x(x-1) + x(x-1)^2]\, dx$$

$$= \int_0^1 (x^3 - x^2)\, dx = \left(\frac{x^4}{4} - \frac{x^3}{3}\right)_0^1 = -\frac{1}{12}$$

$$\therefore \quad I_1 + I_2 + I_3 + I_4 = -\frac{1}{12} + \frac{1}{4} + \frac{1}{4} - \frac{1}{12} = \frac{1}{3}$$

Secondly, we will evaluate R.H.S. of (1)

$$\iint_S \text{curl}\,\mathbf{F} \cdot \mathbf{n}\, ds = \iint_R (y^2 - x)\, dx\, dy$$

$$= \iint_{\substack{\text{over}\\ \Delta ABC}} + \iint_{\substack{\text{over}\\ \Delta ACD}} = J_1 + J_2 \qquad \text{...(3)}$$

$$J_1 = \iint_{\Delta ABC}$$

$$= \int_0^1 \int_{y-1}^{1-y} (y^2 - x)\, dx\, dy$$

$$= \int_0^1 \left[y^2 x - \frac{x^2}{2} \right]_{y-1}^{1-y} dy$$

$$= \int_0^1 (2y^2 - 2y^3)\, dy = \left(\frac{2y^3}{3} - \frac{y^4}{2} \right)_0^1 = \frac{1}{6}$$

$$J_2 = \iint_{\Delta ACD}$$

$$= \int_{-1}^{0} \int_{-1-y}^{1+y} (y^2 - x)\, dx\, dy$$

$$= \int_{-1}^{0} \left[y^2 x - \frac{x^2}{2} \right]_{-1-y}^{1+y} dy$$

$$= \int_{-1}^{0} (2y^2 + 2y^3)\, dy = \left(\frac{2y^3}{3} + \frac{y^4}{2} \right)_{-1}^{0} = \frac{2}{3} - \frac{1}{2} = \frac{1}{6}$$

$$J_1 + J_2 = \frac{1}{6} + \frac{1}{6} = \frac{1}{3}$$

Hence, equation (1) is satisfied and verified.

Example 81. *Evaluate* $\int_C [f_1(x)\, dx + f_2(y)\, dy + f_3(z)\, dz]$ *by using Stoke's theorem where C is any closed curve.*

Solution. Assume any open surface S where boundary is C.

Let $\mathbf{f}(x, y, z) = f_1(x)\,\mathbf{i} + f_2(y)\,\mathbf{j} + f_3(z)\,\mathbf{k}$

$$\mathbf{f} \cdot d\mathbf{r} = [f_1(x)\,\mathbf{i} + f_2(y)\,\mathbf{j} + f_3(z)\,\mathbf{k}] \cdot (dx\mathbf{i} + dy\mathbf{j} + dz\mathbf{k})$$

$$= f_1(x)\, dx + f_2(y)\, dy + f_3(z)\, dz$$

$$\text{curl}\,\mathbf{f} = \begin{vmatrix} \mathbf{i} & \mathbf{j} & \mathbf{k} \\ \frac{\partial}{\partial x} & \frac{\partial}{\partial y} & \frac{\partial}{\partial z} \\ f_1(x) & f_2(y) & f_3(z) \end{vmatrix} = \mathbf{0}, \text{ by actual expansion.}$$

Therefore, by Stoke's theorem,

$$\int_C \mathbf{f} \cdot d\mathbf{r} = \int\int_S \text{curl}\, \mathbf{f} \cdot \mathbf{n}\, ds = 0 \text{ since curl } \mathbf{f} = \mathbf{0}$$

$$\therefore \quad \int_C (f_1\, dx + f_2\, dy + f_3\, dz) = 0$$

Example 82. *Evaluate (i)* $\int_C (\cos x\, dx + e^y\, dy - z^2\, dz)$ *where C is the circle* $x^2 + y^2 + z^2 = a^2;\ z = 0.$

(ii) $\int_C (e^x\, dx - \cos y\, dy + \sin z\, dz)$ *where C is the intersection of* $x^2 + y^2 = 9$ *and* $z = 5.$

(iii) $\int_C [(x^2 - x)\, dx + (y + \sin y)\, dy + e^z\, dz]$ *where C is the intersection of the plane* $x + y + z = 1$ *with coordinate planes.*

Solution. In all the above problems, using the result of the previous problem, the value of the integral is zero.

Example 83. *Evaluate* $\int_C (yz\, dx + zx dy + xy\, dz)$, *by using Stoke's theorem, where (i) C is the any closed curve. (ii) C is the curve of the intersection of* $x^2 + y^2 = 4$ *and* $x = y^2$.

Solution. This is same as example 19 in the chapter. We will see a different approach now.

Let $f(x, y, z) = yz\mathbf{i} + zx\mathbf{j} + xy\mathbf{k}$

$$\mathbf{f} \cdot d\mathbf{r} = (yz\mathbf{i} + zx\mathbf{j} + xy\mathbf{k}) \cdot (dx\mathbf{i} + dy\mathbf{j} + dz\mathbf{k})$$
$$= yz\, dx + zx\, dy + xy\, dz$$

$$\text{curl}\, \mathbf{f} = \begin{vmatrix} \mathbf{i} & \mathbf{j} & \mathbf{k} \\ \frac{\partial}{\partial x} & \frac{\partial}{\partial y} & \frac{\partial}{\partial z} \\ yz & zx & xy \end{vmatrix} = (x - x)\,\mathbf{i} - (y - y)\,\mathbf{j} + (z - z)\,\mathbf{k} = \mathbf{0}$$

$$\int_C (yz\, dx + zx\, dy + xy\, dz) = \int_C \mathbf{f} \cdot d\mathbf{r} = \int\int_S \text{curl}\, \mathbf{f} \cdot \mathbf{n}\, ds$$
$$= 0 \text{ since curl } \mathbf{f} = \mathbf{0}$$

where S is **any** open surface with the boundary C.

Hence, the result is zero in both the problems.

Example 84. *Verify Stoke's theorem if* $\mathbf{F} = y^2\mathbf{i} - x^2\mathbf{j} + (x + z)\,\mathbf{k}$ *where C is the boundary of the triangle formed by the vertices O (0, 0, 0), A (a, 0, 0), B (a, a, 0) and S is any open surface with this triangle as its boundary.*

Solution. Triangle OAB is on the plane $z = 0$, that is on xy plane.

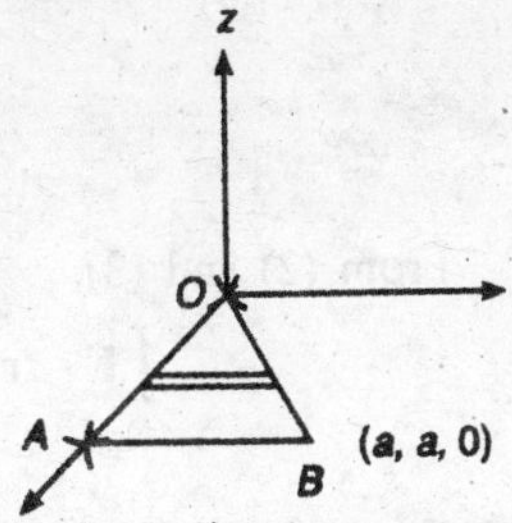

$$\mathbf{F} \cdot d\mathbf{r} = y^2\,dx - x^2\,dy + (x+z)\,dz$$

$$\int_C \mathbf{F} \cdot d\mathbf{r} = \int_{OA} + \int_{AB} + \int_{BO}$$

$$= I_1 + I_2 + I_3 \qquad ...(1)$$

$$I_1 = \int_{OA} [y^2\,dx - x^2\,dy + (x+z)\,dz]$$

$= 0$ since on OA, $y = 0$, $z = 0$, $dy = 0$, $dz = 0$

$$I_2 = \int_{AB} [y^2\,dx - x^2\,dy + (x+z)\,dz]$$

$= -\int_0^a a^2\,dy = -a^3$ since on AB, $x = a$, $z = 0$, $dx = 0$, $dz = 0$

$$I_3 = \int_{BO} y^2\,dx - x^2\,dy + (x+z)\,dz$$

$$= \int_a^0 (x^2\,dx - x^2\,dx) = 0$$

since on BO, $x = y$, $z = 0$, $dx = dy$, $dz = 0$

Therefore, $I_1 + I_2 + I_3 = -a^3$ ...(2)

$$\text{curl } \mathbf{F} = \begin{vmatrix} \mathbf{i} & \mathbf{j} & \mathbf{k} \\ \dfrac{\partial}{\partial x} & \dfrac{\partial}{\partial y} & \dfrac{\partial}{\partial z} \\ y^2 & -x^2 & x+z \end{vmatrix}$$

$$= -\mathbf{j} - 2(x+y)\,\mathbf{k}$$

$$\text{curl } \mathbf{F} \cdot \mathbf{n}\,ds = (-\mathbf{j} - 2(x+y)\,\mathbf{k}) \cdot \mathbf{k}\,dx\,dy$$

$$= -2(x+y)\,dx\,dy$$

Take the surface of triangle itself as S with the sides of triangle as C.

Therefore $\iint_S = \iint_{\Delta OAB} \text{curl } \mathbf{F} \cdot \mathbf{n}\,ds$

$$= -2\int_0^a \int_0^x (x+y)\,dx\,dy$$

$$= -2\int_0^a \left(xy + \frac{y^2}{2} \right)_0^x dx$$

$$= -2\int_0^a \left(\frac{3}{2}x^2 \right) dx$$

$$= -3\left[\frac{x^3}{3}\right]_0^a = -a^3 \qquad \text{...(3)}$$

From (2) and (3),

$$\int_C \mathbf{F}\cdot d\mathbf{r} = \int\int_S \text{curl}\,\mathbf{F}\cdot\mathbf{n}\,ds = -a^3$$

Example 85. *Verify Stoke's theorem for* $\mathbf{F} = -y\mathbf{i} + 2yz\mathbf{j} + y^2\mathbf{k}$ *where S is the upper half of the sphere* $x^2 + y^2 + z^2 = 1$ *and C is the circle* $x^2 + y^2 = 1,\ z = 0.$

Solution. If $\mathbf{n}$ is the unit normal vector, then

$$\mathbf{n} = \frac{\nabla\phi}{|\nabla\phi|} \text{ where } \phi\,(x,\ y,\ z) = x^2 + y^2 + z^2 - 1 = 0 \text{ is the surface.}$$

$$\nabla\phi = 2\,(x\mathbf{i} + y\mathbf{j} + z\mathbf{k})$$

$$|\nabla\phi| = 2\sqrt{x^2 + y^2 + z^2} = 2\sqrt{1} = 2$$

$$\therefore \quad \mathbf{n} = x\mathbf{i} + y\mathbf{j} + z\mathbf{k}$$

(Or, if $P\,(x,\ y,\ z)$ is any point on $x^2 + y^2 + z^2 - 1 = 0$,

$$\mathbf{n} = \frac{\overrightarrow{OP}}{|OP|} = \frac{x\mathbf{i} + y\mathbf{j} + z\mathbf{k}}{1} \text{ since } OP = \text{radius} = 1)$$

$$\mathbf{F}\cdot d\mathbf{r} = F_1 dx + F_2 dy + F_3 dz = -y\,dx + 2yz\,dy + y^2\,dz$$

on C, $x^2 + y^2 = 1$ and $z = 0$; $\therefore\ dz = 0$

$$\int_C \mathbf{F}\cdot d\mathbf{r} = \int_C (-y\,dx + 0\cdot dy + 0) = -\int_C y\,dx$$

$$= -\int_0^{2\pi} \sin\theta\,d\,(\cos\theta), \text{ taking } x = \cos\theta,\ y = \sin\theta \text{ on } x^2 + y^2 = 1$$

$$= \int_0^{2\pi} \sin^2\theta\,d\theta$$

$$= \int_0^{2\pi} \frac{1 - \cos 2\theta}{2}\,d\theta$$

$$= \frac{1}{2}\left[\theta - \frac{\sin 2\theta}{2}\right]_0^{2\pi} = \pi \qquad \text{...(1)}$$

Projecting dS on xy plane we get $dx\,dy = |\mathbf{n}\cdot\mathbf{k}|\,dS$

$$\text{curl } \mathbf{F} = \begin{vmatrix} \mathbf{i} & \mathbf{j} & \mathbf{k} \\ \dfrac{\partial}{\partial x} & \dfrac{\partial}{\partial y} & \dfrac{\partial}{\partial z} \\ -y & 2yz & y^2 \end{vmatrix} = \mathbf{k}, \text{ also } \mathbf{n}\cdot\mathbf{k} = z$$

$$\iint_S \text{curl } \mathbf{F}\cdot\mathbf{n}\, ds = \iint_S \mathbf{k}\cdot\mathbf{n}\, ds = \iint_R dx\, dy$$

$$= \text{area of the circle } x^2 + y^2 = 1 \text{ on } z = 0$$

$$= \pi \qquad \text{...(2)}$$

From (1) and (2),

$$\int_C \mathbf{F}\cdot d\mathbf{r} = \iint_S \text{curl } \mathbf{F}\cdot\mathbf{n}\, ds = \pi$$

Example 86. *Evaluate* $\iint_S \mathbf{F}\cdot\mathbf{n}\, ds$ *where* $\mathbf{F} = (x + y^2)\,\mathbf{i} - 2x\,\mathbf{j} + 2yz\,\mathbf{k}$ *and S is the surface of the plane* $2x + y + 2z = 6$ *in the first octant.*

Solution. Equation of surface is $\phi(x, y, z) = 2x + y + 2z - 6 = 0$

$$\nabla\phi = 2\mathbf{i} + \mathbf{j} + 2\mathbf{k}$$

$$\mathbf{n} = \frac{\nabla\phi}{|\nabla\phi|} = \frac{2\mathbf{i} + \mathbf{j} + 2\mathbf{k}}{\sqrt{9}} = \frac{2\mathbf{i} + \mathbf{j} + 2\mathbf{k}}{3};\ \mathbf{n}\cdot\mathbf{k} = \frac{2}{3}$$

$$\mathbf{F}\cdot\mathbf{n} = \frac{2}{3}(x + y^2) - \frac{2}{3}x + \frac{4}{3}yz$$

$$= \frac{2}{3}[y^2 + 2yz]$$

$$= \frac{2}{3}\left[y^2 + 2y\left(\frac{6 - 2x - y}{2}\right)\right] \quad \text{since on } S,\ 2x + y + 2z = 6$$

$$= \frac{2}{3}[y^2 + 6y - 2xy - y^2]$$

$$= \frac{4}{3}[3y - xy]$$

$$\iint_S \mathbf{F}\cdot\mathbf{n}\, ds = \frac{4}{3}\iint_R (3y - xy)\frac{dx\, dy}{|\mathbf{n}\cdot\mathbf{k}|} = \frac{4}{3}\iint_R (3y - xy)\frac{3}{2}\, dx\, dy$$

$$= 2\iint_R (3y - xy)\, dx\, dy$$

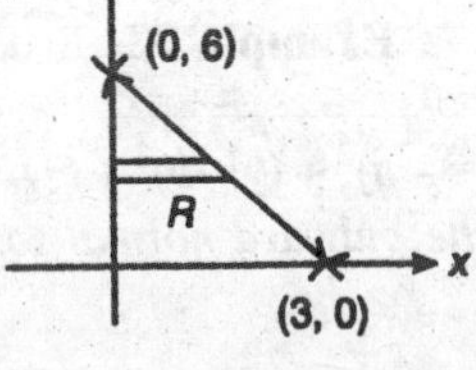

where R is the projection of S on xy plane, *i.e.*, the triangle on $z = 0$ enclosed by $x = 0$, $y = 0$ and $2x + y = 6$.

$$= 2\int_0^6 \int_0^{\frac{6-y}{2}} (3y - xy)\, dx\, dy$$

$$= 2\int_0^6 \left(3xy - \frac{x^2}{2}y\right)_0^{\frac{6-y}{2}} dy$$

$$= 2\int_0^6 y\left[3\left(\frac{6-y}{2}\right) - \frac{1}{2}\left(\frac{6-y}{2}\right)^2\right] dy$$

$$= 2\int_0^6 \left\{\frac{y(6-y)}{2}\left[3 - \frac{6-y}{4}\right]\right\} dy$$

$$= \frac{1}{4}\int_0^6 (36y - y^3)\, dy = \frac{1}{4}\left[18y^2 - \frac{y^4}{4}\right]_0^6 = 81$$

Example 87. *If* $\mathbf{F} = xy^2\boldsymbol{i} + yz^2\boldsymbol{j} + zx^2\boldsymbol{k}$ *find* $\int_S\int \boldsymbol{F}\,.\,\boldsymbol{n}\, ds$ *over the sphere* $x^2 + y^2 + z^2 = 1$

Solution. $\text{div}\ \overrightarrow{F} = y^2 + z^2 + x^2$

$$\int_S\int \boldsymbol{F}\,.\,\boldsymbol{n}\, ds = \iiint_V \text{div}\,\boldsymbol{F}\, dv$$

$$= \iiint_V (x^2 + y^2 + z^2)\, dxdydz$$

$$= \int_0^{2\pi}\int_0^{\pi}\int_0^1 r^2\,.\,r^2 \sin\theta\, dv\, d\theta\, d\phi$$

by putting $x = r\sin\theta\cos\phi$

$y = r\sin\theta\sin\phi$

$z = r\cos\theta$

$$= \left(\frac{r^5}{5}\right)_0^1 (-\cos\theta)_0^{\pi}\, (\phi)_0^{2\pi}$$

$$= \frac{1}{5} \times 2 \times 2\pi = \frac{4}{5}\pi$$

Example 88. *Evaluate* $\iint_S (lx^2 + my^2 + nz^2)\, ds$ *taken over the sphere* $(x-a)^2 + (y-b)^2 + (z-c)^2 = \rho^2$ *where l, m, n are the direction cosines of the outward normal to the sphere.*

Solution. Let $F = x^2 i + y^2 j + z^2 k$ and $n = li + mj + nk$

$$I = \iint_S (lx^2 + my^2 + nz^2)\, ds$$

$$= \iint_S F \, . \, n \, ds$$

$$= \iiint_V \text{div } F \, dv$$

$$= 2 \iiint_V (x + y + z)\, dv \qquad \ldots (i)$$

The centre of the sphere is (a, b, c). Any point on the sphere is $x = a + r\sin\theta\cos\phi$, $y = b + r\sin\theta\sin\phi$, $z = c + r\cos\theta$.

$$dx\, dy\, dz = r^2 \sin\theta \, dr\, d\theta\, d\phi$$

$$I = 2\int_0^{2\pi}\int_0^{\pi}\int_0^{\rho} [(a + b + c) + r\sin\theta\cos\phi + r\sin\theta\sin\phi + r\cos\theta] \times r^2 \sin\theta \, dx\, d\theta\, d\phi$$

$$= 2\,(a + b + c)\left(\frac{r^3}{3}\right)_0^{\rho} (-\cos\theta)_0^{\pi}\, (\phi)_0^{2\pi}$$

$$+ 2\left(\frac{r^4}{4}\right)_0^{\rho}\left(\int_0^{\pi} \sin^2\theta\, d\theta\right)(\sin\phi)_0^{2\pi}$$

$$+ 2\left(\frac{r^4}{4}\right)_0^{\rho}\left(\int_0^{\pi} \sin^2\theta\, d\theta\right)(-\cos\phi)_0^{2\pi} + \left(\frac{r^4}{4}\right)_0^{\rho}\left(-\frac{1}{2}\cos 2\theta\right)_0^{\pi}[\phi]_0^{2\pi}$$

$$= \frac{2}{3}(a + b + c) \, . \, \rho^3 \times 2 \times 2\pi = \frac{8}{3}\pi\,(a + b + c)\,\rho^3.$$

Example 89. *Verify Gauss theorem for* $\mathbf{F} = 4x\mathbf{i} - 2y^2\mathbf{j} + z^2\mathbf{k}$ *taken over the region bounded by* $x^2 + y^2 = 4$, $z = 0$ *and* $z = 3$.

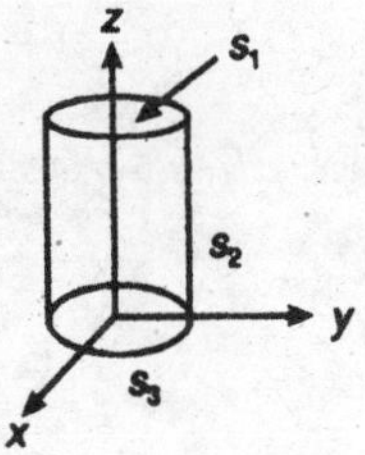

$$\text{div } \mathbf{F} = 4 - 4y + 2z$$

$$\iiint_V \text{div } F\, dv$$

$$= \int_0^3 \int_{-2}^{2} \int_{-\sqrt{4-x^2}}^{\sqrt{4-x^2}} (4 - 4y + 2z)\, dx\, dy\, dz$$

$$= \int_{-2}^{2} \int_{-\sqrt{4-x^2}}^{\sqrt{4-x^2}} (4z - 4yz + z^2)_0^3\, dxdy, \text{ integrating w.r.t.z}$$

$$= \int_{-2}^{2} \int_{-\sqrt{4-x^2}}^{\sqrt{4-x^2}} (12 - 12y + 9)\, dxdy$$

$$= 21 \int_{-2}^{2} \int_{-\sqrt{4-x^2}}^{\sqrt{4-x^2}} dxdy - 12 \int_{-2}^{2} \int_{-\sqrt{(4-x^2)}}^{\sqrt{4-x^2}} y\ dxdy$$

$$= 42 \int_{-2}^{2} \sqrt{4-x^2}\, dx - 0$$

$$= 42 \times 2 \int_{0}^{2} \sqrt{4-x^2}\, dx$$

$$= 84 \left[2 \sin^{-1}\left(\frac{x}{2}\right) + \frac{x}{2}\sqrt{4-x^2} \right]_0^2$$

$$= 168 \times \left[\frac{\pi}{2}\right] = 84\,\pi \qquad \ldots (1)$$

$$\iint_S \boldsymbol{F} \cdot \boldsymbol{n}\, ds = \iint_{S_1} + \iint_{S_2} + \iint_{S_3} \qquad \ldots (2)$$

on S_1, $z = 3$, $\boldsymbol{n} = \boldsymbol{k}$ $\quad \boldsymbol{F} \cdot \boldsymbol{n} = z^2 = 9$

$$\int_{S_1}\int \boldsymbol{F} \cdot \boldsymbol{n}\, ds = 9 \int_{S_1}\int dxdy = 9 \times \pi\,(2)^2 = 36\,\pi$$

on S_3, $z = 0$, $\boldsymbol{n} = -\boldsymbol{k}$, $\boldsymbol{F} \cdot \boldsymbol{n} = 0 \;\therefore\; \iint_{S_3} \boldsymbol{F} \cdot \boldsymbol{n}\, ds = 0.$

on S_2, $x^2 + y^2 = 4$, $\therefore\ x = 2\cos\theta, y = 2\sin\theta, ds = (2d\theta)\, dz = 2d\theta dz$

$$\boldsymbol{n} = \frac{\nabla\phi}{|\nabla\phi|} = \frac{2x\,\boldsymbol{i} + 2y\,\boldsymbol{j}}{2\sqrt{x^2+y^2}} = \frac{2x\boldsymbol{i} + 2y\boldsymbol{j}}{2 \times 2} = \frac{x}{2}\boldsymbol{i} + \frac{y}{2}\boldsymbol{j}$$

$\boldsymbol{F} \cdot \boldsymbol{n} = 2x^2 - y^3$

$$\int_{S_2}\int \boldsymbol{F} \cdot \boldsymbol{n}\, ds = \iint_{S_2} (2x^2 - y^3)\ 2d\theta dz \quad \text{where } x = 2\cos\theta, y = 2\sin\theta$$

$$= 2 \int_{S_2}\int (8\cos^2\theta - 8\sin^3\theta)\, d\theta\, dz$$

$$= 2 \left(\int_0^3 dz\right) \int_0^{2\pi} (8\cos^2\theta - 8\sin^3\theta) d\theta$$

$$= 16 \times 3 \times \int_0^{2\pi} \left[\frac{1+\cos 2\theta}{2} - \left(\frac{3}{4}\sin\theta - \frac{1}{4}\sin 3\theta \right) \right] d\theta$$

$$= 48 \left[\frac{1}{2}\theta + \frac{\sin 2\theta}{4} + \frac{3}{4}\cos\theta - \frac{1}{4}\frac{\cos 3\theta}{3} \right]_0^{2\pi}$$

$$= 48\,\pi$$

$$\therefore \quad \int_S\int \boldsymbol{F} \cdot \boldsymbol{n}\, ds = 36\,\pi + 0 + 48\,\pi = 84\,\pi = \iiint_V \text{div}\, \boldsymbol{F}\, dv$$

Hence the theorem is verified

Example 90. *Verify Green's theorem for* $\int_C \left[(xy + y^2)\,dx + x^2\,dy\right]$ *where C is the boundary of the common area between* $y = x^2$ *and* $y = x$.

Solution. Here $M = xy + y^2$

$$N = x^2$$

We have to verify

$$\int_C [Mdx + Ndy] = \iint_R \left(\frac{\partial N}{\partial x} - \frac{\partial M}{\partial y}\right) dxdy$$

i.e. $$\int_C [(xy + y^2)\,dx + x^2\,dy] = \int_R\int (x - 2y)\,dxdy$$

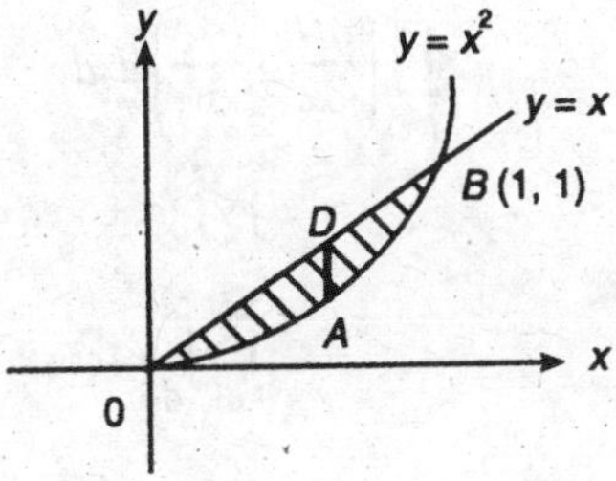

$$\text{R.H.S.} = \int_0^1 \int_{x^2}^{x} (x - 2y)\,dy\,dx$$

$$= \int_0^1 [xy - y^2]_{x^2}^{x}\,dx$$

$$= \int_0^1 (x^4 - x^3)\,dx$$

$$= \left(\frac{x^5}{5} - \frac{x^4}{4}\right)_0^1$$

$$= \frac{1}{5} - \frac{1}{4}$$

$$= -\frac{1}{20}$$

$$\text{LHS} = \int_C [(xy + y^2)\,dx + x^2\,dy]$$

$$= \underset{\substack{OAB \\ y = x^2}}{\int} + \underset{\substack{BDO \\ y = x}}{\int}$$

$$= \int_0^1 [(x^3 + x^4)\,dx + 2x^3\,dx] + \int_1^0 (2x^2\,dx + x^2 dx)$$

$$= \int_0^1 (3x^3 + x^4)\,dx + \int_1^0 3x^2\,dx$$

$$= \left(3\frac{x^4}{4} + \frac{x^5}{5}\right)_0^1 + (x^3)_1^0$$

$$= \frac{3}{4} + \frac{1}{5} - 1$$

$$= \frac{19}{20} - 1 = -\frac{1}{20}$$

Hence $\quad \text{LHS} = -\dfrac{1}{20} = \text{R.H.S.}$

$\therefore$ The theorem is verified to be true.

Example 91. *If C is the closed curve consisting of the upper semi circle $x^2 + y^2 = a^2$ together with the diameter on the x-axis from $-a$ to a, verify Green's theorem for* $\int_C [(2x^2 - y^2)\, dx + (x^2 + y^2)\, dy]$

Solution. Here $M = 2x^2 - y^2$, $N = x^2 + y^2$

$$\frac{\partial N}{\partial x} - \frac{\partial M}{\partial y} = 2x + 2y$$

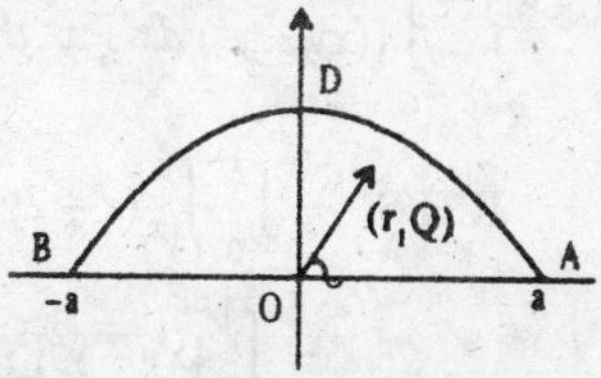

$$\iint_R \left(\frac{\partial N}{\partial x} - \frac{\partial M}{\partial y}\right) dxdy$$

$$= 2\int_R \int (x + y)\, dx\, dy$$

$$= 2\int_0^{\pi} \int_0^{a} r\,(\cos\theta + \sin\theta)\,(dr\ d\theta), \text{ changing to polar system}$$

$$= 2\left(\int_0^a r^2 dr\right)\left[\int_0^{\pi} (\cos\theta + \sin\theta)\, d\theta\right]$$

$$= 2.\left(\frac{r^3}{3}\right)_0^a [\sin\theta - \cos\theta]_0^{\pi} = \frac{2}{3}a^3 \times 2 = \frac{4}{3}a^3$$

$$\int_C Mdx + Ndy = \int_{BA} (Mdx + Ndy) + \int_{ADB} (Mdx + Ndx) = I_1 + I_2$$

on BA, $y = 0$; $\quad \therefore \quad I_1 = \int_{\substack{BA \\ y=0}} 2x^2\, dx = \left(\dfrac{2x^3}{3}\right)_{-a}^{a} = \dfrac{4}{3}a^3$

on ADB, $x = a\cos\theta$, $y = a\sin\theta$;

$$\therefore \quad I_2 = a^3 \int_{\theta=0}^{\pi} \left[(2\cos^2\theta - \sin^2\theta)(-\sin\theta\, d\theta) + \cos\theta\, d\theta\right]$$

$$= a^3 \int_0^{\pi} (3\sin^3\theta - 2\sin\theta + \cos\theta)\, d\theta$$

$$= a^3\left[2\int_0^{\pi/2} 3\sin^3\theta\, d\theta + (2\cos\theta + \sin\theta)_0^{\pi}\right]$$

$$= a^3\left[2 \times 3 \times \frac{2}{3} + (-4)\right] = 0$$

$$I_1 + I_2 = \frac{4}{3}a^3 + 0 = \frac{4}{3}a^3.$$

Hence $\int_C (Mdx + Ndy) = \frac{4}{3}a^3 = \iint_C \left(\frac{\partial N}{\partial x} - \frac{\partial M}{\partial y}\right) dxdy$

Example 92. *Evaluate* $\int_C [(x+y)\,dx + (2x-z)\,dy + (y+z)\,dz]$ *where C is the boundary of the triangle whose vertices are (2, 0, 0), (0, 3, 0) and (0, 0, 6)*

Solution. Let $\boldsymbol{F} = (x+y)\,\boldsymbol{i} + (2x-z)\,\boldsymbol{j} + (y+z)\,\boldsymbol{k}$

Then curl $\boldsymbol{F} = \begin{vmatrix} \boldsymbol{i} & \boldsymbol{j} & \boldsymbol{k} \\ \frac{\partial}{\partial x} & \frac{\partial}{\partial y} & \frac{\partial}{\partial z} \\ x+y & 2x-z & y+z \end{vmatrix}$

$= 2\boldsymbol{i} + \boldsymbol{k}$

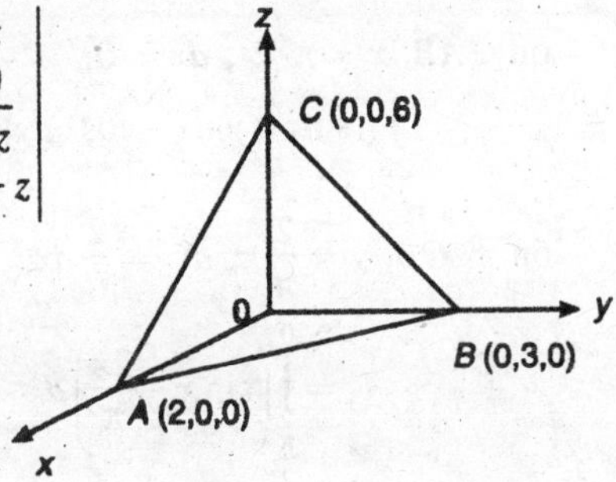

Equation of the plane *ABC* is

$$\frac{x}{2} + \frac{y}{3} + \frac{z}{6} = 1$$

i.e.

$$\phi = 3x + 2y + z - 6 = 0$$

$$\nabla\phi = 3\boldsymbol{i} + 2\boldsymbol{j} + \boldsymbol{k}$$

$$\boldsymbol{n} = \frac{\nabla\phi}{|\nabla\phi|} = \frac{3\boldsymbol{i} + 2\boldsymbol{j} + \boldsymbol{k}}{\sqrt{14}}$$

$$\text{Curl } \boldsymbol{F}.\boldsymbol{n} = \frac{1}{\sqrt{14}}(6 + 0 + 1) = \frac{7}{\sqrt{14}}$$

$$\int_C [(x+y)\,dx + (2x-z)\,dy + (y+z)\,dz]$$

$$= \int_C \boldsymbol{F}.\,d\boldsymbol{r}$$

$$= \iint_S \text{curl}\,\boldsymbol{F}\,.\,\boldsymbol{n}\,ds$$

$$= \int_S\!\!\int \frac{7}{\sqrt{14}}.(\sqrt{14}\,dx\,dy) \text{ since } dxdy = \boldsymbol{K}\,.\,\boldsymbol{n}\,ds = \frac{1}{\sqrt{14}}ds$$

$$= 7 \times \text{area of } \Delta\, OAB$$

$$= 7 \times \frac{2 \times 3}{2} = 21$$

Example 93. *Evaluate* $\int_C [(\sin x - y)\,dx - \cos x dy]$ *where C is the triangle whose vertices are (0, 0),* $\left(\frac{\pi}{2}, 0\right)$ *and* $\left(\frac{\pi}{2}, 1\right)$

Solution. $\mathrm{I} = \int_C [(\sin x - y)\,dx - \cos x dy]$

$$= \int_{OA} + \int_{AB} + \int_{BO} = I_1 + I_2 + I_3$$

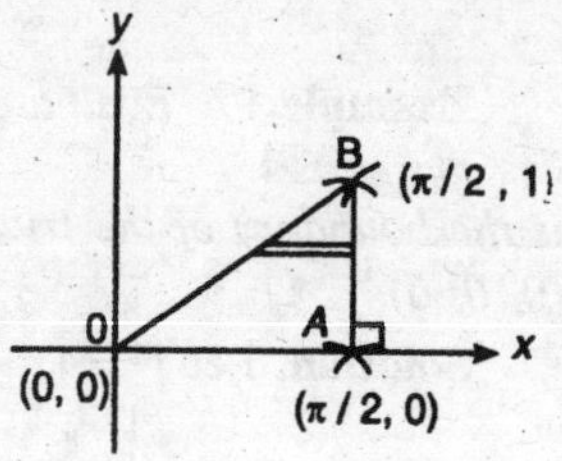

On OA, $y = 0$;

$$\int_{OA} [(\sin x - y)\, dx - \cos x dy]$$

$$= \int_0^{\pi/2} \sin x\, dx = (-\cos x)_0^{\pi/2} = 1$$

on AB, $x = \pi/2$; $dx = 0$;

$$\int_{AB} [(\sin x - y)dx - \cos x\, dy] = 0$$

on OB, $y = \frac{2}{\pi}x$; $dy = \frac{2}{\pi}dx$;

$$I_3 = \int_{\frac{\pi}{2}}^{0}\left(\sin x - \frac{2x}{\pi}\right)dx - \frac{2}{\pi}\int_{\frac{\pi}{2}}^{0} \cos x\, dx$$

$$= \left(\frac{x^2}{\pi} + \cos x\right)^{\pi/2} + \frac{2}{\pi}(\sin x)_0^{\pi/2}$$

$$= \frac{\pi}{4} - 1 + \frac{2}{\pi}$$

$$\therefore \quad I = I_1 + I_2 + I_3 = 1 + 0 + \frac{\pi}{4} + \frac{2}{\pi} - 1 = \frac{\pi}{4} + \frac{2}{\pi}$$

Example 94. *Verify Gauss divergence theorem for* $F = (2xy + z)i + y^2 j - (x + 3y)k$ *when taken over the region bounded by* $x = 0$, $y = 0$, $z = 0$ *and* $2x + 2y + z = 6$

Solution. div $F = \Sigma \frac{\partial}{\partial x} F_1 = 2y + 2y = 4y$

$$\iiint_V \text{div F dv}$$

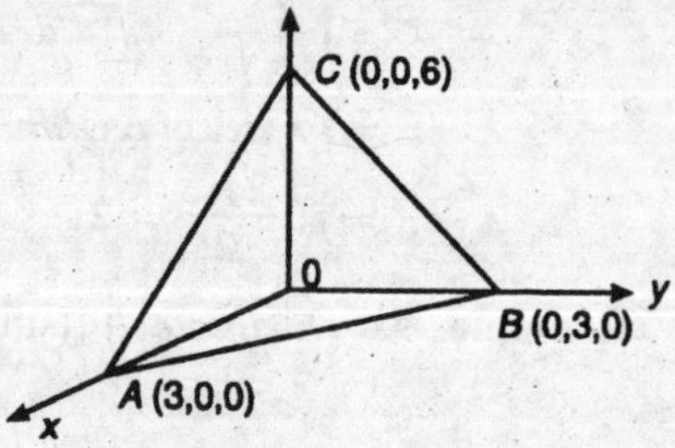

$$= 4\iiint_V y\, dz dy dx$$

$$= 4\int_0^3\int_0^{3-x}\int_0^{6-2x-2y} y\, dz\, dy\, dx$$

$$= 4\int_0^3\int_0^{3-x} y\,(z)_0^{6-2x-2y}\, dy\, dx$$

$$= 4\int_0^3\int_0^{3-x} y(6-2x-2y)\,dydx$$

$$= 8\int_0^3\int_0^{3-x} (3y-xy-y^2)\,dydx$$

$$= 8\int_0^3\left[(3-x)\frac{y^2}{2}-\frac{y^3}{3}\right]_0^{3-x} dx$$

$$= 8\int_0^3\left[\frac{1}{2}(3-x)^3-\frac{1}{3}(3-x)^3\right]dx$$

$$= \frac{4}{3}\int_0^3 (3-x)^3\,dx$$

$$= \frac{4}{3}\left[\frac{(3-x)^4}{(-4)}\right]_0^3$$

$$= -\frac{1}{3}(-81) = 27 \qquad \text{... (1)}$$

$$\iint_S \boldsymbol{F}\cdot\boldsymbol{n}\,ds = \iint_{BOC}+\iint_{COA}+\iint_{AOB}+\iint_{ABC} = I_1+I_2+I_3+I_4 \qquad \text{... (2)}$$

On BOC, $x = 0$, $\mathbf{n} = -i$, $ds = dy\,dz$; $\mathbf{F}\cdot\mathbf{n} = -z$

$$\therefore \qquad I_1 = -\iint_{BOC} z\,dy\,dz = -\int_0^3\int_0^{6-2y} z\,dz\,dy$$

$$= -\frac{1}{2}\int_0^3 (Z^2)_0^{6-2y}\,dy$$

$$= -\frac{1}{2}\int_0^3 (6-2y)^2\,dy$$

$$= -\frac{1}{2}\left[\frac{(6-2y)^3}{(-6)}\right]_0^3$$

$$= \frac{1}{12}(-216) = -18$$

On AOC, $y = 0$, $ds = dxdz$, $\boldsymbol{n} = -\boldsymbol{j}$, $\boldsymbol{F}\cdot\boldsymbol{n} = -y^2 = 0$

$$\therefore \qquad I_2 = \iint_{OAC} 0\,dxdz = 0$$

On OAB, $z = 0$, $\boldsymbol{n} = -\boldsymbol{k}$, $ds = dx\,dy$, $\boldsymbol{F}\cdot\boldsymbol{n} = x+3y$

$$I_3 = \iint_{z=0} (x+3y)\,dydx = \int_0^3\int_0^{3-x} (x+3y)\,dydx$$

$$= \int_0^3 \left(xy+\frac{3}{2}y^2\right)_0^{3-x} dx$$

$$= \int_0^3 (3-x)\,[x + \frac{3}{2}(3-x)]\,dx$$

$$= \frac{1}{2}\int_0^3 (3-x)(9-x)\,dx$$

$$= \frac{1}{2}\int_0^3 (27 + x^2 - 12x)\,dx$$

$$= \frac{1}{2}\left[27x + \frac{x^3}{3} - 6x^2\right]_0^3$$

$$= \frac{1}{2}[81 + 9 - 54] = 18$$

On ABC, $\boldsymbol{n} = \dfrac{\nabla\phi}{|\nabla\phi|} = \dfrac{2i + 2j + k}{\sqrt{9}} = \dfrac{2}{3}i + \dfrac{2}{3}j + \dfrac{1}{3}k$

$$\bar{F}\,.\,\bar{n}\,ds = \frac{1}{3}[2(2xy + z) + 2\,(y^2) - (x + 3y)]\,ds$$

$$= \frac{1}{3}(4xy + 2z + 2y^2 - x - 3y)\,ds$$

$$= \frac{1}{3}[4xy + 2\,(6 - 2x - 2y) + 2y^2 - x - 3y]\,ds$$

$$= \frac{1}{3}[4xy + 2y^2 - 5x - 7y + 12]\,ds$$

But $k\,.\,\bar{n}\,ds = dxdy \quad \dfrac{1}{3}ds = dxdy$

$$\therefore \quad \mathbf{F}\cdot\mathbf{n} = [4xy + 2y^2 - 5x - 7y + 12]\,dxdy$$

$$I_4 = \iint_{ABC} \boldsymbol{F}\,.\,\boldsymbol{n}\,ds = \iint_{OAB} (4xy + 2y^2 - 5x - 7y + 12)\,dxdy$$

$$= \int_0^3\int_0^{3-x} (4xy + 2y^2 - 5x - 7y + 12)\,dxdy$$

$$= \int_0^3 \left[2xy^2 + \frac{2}{3}y^3 - 5xy - \frac{7y^2}{2} + 12y\right]_0^{3-x} dx$$

$$= \int_0^3 2x\,(3-x)^3 + \frac{2}{3}(3-x)^2 - 5x\,(3-x) - \frac{7}{2}(3-x)^2 + 12\,(3-x)]\,dx$$

$$= \int_0^3 (3-x)\,[2x\,(3-x) + \frac{2}{3}(3-x)^2 - 5x - \frac{7}{2}(3-x) + 12]\,dx$$

$$= \int_0^3 (3-x)\,[-\frac{4}{3}x^2 + \frac{1}{2}x + \frac{15}{2}]\,dx$$

$$= \int_0^3 \left(\frac{4}{3}x^3 - \frac{9}{2}x^2 - 6x + \frac{45}{2} \right) dx$$

$$= \left[\frac{x^4}{3} - \frac{3}{2}x^3 - 3x^2 + \frac{45}{2}x \right]_0^3$$

$$= 27 - \frac{81}{2} - 27 + \frac{135}{2} = 27$$

$$\therefore \quad \iiint_V \text{div}\, \boldsymbol{F}\, dv = 27 = \iint_V \boldsymbol{F} \, . \, \boldsymbol{n} \, ds$$

since $\iint_S \boldsymbol{F} \, . \, \boldsymbol{n} \, ds = I_1 + I_2 + I_3 + I_4 = -18 + 0 + 18 + 27 = 27$

Hence the theorem is verified.

Example 95. Prove that the directional derivative of $x + xy^2 + yz^3$ at (0, 1, 1) in the direction (i) whose direction cosines are $\left(\frac{2}{3}, \frac{2}{3}, -\frac{1}{3}\right)$ is 1 (ii) $2\mathbf{i} + 2\mathbf{j} - \mathbf{k}$ is also 1.

Solution. (*i*) The directional derivative of $f(x, y, z)$ in the direction whose direction cosines are l, m, n is $l\frac{\partial f}{\partial x} + m\frac{\partial f}{\partial y} + n\frac{\partial f}{\partial z}$.

Hence in our case, the directional derivative is

$$= \frac{2}{3}\frac{\partial f}{\partial x} + \frac{2}{3}\frac{\partial f}{\partial y} - \frac{1}{3}\frac{\partial f}{\partial z}$$

$$= \frac{2}{3}(1 + y^2) + \frac{2}{3}(2xy + z^3) - \frac{1}{3}(3yz^2)$$

At (0, 1, 1), directional derivative

$$= \frac{2}{3}(2) + \frac{2}{3}(1) - \frac{1}{3}(3) = 1$$

(*ii*) Direction cosines of $2\mathbf{i} + 2\mathbf{j} - \mathbf{k}$ are $\left(\frac{2}{3}, \frac{2}{3}, -\frac{1}{3}\right)$.

Hence the answer in this case is also 1.

Example 96. *If* $\mathbf{r} = x\mathbf{i} + y\mathbf{j} + z\mathbf{k}$, *and* $r = |\mathbf{r}|$, *prove* (*i*) $\nabla (f(r)) = f'(r)\,\hat{\mathbf{r}}$ *where* $\hat{\mathbf{r}}$ *is the unit vector in the direction of* $\mathbf{r}$. *Hence deduce the value of* (*ii*) $\nabla\left(\frac{1}{r}\right)$ (*iii*) $\nabla (\log r)$

Solution. $\quad \nabla \phi = \sum \mathbf{i} \frac{\partial \phi}{\partial x}$

(*i*) $\therefore \quad \nabla f(r) = \sum \mathbf{i} \frac{\partial f(r)}{\partial x}$

$$= \sum \mathbf{i} \frac{\partial}{\partial r} f(r). \frac{\partial r}{\partial x}$$

$$= f'(r) \sum \mathbf{i} \left(\frac{x}{r}\right) \quad \because \frac{\partial r}{\partial x} = \frac{x}{r}$$

$$= f'(r) \frac{1}{r} \cdot \sum x \mathbf{i}$$

$$= f'(r) \cdot \frac{1}{r} \mathbf{r}$$

$$= f'(r) \hat{\mathbf{r}} \text{ where } \hat{\mathbf{r}} = \text{unit vector} = \frac{\mathbf{r}}{r}.$$

(ii) $\nabla \left(\frac{1}{r}\right) = \left[\frac{d}{dr}\left(\frac{1}{r}\right)\right] \cdot \left(\frac{\mathbf{r}}{r}\right) = -\frac{1}{r^2} \cdot \frac{\mathbf{r}}{r} = -\frac{\mathbf{r}}{r^3}.$

(iii) $\nabla (\log r) = \frac{1}{r} \cdot \frac{\mathbf{r}}{r} = \frac{\mathbf{r}}{r^2}.$

Example 97. *Find the equation of the tangent plane to $x^2 + 2y^2 + 3z^2 = 6$ at $(1, -1, 1)$.*

Solution. $\nabla \phi = \sum \mathbf{i} \frac{\partial \phi}{\partial x} = 2x\mathbf{i} + 4y\mathbf{j} + 6z\mathbf{k}$

$(\nabla \phi)_{(1, -1, 1)} = 2\mathbf{i} - 4\mathbf{j} + 6\mathbf{k}$ is along normal to the tangent plane. Also the plane passes through $(1, -1, 1)$. Hence, the equation of the tangent plane is (The direction ratios of the normal are $2, -4, 6$).

$2(x-1) - 4(y+1) + 6(z-1) = 0$ *i.e.*, $x - 2y + 3z = 6$.

Example 98. *Find the angle between the normals to the surface $xy - z^2 = 0$ at points $(1, 4, -2)$ and $(-3, -3, 3)$.*

Solution. $\nabla \phi = \sum \mathbf{i} \frac{\partial \phi}{\partial x} = y\mathbf{i} + x\mathbf{j} - 2z\mathbf{k}$

$$(\nabla \phi)_{(1, 4, -2)} = 4\mathbf{i} + 1\mathbf{j} + 4\mathbf{k}$$

$$(\nabla \phi)_{(-3, -3, 3)} = -3\mathbf{i} - 3\mathbf{j} - 6\mathbf{k}$$

We want angle between these two normal vectors.

$$\therefore \quad \cos\theta = \frac{(4\mathbf{i} + \mathbf{j} + 4\mathbf{k}) \cdot (-3\mathbf{i} - 3\mathbf{j} - 6\mathbf{k})}{\sqrt{(16 + 1 + 16)(9 + 9 + 36)}}$$

$$= \frac{-13}{3\sqrt{22}}$$

$$\theta = \cos^{-1}\left(\frac{-13}{3\sqrt{22}}\right).$$

Example 99. *Prove that the surfaces $5x^2 - 2yz - 9x = 0$ and $4x^2y + z^3 - 4 = 0$ cut orthogonally at $(1, -1, 2)$.*

Solution. $\nabla(5x^2 - 2yz - 9x) = (10x - 9)\,\mathbf{i} + (-2z)\,\mathbf{j} + (-2y)\,\mathbf{k}$

$$\nabla\phi_1 \text{ at } (1, -1, 2) \text{ is } \mathbf{i} - 4\mathbf{j} + 2\mathbf{k}$$

$$\nabla(4x^2y + z^3 - 4) = (8xy)\,\mathbf{i} + (4x^2)\,\mathbf{j} + (3z^2)\,\mathbf{k}$$

$$\nabla\phi_2 \text{ at } (1, -1, 2) = -8\mathbf{i} + 4\mathbf{j} + 12\mathbf{k}$$

$$\nabla\phi_1 \cdot \nabla\phi_2 = (\mathbf{i} - 4\mathbf{j} + 2\mathbf{k}) \cdot (-8\mathbf{i} + 4\mathbf{j} + 12\mathbf{k})$$

$$= -8 - 16 + 24 = 0$$

$\therefore$ The surfaces cut orthogonally at $(1, -1, 2)$.

Example 100. *If* $\nabla\phi = 5r^3\mathbf{r}$, *find* ϕ.

Solution. We know $\nabla\phi(r) = \phi'(r)\dfrac{\mathbf{r}}{r} = 5r^3\mathbf{r}$

$$\therefore \quad \frac{\phi'(r)}{r} = 5r^3 \quad \therefore \quad \phi'(r) = 5r^4$$

On integrating, $\phi(r) = r^5 + C$.

Example 101. *If* $\nabla\phi = (6r - 8r^2)\,\mathbf{r}$ *and* $\phi(2) = 4$, *find* ϕ.

Solution. $\nabla\phi(r) = \phi'(r)\dfrac{\mathbf{r}}{r} = (6r - 8r^2)\,\mathbf{r}$

$$\therefore \quad \frac{\phi'(r)}{r} = 6r - 8r^2$$

$$\phi'(r) = 6r^2 - 8r^3$$

Integrating, $\phi(r) = 2r^3 - 2r^4 + C$

$$\phi(2) = 16 - 32 + C = 4 \quad \therefore \quad C = 20$$

$$\therefore \quad \phi(r) = 2(r^3 - r^4 + 10).$$

Example 102. *If* $\nabla\phi = (y + y^2 + z^2)\,\mathbf{i} + (x + z + 2xy)\,\mathbf{j} + (y + 2zx)\,\mathbf{k}$ *and if* $\phi(1, 1, 1) = 3$, *find* $\phi(x, y, z)$.

Solution. $\nabla\phi = \sum \mathbf{i}\dfrac{\partial\phi}{\partial x} = (y + y^2 + z^2)\,\mathbf{i} + (x + z + 2xy)\,\mathbf{j} + (y + 2zx)\,\mathbf{k}$

$$\therefore \quad \frac{\partial\phi}{\partial x} = y + y^2 + z^2 \Rightarrow \phi = xy + xy^2 + xz^2 + f_1(y, z) \quad \text{...(1)}$$

$$\frac{\partial\phi}{\partial y} = x + z + 2xy \Rightarrow \phi = xy + zy + xy^2 + f_2(x, z) \quad \text{...(2)}$$

$$\frac{\partial\phi}{\partial z} = y + 2zx \Rightarrow \phi = yz + z^2x + f_3(x, y) \quad \text{...(3)}$$

From (1), (2), (3), we infer,

$$\phi(x, y, z) = xy + xy^2 + xz^2 + yz + C$$

$$\phi(1, 1, 1) = 4 + C = 3 \quad \therefore \quad C = -1$$

Hence $\phi(x, y, z) = xy + xy^2 + xz^2 + yz - 1$.

Example 103. Prove $\nabla^2 f(r) = f''(r) + \frac{2}{r} f'(r)$ *where* $r = |\mathbf{r}|$. *Hence show that, if* $\nabla^2 f(r) = 0$, *then* $f(r) = \frac{\alpha}{r} + \beta$.

Solution. $\nabla f(r) = f'(r) \cdot \frac{\mathbf{r}}{r}$

$$\nabla^2 f(r) = \nabla \cdot \nabla f(r) = \text{div}\left[\frac{f'(r)}{r}\mathbf{r}\right]$$

$$= \nabla\left[\frac{f'(r)}{r}\right] \cdot \mathbf{r} + \frac{f'(r)}{r} \text{div}\, \mathbf{r}$$

$$= \left[\sum i \frac{\partial}{\partial x}\left(\frac{f'(r)}{r}\right)\right] \cdot \mathbf{r} + \frac{3f'(r)}{r}$$

$$= \left\{\sum i \left[\frac{r f''(r) - f'(r)}{r^2} \times \frac{x}{r}\right]\right\} \cdot \mathbf{r} + \frac{3}{r} f'(r)$$

$$= \frac{r f''(r) - f'(r)}{r^3}\left(\sum x i\right) \cdot \mathbf{r} + \frac{3}{r} f'(r)$$

$$= \left\{\frac{r f''(r) - f'(r)}{r^3}\right\} \cdot r^2 + \frac{3}{r} f'(r)$$

$$= f''(r) - \frac{f'(r)}{r} + \frac{3}{r} f'(r)$$

$$= f''(r) + \frac{2}{r} f'(r)$$

If $\nabla^2 f(r) = 0$ then $f''(r) = -\frac{2}{r} f'(r)$

i.e., $\frac{f''(r)}{f'(r)} = -\frac{2}{r}$

Integrating, $\log [f'(r)] = -2 \log r + \log C$

$$= \log \frac{C}{r^2}$$

$$f'(r) = \frac{C}{r^2}$$

Integrating, $f(r) = -\frac{C}{r} + k = \frac{\alpha}{r} + \beta.$

Example 104. *Show that the line integration of $f = (3x^2 + 6xy)\,\boldsymbol{i} + (3x^2 - y^2)\,\boldsymbol{j}$ is independent of the path of integration. Find $\int_c$ **F**. dr where C is the straight line joining (0, 0) and (1, 2).*

Solution.

$$\text{curl } \mathbf{F} = \begin{vmatrix} \mathbf{i} & \mathbf{j} & \mathbf{k} \\ \frac{\partial}{\partial x} & \frac{\partial}{\partial y} & \frac{\partial}{\partial z} \\ 3x^2 + 6xy & 3x^2 - y^2 & 0 \end{vmatrix}$$

$$= (0)\,\mathbf{i} - (0)\,\mathbf{j} + (6x - 6x)\,\mathbf{k} = \mathbf{0}$$

Therefore, the integration is independent of the path.

The equation of the line is $y = 2x$.

$$\int_C \mathbf{F} \cdot d\mathbf{r} = \int_C (F_1\,dx + F_2\,dy + F_3\,dz)$$

$$= \int_{x=0}^{x=1} (3x^2 + 6xy)\,dx + (3x^2 - y^2)\,dy$$

$$= \int_0^1 [3x^2 + 12x^2 + (3x^2 - 4x^2)\,2]\,dx$$

$$= \int_0^1 13x^2\,dx = \frac{13}{3}$$

Example. 105. *Prove that the vector field*

$\mathbf{F} = (y + y^2 + z^2)\,\mathbf{i} + (x + z + 2xy)\,\mathbf{j} + (y + 2zx)\,\mathbf{k}$

is conservative and find its scalar potential.

Solution. $$\text{curl } \mathbf{F} = \begin{vmatrix} \mathbf{i} & \mathbf{j} & \mathbf{k} \\ \frac{\partial}{\partial x} & \frac{\partial}{\partial y} & \frac{\partial}{\partial z} \\ y + z^2 + y^2 & x + z + 2xy & y + 2zx \end{vmatrix}$$

$$= (1 - 1)\,\mathbf{i} - (2z - 2z)\,\mathbf{j} + (1 + 2y - 1 - 2y)\,\mathbf{k} = \mathbf{0}$$

Therefore, **F** is conservative field.

A scalar potential ϕ exists such that $\nabla\phi = \mathbf{F}$

$$\therefore \quad \frac{\partial \phi}{\partial x} = y + y^2 + z^2 \quad \therefore \quad \phi = xy + xy^2 + xz^2 + f_1(y, z)$$

$$\frac{\partial \phi}{\partial y} = x + z + 2xy\,;\ \phi = xy + zy + xy^2 + f_2(z, x)$$

$$\frac{\partial\phi}{\partial z} = y + 2zx \; ; \; \phi = yz + z^2x + f_3(x, y) \;\; \therefore \;\; \phi = xy + xy^2 + xz^2 + yz + C$$

Example. 106. *Find the area of the surface of the hemisphere* $x^2 + y^2 + z^2 = a^2 \; ; z \geq 0$

Solution. Surface area $= \iint_S dS = \iint_R \frac{dx\,dy}{\mathbf{n}\cdot\mathbf{k}} = \iint_R \frac{a\,dx\,dy}{z}$

since $\mathbf{n}\cdot\mathbf{k} = \frac{z}{a}$ (refer previous example)

$$= \iint_R \frac{a\,dx\,dy}{\sqrt{a^2 - x^2 - y^2}}$$

$$= \iint_R \frac{a\,r\,dr\,d\theta}{\sqrt{a^2 - r^2}}$$

$$= \left(\int_0^{2\pi} d\theta\right)\left(a\int_0^a \frac{r\,dr}{\sqrt{a^2 - r^2}}\right)$$

$$= 2\pi\,a \int_0^a \frac{r\,dr}{\sqrt{a^2 - r^2}}$$

$$= 2\pi\,a \int_0^{\pi/2} a \sin\theta\,d\theta \qquad \text{(Put } r = a\sin\theta\text{)}$$

$$= 2\pi\,a^2 \times 1 = 2\pi\,a^2$$

Example. 107. *Evaluate* $\int_C \mathbf{F}\cdot d\mathbf{r}$ *where* $\mathbf{F} = xz\,\mathbf{i} + yz\,\mathbf{j} + z^2\,\mathbf{k}$ *where c is*

(*i*) $x = t, y = t^2, z = t^3$ *from (0, 0, 0) to (1, 1, 1).*

(*ii*) rectilinear curve obtained by joining $O\,(0, 0, 0)$ $A\,(1, 0, 0)$, $B\,(1, 1, 0)$, and $C\,(1, 1, 1)$.

(*iii*) the straight line joining the points $C\,(0, 0, 0)$ and $A\,(1, 1, 1)$.

Solution. (*i*) $\mathbf{r} = x\,\mathbf{i} + y\,\mathbf{j} + z\,\mathbf{k} = t\,\mathbf{i} + t^2\,\mathbf{j} + t^3\,\mathbf{k}$

$$d\mathbf{r} = (\mathbf{i} + 2t\,\mathbf{j} + 3t^2\,\mathbf{k})\,dt$$

$$\mathbf{F}\cdot d\mathbf{r} = (xz\,\mathbf{i} + yz\,\mathbf{j} + z^2\,\mathbf{k})\cdot(\mathbf{i} + 2t\,\mathbf{j} + 3t^2\,\mathbf{k})\,dt$$

$$= (xz + 2yzt + 3t^2z^2)\,dt$$

$$= (t^4 + 2t^6 + 3t^8)\,dt$$

$$\int_C \mathbf{F}\cdot d\mathbf{r} = \int_0^1 (t^4 + 2t^6 + 3t^8)\,dt = \left(\frac{t^5}{5} + \frac{2}{7}t^7 + \frac{t^9}{3}\right)_0^1$$

$$= \frac{1}{5} + \frac{2}{7} + \frac{1}{3} = \frac{86}{105}$$

(*ii*) On OA, $x = t, y = 0, z = 0$ where t varies from 0 to 1.

$$\mathbf{r} = x\mathbf{i}, \ (x \text{ varies from } 0 \text{ to } 1)$$

$$\int_{OA} \mathbf{F} \cdot d\mathbf{r} = \int_{OA} (F_1\,dx + F_2\,dy + F_3\,dz) = \int_{OA} 0\,dx = 0$$

On $AB, x = 1, y = y, z = 0,$ (y varies from 0 to 1)

$$\int_{AB} \mathbf{F} \cdot d\mathbf{r} = \int_{AB} (F_1 dx + F_2 dy + F_3 dz) = \int_0^1 0\,dy = 0$$

since $F_2 = yz = 0$ on $z = 0$.

On $BC, x = 1, y = 1, z = z$ (z varies from 0 to 1)

$$\int_{BC} \mathbf{F} \cdot d\mathbf{r} = \int_{BC} (F_1\,dx + F_2\,dy + F_3\,dz)$$

$$= \int_0^1 F_3\,dz \qquad \because dx = 0, dy = 0$$

$$= \int_0^1 z^2\,dz = \left(\frac{z^3}{3}\right)_0^1 = 1/3$$

$$\therefore \int_{OABC} \mathbf{F} \cdot d\mathbf{r} = 0 + 0 + \frac{1}{3} = \frac{1}{3}$$

(*iii*) On line joining (0, 0, 0) to (1, 1, 1), the equations of straight line are $\dfrac{x}{1} = \dfrac{y}{1} = \dfrac{z}{1} = t$

$$\therefore \int_{OA} \mathbf{F} \cdot d\mathbf{r} = \int_{OA} (F_1\,dx + F_2\,dy + F_3\,dz)$$

$$= \int_0^1 (t^2 + t^2 + t^2)\,dt = \left(t^3\right)_0^1 = 1$$

Example. 108. Evaluate $\iint_S \mathbf{F} \cdot d\mathbf{S}$ where $\mathbf{F} = 4y\,\mathbf{i} + 18z\,\mathbf{j} - x\,\mathbf{k}$ and S is the plane saurface of the portion of the plane $3x + 2y + 6z = 6$ contained in the first octant.

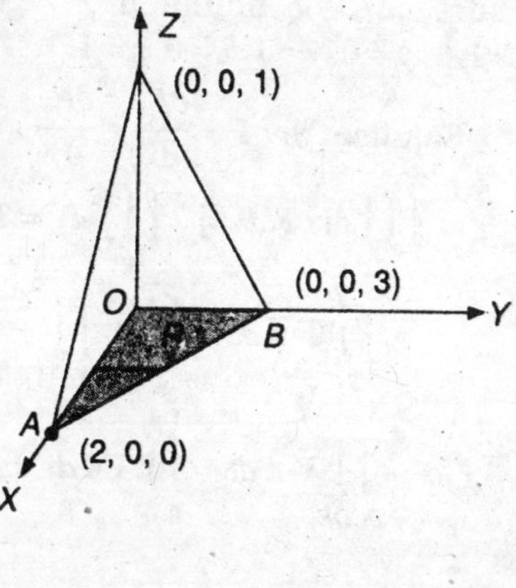

Solution. Let the plane ABC be projected on $z = 0$ plane.

The unit vector normal to the plane ABC is

$$\frac{\nabla\phi}{|\nabla\phi|}$$

$$\mathbf{n} = \frac{3\mathbf{i}+2\mathbf{j}+6\mathbf{k}}{\sqrt{9+4+36}} = \frac{3}{7}\mathbf{i}+\frac{2}{7}\mathbf{j}+\frac{6}{7}\mathbf{k}$$

$$\therefore \quad \mathbf{n}\cdot\mathbf{k} = \frac{6}{7}$$

$$\iint_S \mathbf{F}\cdot\mathbf{n}\, dS = \iint_S \left(\frac{3}{7}\times 4y + 18\left(\frac{2}{7}\right)z - x\times\frac{6}{7}\right) dS$$

$$= \iint_R \left(\frac{12}{7}y + \frac{36}{7}z - \frac{6x}{7}\right)\left(\frac{dxdy}{|\mathbf{n}\cdot\mathbf{k}|}\right) \qquad \text{since } dS = \frac{dxdy}{|\mathbf{n}\cdot\mathbf{k}|}$$

$$= \iint (2y+6z-x)\, dx\, dy$$

$$= \iint (6-4x)\, dxdy \qquad \text{using } 6z = 6-3x-2y$$

$$= \int_0^2 \int_0^{\frac{6-3x}{2}} (6-4x)\, dx\, dy$$

$$= \int_0^2 (6-4x)\left(\frac{6-3x}{2}\right) dx$$

$$= 3\int_0^2 (3-2x)(2-x)\, dx$$

$$= 3\int_0^2 (6-7x+2x^2)\, dx$$

$$= 3\left[6x - \frac{7x^2}{2} + \frac{2}{3}x^3\right]_0^2 = 3\left[12 - 14 + \frac{16}{3}\right] = 10$$

Example. 109. Verify Gauss Theorem taking $\mathbf{F} = (x+y)\,\mathbf{i} + x\,\mathbf{j} + z\,\mathbf{k}$ and the surface as the surface bounded by $x = 0$, $x = 1$, $y = 0$, $y = 1$, $z = 0$, $z = 1$.

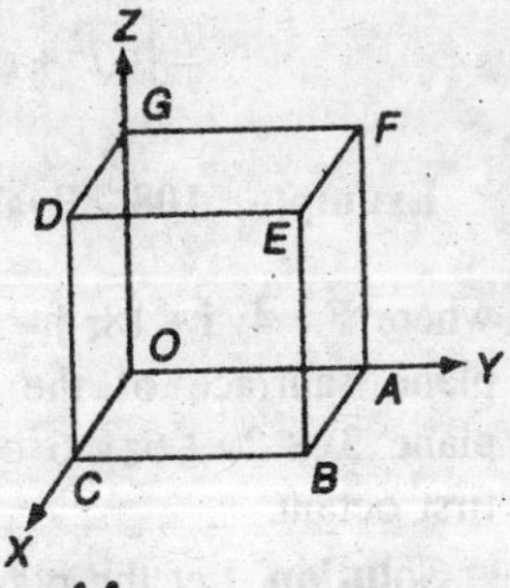

Solution. div $\mathbf{F} = \frac{\partial F_1}{\partial x} + \frac{\partial F_2}{\partial y} + \frac{\partial F_3}{\partial z} = 1+0+1 = 2$

$$\iiint_V \text{div}\,\mathbf{F}\, dV = 2\iiint_V dV = 2V = 2\times 1 = 2$$

$$\iint_S \mathbf{F}\cdot\bar{\mathbf{n}}\, dS = \iint_{BCDE} + \iint_{OGFA} + \iint_{OCDG} + \iint_{ABEF} + \iint_{OABC} + \iint_{EDGF} \quad \ldots(1)$$

$$\iint_{BCDE} \mathbf{F}\cdot\mathbf{n}\, dS = \int_0^1\int_0^1 \mathbf{F}\cdot\mathbf{i}\, dy\, dz = \int_0^1\int_0^1 (1+y)\, dy\, dz = \frac{3}{2}$$

$$\iint_{OGFA} \mathbf{F}\cdot\mathbf{n}\, dS = \int_0^1\int_0^1 \mathbf{F}\cdot(-\mathbf{i})\, dydz = -\int_0^1\int_0^1 y\, dy\, dz = -\frac{1}{2}$$

On $\quad OCDG, y = 0, \mathbf{n} = -\mathbf{j}$.

$$\iint_{OCDG} \mathbf{F}\cdot\mathbf{n}\, dS = \iint_{OCDG} \mathbf{F}\cdot(-\mathbf{j})\, dx\, dz = -\int_0^1\int_0^1 x\, dx\, dz = -\frac{1}{2}$$

$$\iint_{BEFA} \mathbf{F}\cdot\mathbf{n}\, dS = \iint_{BEFA} \mathbf{F}\cdot\mathbf{j}\, dxdz = \int_0^1\int_0^1 x\, dx\, dz = \frac{1}{2}$$

$$\iint_{OABC} \mathbf{F}\cdot\mathbf{n}\, dS = \iint_{OABC} \mathbf{F}\cdot(-\mathbf{k})\, dS = -\int_0^1\int_0^1 z\, dx\, dy = 0$$

$$\iint_{DGFE} \mathbf{F}\cdot\mathbf{n}\, dS = \iint_{DGEF} \mathbf{F}\cdot\mathbf{k}\, dx\, dy = \int_0^1\int_0^1 z\, dx\, dy = 1$$

Adding the integrals, $\iint_S \mathbf{F}\cdot\mathbf{n}\, dS = \frac{3}{2} - \frac{1}{2} - \frac{1}{2} + \frac{1}{2} + 1 = 2$

Hence, $\iint_S \mathbf{F}\cdot\mathbf{n}\, dS = \iiint_V \text{div}\,\mathbf{F}\, dV$ is verified.

Exmple. 110. If $\mathbf{F} = z\mathbf{k}$, show that

$$\iint_S \mathbf{F}\cdot\mathbf{n}\, dS = \iint_R \sqrt{a^2 - x^2 - y^2}\, dx\, dy = \frac{2}{3}\pi a^3$$

where S is the surface of the upper hemisphere $x^2 + y^2 + z^2 = a^2, z \geq 0$ and R is the region $x^2 + y^2 \leq a^2, z = 0$.

Solution. On S, the normal at any point $P(x, y, z)$ to the surface is along the radius (outward). Unit normal at $P(x, y, z)$

$$\frac{\mathbf{OP}}{|\mathbf{OP}|} = \frac{x\,\mathbf{i} + y\,\mathbf{j} + z\,\mathbf{k}}{a}$$

$$\therefore \text{ On } \quad S, \mathbf{n} = \frac{x\,\mathbf{i} + y\,\mathbf{j} + z\,\mathbf{k}}{a}$$

$$\mathbf{n}\cdot\mathbf{k} = \frac{z}{a}$$

$$dS = \frac{dx\, dy}{\mathbf{n}\cdot\mathbf{k}} = \frac{a}{z}\, dxdy\, ; \mathbf{F}\cdot\mathbf{n} = \frac{z^2}{a}$$

$$\iint_S \mathbf{F}\cdot\mathbf{n}\, dS = \iint_R \frac{z^2}{a}\cdot\frac{a}{z}\, dxdy = \iint_R z\, dx\, dy = \iint_R \sqrt{a^2 - x^2 - y^2}\, dxdy$$

On R, any point (x, y) is such that $x = r\cos\theta, y = r\sin\theta$

$\therefore\ dxdy = r\, dr\, d\theta \ \therefore\ x^2 + y^2 = r^2$

$$\therefore \iint_R \sqrt{a^2 - x^2 - y^2}\, dx dy = \int_0^{2\pi}\int_0^{a} \sqrt{a^2 - r^2}\; r\, dr\, d\theta$$

$$= \left(\int_0^{2\pi} d\theta\right)\int_0^{a} \sqrt{a^2 - r^2}\; r\, dr$$

$$= 2\pi \times \int_0^{a} \sqrt{u^2}\; u\, du$$

$$= 2\pi \times \left(\frac{u^3}{3}\right)_0^a = \frac{2}{3}\pi a^3$$

FOURIER SERIES AND ITS APPLICATIONS

CHAPTER 1

FOURIER SERIES

Periodic functions. Let $f(x)$ be a function of x, defined as follows. For each positive integer x, $f(x)$ is the remainder obtained when x is divided by 3. We can easily make a list of the first few values of $f(x)$: $f(1) = 1$, $f(2) = 2$, $f(3) = 0$; $f(4) = 1$, $f(5) = 2$, $f(6) = 0$; $f(7) = 1$, $f(8) = 2$, $f(9) = 0$; $f(10) = 1$, $f(11) = 2$, $f(12) = 0$.

From an examination of this table, it is found that for any positive integer x, $f(x+3) = f(x)$; further, if y is any positive integer less than 3,

$f(x+y) \neq f(x)$. We say that $f(x)$ is a periodic function with the period 3.

Definition. A function $f(x)$ is said to be periodic, if and only if $f(x+p) = f(x)$ is true for some value of p and every value of x. The smallest value of p for which this equation is true for every value of x will be called the *period of the function.*

When n is any integer, $\sin(x + 2n\pi) = \sin x$, for all x; hence $\sin x$ is periodic. Taking $n = 1$, $\sin(x + 2\pi) = \sin x$ for every x. However, if a is any positive number smaller than 2π, it can be shown that $\sin(x + a) = \sin x$ is *not* true for *all* values of x. Therefore, 2π is the smallest number p such that $\sin(x + p) = \sin x$ for every x. Accordingly, 2π is the period of $\sin x$.

Similarly, $\cos x$ is a periodic function with the period 2π. Now $\tan(x + \pi) = \tan x$, for all x. Hence $\tan x$ is periodic; its period is π.

The graph of a periodic function of period p is obtained by periodic repetition of its graph in any interval of length p.

Note that $f(x) = k$, where k is a constant, is a periodic function in the sense of the definition, since $f(x+p) = f(x)$. Again, if $f(x)$ and $F(x)$ have the same period p, then the function $\phi(x) = af(x) + bF(x)$ has the period p.

Exercises 1(a)

1. Find the smallest period p of the following functions:

(*i*) $\sin 2x$, (*ii*) $\cos 3x$, (*iii*) $\sin nx$, (*iv*) $\cos 2\pi x$, (*v*) $\sin(2\pi x/k)$.

2. Plot graphs of the following functions:

(*i*) $\sin x$, (*ii*) $\frac{\pi}{4} + \cos x$, (*iii*) $\sin x + \frac{1}{3}\sin 3x$, (*iv*) $\sin 2\pi x$,

(*v*) $f(x) = x$, when $-\pi < x < \pi$ and $f(x+2\pi) = f(x)$,

(*vi*) $f(x) = x^2$, when $-\pi < x < \pi$ and $f(x+2\pi) = f(x)$,

(*vii*) $f(x) = \begin{cases} -1 \text{ when } -\pi < x < 0 \\ +1 \text{ when } 0 < x < \pi \end{cases}$ and $f(x+2\pi) = f(x)$,

(*viii*) $f(x) = \begin{cases} 0 \text{ when } -\pi < x < 0 \\ x \text{ when } 0 < x < \pi \end{cases}$ and $f(x+2\pi) = f(x)$.

3. If p is a period of $f(x)$, show that np is also a period, where n is any integer, (positive or negative)

4. Show that the function $f(x) = $ constant, is a periodic function of period p, for any value of p.

5. If $f(x)$ and $F(x)$ have the period p, show that $\phi = af + bF$ where a and b are constants, has the same period p.

Bounds of a function. If for all values of x in a given interval, $f(x)$ is never greater than some fixed number M, the number M is called an *upper bound* for $f(x)$ in that interval; if $f(x)$ is never less than some number m, then m is called a *lower bound*. One or both of these bounds may not exist. Thus, in the interval $(0, 2\pi)$, $y = \sin x$ has an upper bound $+1$ and a lower bound -1. $y = \frac{1}{2} + \cos x$, has an upper bound $\frac{3}{2}$ and a lower bound $-\frac{1}{2}\cdot$ The function $y = \frac{1}{x}$ in the interval (0, 1) has a lower bound unity, but has no upper bound. The function $y = x - \frac{1}{x}$ has neither an upper nor a lower bound in the interval $(0, \infty)$.

If a function has an upper bound and a lower bound in a given interval, then that function is said to be *bounded*. We can put the idea differently: for all values of x in a given interval, if $|f(x)| \le a$, where a is some fixed number, then the function is bounded in that interval.

Continuity of a function. It is sometimes important to know the behaviour of a function $f(x)$ when $x \to a$ in such a way that x always remains greater than a, that is when x is allowed to approach a from the right side only. In such a case, the following notation is adopted: we write $x \to a+$, and the limit is called the *right-hand limit*. If the limit is b, we give the symbol

$$\lim_{x \to a+} f(x) = b, \text{ or } f(a+) = b.$$

The left-hand limit is defined similarly, and is denoted by either

$$\lim_{x \to a-} f(x), \text{ or } f(a-)$$

Let $f(x) = \frac{3}{2} + \frac{x-a}{2|x-a|}$.

Here $f(a-) = 1$, and $f(a+) = 2$. The symbols $f(a-)$ and $f(a+)$ should not be confused with the symbol $f(a)$, which means the value of $f(x)$ calculated at the point $x = a$. In the present example, $f(a) = \frac{3}{2} + \frac{1}{2} \times \frac{0}{0}$; hence $f(a)$ *does not exist*.

Or again, if $f(x) = 2^{\frac{1}{x-1}}$, $f(1-) = 0$, $f(1+) = \infty$, but $f(1)$ is not defined.

If $f(a-) = f(a+) = b$, then we call b the limit of $f(x)$ as $x \to a$ and write $\lim_{x \to a} f(x) = b$.

If $f(x) = \frac{x^2 - a^2}{x - a}$, $f(a-) = 2a$ and $f(a+) = 2a$; thus, $\lim_{x \to a} f(x) = 2a$; note that $f(a) = \frac{0}{0}$ and hence is not defined.

We have already studied that a function $f(x)$ is said to be continuous at $x = a$, if $\lim_{x \to a} f(x) = f(a)$. With our present knowledge, we can say that $f(x)$ is continuous at $x = a$, if $f(a-) = f(a) = f(a+)$. Thus, $f(x) = \frac{x^2 - a^2}{x - a}$, when $x \neq a$, and $f(x) = 2a$, when $x = a$ is continuous at $x = a$, even though the function $f(x) = \frac{x^2 - a^2}{x - a}$ as such is not continuous at $x = a$.

From the above, it is clear that $f(x)$ is discontinuous at $x = a$, if $f(a-) \neq f(a+)$. The discontinuity at $x = a$ is called *infinite* whenever $f(x)$ becomes infinite as $x \to a$. If the function is discontinuous at point $x = a$ but is bounded in its vicinity, the discontinuity is called *finite* or *ordinary*. If $f(a+)$ and $f(a-)$ exist and are unequal, their difference $f(a+) - f(a-)$ is called the *jump of the function* at $x = a$. Thus the function $\frac{3}{2} + \frac{x-a}{2|x-a|}$ has a jump of one unit at $x = a$. On the other hand, the function $2^{\frac{1}{x-1}}$ has an infinite discontinuity at $x = 1$. A function which is continuous everywhere except for a finite number of jumps in a given interval is called *piece-wise continuous or sectionally continuous*, in that interval.

Exercises 1(b)

1. If $f(x) = x^2$, find $f(3-)$, $f(3+)$. What is $\lim_{x \to 3} x^3$?

2. What is $\lim_{x \to \pi/3} \cos x$? $\cos\left(\frac{\pi}{3}+\right)$? $\cos\left(\frac{\pi}{3}-\right)$?

3. Given that $f(x) = \dfrac{1}{x-1}$, find $f(1+)$ and $f(1-)$

4. Find $f(0+)$ and $f(0-)$, if $f(x) = \dfrac{e^{1/x}}{1+e^{1/x}}$; what is $\lim_{x \to 0} f(x)$?

5. If $f(x) = \dfrac{1}{1-e^{-1/x}}$, find $f(0+)$ and $f(0-)$. What is $f(0)$?

Is the function continuous at $x = 0$? What is the jump of the function at $x = 0$? Find $f(0)$.

6. $f(x) = 1,\ 0 < x < \pi$

$f(x) = 2,\ \pi < x < 2\pi$. Find $f(\pi -)$ and $f(\pi +)$. Evaluate the jumps of the function at $x = \pi$.

Fourier Series. Periodic functions occur frequently in engineering problems. Such periodic functions are often complicated; it is therefore desirable to represent these in terms of the simple periodic functions of sine and cosine. A development of a given periodic function into a series of sines and cosines was effected by the French physicist and mathematician Joseph Fourier (1768–1830); the series of sines and cosines is known after him.

Below we prove a few results which we need in deriving Fourier series. Given that m and n are positive integers or zero:

1. If $n \neq 0$,

$$\int_c^{c+2\pi} \sin nx\,dx = \left[-\frac{\cos nx}{n}\right]_c^{c+2\pi} = -\frac{\cos n(c+2\pi)}{n} + \frac{\cos nc}{n}$$

$$= \frac{\cos nc - \cos(nc + 2n\pi)}{n} = \frac{\cos nc - \cos nc}{n} = 0.$$

Note. if $n = 0$, $\displaystyle\int_c^{c+2\pi} \sin nx\,dx = \int_c^{c+2\pi} 0 \cdot dx = 0.$

2. If $n \neq 0$, $\displaystyle\int_c^{c+2\pi} \cos nx\,dx = \left[\frac{\sin nx}{n}\right]_c^{c+2\pi} = 0.$

3. $\displaystyle\int_c^{c+2\pi} \sin mx \cos nx\,dx$

$$= \frac{1}{2}\int_c^{c+2\pi} [\sin(m+n)x + \sin(m-n)x]\,dx = 0, \text{ as in (1).}$$

4. If $m \neq n$, $\int_c^{c+2\pi} \sin mx \sin nx \, dx$

$$= \frac{1}{2}\int_c^{c+2\pi} [\cos (m-n)x - \cos (m+n)x] \, dx = 0, \text{ as in (2).}$$

5. If $m \neq n$, $\int_c^{c+2\pi} \cos mx \cos nx \, dx$

$$= \frac{1}{2}\int_c^{c+2\pi} [\cos (m+n)x + \cos (m-n)x] \, dx = 0, \text{ as in (2)}$$

6. If $n \neq 0$, $\int_c^{c+2\pi} \sin^2 nx \, dx = \frac{1}{2}\int_c^{c+2\pi} (1 - \cos 2nx) \, dx$

$$= \frac{1}{2}\int_c^{c+2\pi} dx - \frac{1}{2}\int_c^{c+2\pi} \cos 2nx \, dx = \frac{1}{2}[x]_c^{c+2\pi} - \frac{1}{2} \times 0$$

$$= \frac{1}{2}\{(c+2\pi) - c\} = \frac{1}{2} \times 2\pi = \pi.$$

7. If $n \neq 0$, $\int_c^{c+2\pi} \cos^2 nx \, dx = \frac{1}{2}\int_c^{c+2\pi} (1 + \cos 2nx) \, dx = \pi.$

Theorem 1. *If $f(x)$ is a periodic function with period 2π and if $f(x)$ can be represented by a trigonometric series in $(-\infty, \infty)$.*

$$f(x) = \frac{a_0}{2} + \sum_{n=1}^{\infty} (a_n \cos nx + b_n \sin nx), \qquad \text{...}(i)$$

then $$a_n = \frac{1}{\pi}\int_c^{c+2\pi} f(x) \cos nx \, dx, \text{ and } b_n = \frac{1}{\pi}\int_c^{c+2\pi} f(x) \sin nx \, dx.$$

Proof. First of all, let us determine a_0. Integrating both the sides of (i) from c to $c + 2\pi$, we get

$$\int_c^{c+2\pi} f(x) \, dx = \int_c^{c+2\pi} \left[\frac{a_0}{2} + \sum_{n=1}^{\infty} (a_n \cos nx + b_n \sin nx)\right] dx.$$

We assume that term by term integration of the series is allowed. [This is possible, if the series is uniformly convergent.] Then we have

$$\int_c^{c+2\pi} f(x)\,dx = \frac{a_0}{2}\int_c^{c+2\pi} dx + \sum_{n=1}^{\infty}\left\{a_n \int_c^{c+2\pi} \cos nx\,dx + b_n \int_c^{c+2\pi} \sin nx\,dx\right\}$$

The first term on the right is $\frac{a_0}{2}[x]_c^{c+2\pi} = \frac{a_0}{2} \times 2\pi = a_0\pi$. Using formulae 2 and 1 on p. 96, the other integrals on the right become zero each. Thus,

$$\int_c^{c+2\pi} f(x)\,dx = a_0\pi, \text{ giving } a_0 = \frac{1}{\pi}\int_c^{c+2\pi} f(x)\,dx.$$

We now determine $a_1, a_2, a_3, \ldots$ by a similar method. Multiplying (*i*) by $\cos mx$, where m is any fixed positive integer, and integrating from c to $c + 2\pi$, we get

$$\int_c^{c+2\pi} f(x) \cos mx\,dx = \int_c^{c+2\pi}\left\{\frac{a_0}{2} + \sum_{n=1}^{\infty} (a_n \cos nx + b_n \sin nx)\right\} \cos mx\,dx \quad ..(ii)$$

Using term by term integration, the right hand side becomes

$$\frac{a_0}{2}\int_c^{c+2\pi} \cos mx\,dx + \sum_{n=1}^{\infty} a_n \int_c^{c+2\pi} \cos nx \cos mx\,dx + \sum_{n=1}^{\infty} b_n \int_c^{c+2\pi} \sin nx \cos mx\,dx.$$

By formula 2 on p. 96, the first of these integrals is zero. Again by formula 3 on p. 96 all the integrals of the last Σ are zero each. The remaining integrals are

$$a_1\int_c^{c+2\pi} \cos x \cos mx\,dx + a_2\int_c^{c+2\pi} \cos 2x \cos mx\,dx + \cdots + a_m\int_c^{c+2\pi} \cos^2 mx\,dx$$

$$+ \cdots + a_n\int_c^{c+2\pi} \cos nx \cos mx\,dx + \cdots$$

Of these, $a_m \int_c^{c+2\pi} \cos^2 mx\,dx = a_m \cdot \pi$, using formula 7, p. 97

Lastly by formula 5, p. 97 each of the remaining integrals is zero. Substituting the above results in (*ii*) above, we have

$$\int_c^{c+2\pi} f(x) \cos mx\,dx = a_m \cdot \pi, \text{ giving } a_m = \frac{1}{\pi}\int_c^{c+2\pi} f(x) \cos mx\,dx.$$

Writing n in the place of m, we get $a_n = \frac{1}{\pi}\int_c^{c+2\pi} f(x) \cos nx\,dx,\ n = 1, 2, 3, \ldots$.

Finally multiplying (*i*) by $\sin mx$, where m is any fixed positive integer, and integrating from c to $c + 2\pi$, we have

$$\int_c^{c+2\pi} f(x) \sin mx \, dx = \int_c^{c+2\pi} \left\{ \frac{a_0}{2} + \sum_{n=1}^{\infty} (a_n \cos nx + b_n \sin nx) \sin mx \, dx \right. \qquad ...(iii)$$

Integrating the R.H.S. term by term, we find that every one of the integrals except $\int_c^{c+2\pi} b_m \sin^2 mx \, dx$ is zero; and, by formula 6 on p. 97 this integral is $b_m \pi$. Hence from (*iii*) above

$$\int_c^{c+2\pi} f(x) \sin mx \, dx = b_m \pi, \text{ giving } b_m = \frac{1}{\pi} \int_c^{c+2\pi} f(x) \sin mx \, dx.$$

Writing n in the place of m, $b_n = \frac{1}{\pi} \int_c^{c+2\pi} f(x) \sin nx \, dx, \; n = 1, 2, 3, \ldots$

Thus we have found the values of the coefficients a_0, a_n and b_n. These relations are known as Euler formulas. Thus, the **Euler formulas** are ;

$$a_0 = \frac{1}{\pi} \int_c^{c+2\pi} f(x) \, dx; \quad a_n = \frac{1}{\pi} \int_c^{c+2\pi} f(x) \cos nx \, dx, \text{ and}$$

$$b_n = \frac{1}{\pi} \int_c^{c+2\pi} f(x) \sin nx \, dx, \; n = 1, 2, \ldots.$$

They are sometimes called *Euler-Fourier* formulas. The R.H.S. of (i), when the coefficients there are given by the Euler formulas, is known as the **Fourier series** of $f(x)$. In most applications, the interval over which $f(x)$ is to be expanded is either $(-\pi, \pi)$ or $(0, 2\pi)$, so that the value of c in the Euler formulas is either $-\pi$ or 0. Actually, the formula for a_0 need not be listed, for it can be obtained from the general expression for a_n by putting $n = 0$; it was to achieve this that we wrote the constant term in the original expansion as $\frac{a_0}{2}$, (and not a_0 as usual).

Note that we have not proved that any given function $f(x)$ has a Fourier expansion that converges to $f(x)$. What we have shown is that *if* a function $f(x)$ has an expansion of the form (i) for which term by term integration is valid, *then* the coefficients in that series will be given by the Euler formulas. Questions concerning the convergence of the Fourier series, and if they converge, the conditions under which they will represent the functions which generated them are many and difficult. But, all the functions that you will have to deal with will be covered by the famous theorem of Dirichlet, (the great German mathematician 1805–1859).

Theorem 2. *If $f(x)$ is a bounded function of period 2π which in any one period has at most a finite number of maxima and minima and a finite number of points of discontinuity, then the Fourier series of $f(x)$ converges to $f(x)$ at all points where $f(x)$ is continuous; also, the series converges to the average value of the right and left-hand limits of $f(x)$ at each point where $f(x)$ is discontinuous.* The conditions given in this theorem are known as the **Dirichlet conditions**. (The proof of this theorem is not expected of you; you have merely to memorise the statement). These conditions are *not necessary.*

Dirichlet's conditions

If a function $f(x)$ is defined in $c \le x \le c + 2\pi$, it can be expanded as a Fourier series of the form $\frac{a_0}{2} + \sum_{n=1}^{\infty} (a_n \cos nx + b_n \sin nx)$, provided the following DIRICHLET'S conditions are satisfied.

(*i*) $f(x)$ is single valued and finite in $(c, \ c + 2\pi)$

(*ii*) $f(x)$ is continuous or piece-wise continuous with finite number of finite discontinuities in $(c, \ c + 2\pi)$

(*iii*) $f(x)$ has a finite number of maxima or minima in $(c, \ c + 2\pi)$

Note 1. These conditions are *not necessary* but *only sufficient* for the existence of Fourier series.

Note 2. If $f(x)$ satisfies Dirichlet's conditions and $f(x)$ is defined in $(-\infty, \infty)$, then $f(x)$ has to be periodic of periodicity 2π for the existence of Fourier series of period 2π.

Note 3. If $f(x)$ satisfies Dirichlet's conditions and $f(x)$ is defined in $(c, \ c + 2\pi)$, then $f(x)$ *need not be periodic* for the existence of Fourier series of period 2π

Note 4. If $x = a$ is a point of discontinuity of $f(x)$, then the value of the Fourier series at $x = a$ is $\frac{1}{2}\,[f(a+) + f(a-)]$

The Dirichlet conditions make it clear that a function need not be continuous in order to possess a valid Fourier expansion. This means that the graph of a function may consist of a number of dis-jointed arcs of different curves, each defined by a different equation, and still be representable by a Fourier series. In using the Euler formulas to find the coefficients in the expansion of such a function, it will be necessary to break up the range of integration $(c, c + 2\pi)$ to correspond to the various segments of the function. Thus, in the above

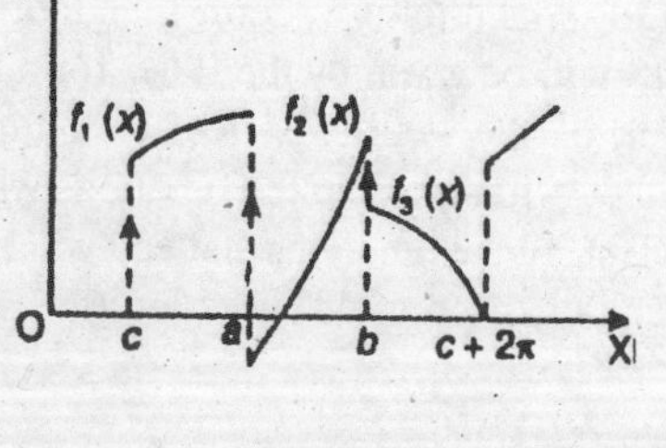

function, $f(x)$ is defined by three expressions $f_1(x), f_2(x)$ and $f_3(x)$ on successive portions of the period interval $c \le x \le c + 2\pi$. Hence it is necessary to write the Euler formulas as

$$a_n = \frac{1}{\pi} \int_c^{c+2\pi} f(x) \cos nx \, dx$$

$$= \frac{1}{\pi} \left\{ \int_c^a f(x) \cos nx \, dx + \int_a^b f(x) \cos nx \, dx + \int_b^{c+2\pi} f(x) \cos nx \, dx \right\}.$$

$$= \frac{1}{\pi} \left\{ \int_c^a f_1(x) \cos nx \, dx + \int_a^b f_2(x) \cos nx \, dx + \int_b^{c+2\pi} f_3(x) \cos nx \, dx \right\}.$$

A similar break up is necessary for evaluating b_n.

According to the above theorem, the Fourier series of the function shown in the figure will converge to the average values, indicated by dots, at the discontinuities at c, a and b, regardless of the definition (or lack of definition) of the function at these points.

The following two standard integrals which you have studied in lower classes occur in almost all problems in this chapter. It is better to remember these results to apply them in problems.

1. $\int e^{ax} \sin bx \, dx = \frac{e^{ax}}{a^2 + b^2} [a \sin bx - b \cos bx].$

2. $\int e^{ax} \cos bx \, dx = \frac{e^{ax}}{a^2 + b^2} [a \cos bx + b \sin bx].$

Bernoulli's generalised formula of integration by parts

This formula is only an extension of the formula of integration by parts.

We know that

$$\int u \, dv = uv - \int v \, du.$$

But in many problems, the integral will be of the form $\int uv \, dx$ and not $\int u \, dv$. Now, we will find a formula to evaluate $\int uv dx$.

In what follows, we write

$$\frac{du}{dx} = u', \ \frac{d^2u}{dx^2} = u'', \text{ etc.}$$

and $$\int v \, dx = v_1, \int v_1 dx = v_2 \text{ etc.}$$

Hence,

$$\int uv \, dx = \int u d(v_1)$$

$= uv_1 - \int v_1\, du$, using integration by parts

$= uv_1 - \int v_1 u' dx$

$= uv_1 - \int u'\, d\,(v_2)$

$= uv_1 - u'v_2 + \int v_2 u''\, dx$

$= uv_1 - u'v_2 + \int u''\, d\,(v_3)$

$= uv_1 - u'v_2 + u''v_3 - \int v_3 u'''\, dx$

....

....

$= uv_1 - u' v_2 + u''v_3 - u'''v_4 + \cdots\cdots$

Thus,

$\int uv\, dx = uv_1 - u' v_2 + u''v_3 - \cdots$ is called Bernoulli's extented formula of integration by parts.

Note 1. Whenever there is a polynomial as a factor in the integrand, take that polynomial as u and the remaining factor as v, since after a certain stage, the derivatives of u will vanish.

To illustrate the *advantage* of this formula, consider

$\int (x^3 + 2x^2 + 5) \cos 2x\, dx$

Here take $x^3 + 2x^2 + 5 = u$ and $\cos 2x = v$.

$$\int (x^3 + 2x^2 + 5) \cos 2x\, dx = \int uv\, dx$$

$$= (x^3 + 2x^2 + 5)\left(\frac{\sin 2x}{2}\right) - (3x^2 + 4x)\left(-\frac{\cos 2x}{4}\right) + (6x + 4)\left(-\frac{\sin 2x}{8}\right) - (6)\left(\frac{\cos 2x}{16}\right) + C$$

Note 2. In the above problem, we take the given integral as $\int uv\, dx$ and not as $\int u\, dv$.

Example 1. *Obtain the Fourier series of periodicity* 2π *for* $f(x) = e^{-x}$ *in the interval* $0 < x < 2\pi$. *Hence deduce the value of* $\sum_{n=2}^{\infty} \frac{(-1)^n}{1+n^2}$. *Further, derive a series for* $\operatorname{cosech} \pi$.

$$\text{Let } e^{-x} = \frac{a_0}{2} + \sum_{n=1}^{\infty} (a_n \cos nx + b_n \sin nx) \qquad \ldots(i)$$

Taking $c = 0$ in the Euler formulas, we have,

$$a_0 = \frac{1}{\pi}\int_0^{2\pi} f(x)\,dx = \frac{1}{\pi}\int_0^{2\pi} e^{-x}\,dx = \frac{1}{\pi}\left[-e^{-x}\right]_0^{2\pi}$$

$$= \frac{1}{\pi}\left[-e^{-2\pi} + 1\right] = \frac{1-e^{-2\pi}}{\pi}$$

For $n = 1, 2, 3, \ldots\ldots$

$$a_n = \frac{1}{\pi}\int_0^{2\pi} f(x)\cos nx\,dx$$

$$= \frac{1}{\pi}\int_0^{2\pi} e^{-x}\cos nx\,dx$$

$$= \frac{1}{\pi}\left[\frac{e^{-x}}{(-1)^2+n^2}\{(-1)\cos nx + n\sin nx\}\right]_0^{2\pi},$$

using the result of the standard integral.

$$= \frac{1}{\pi(1+n^2)}\left[e^{-2\pi}(-1) - (-1)\right]$$

$$= \frac{1}{\pi(1+n^2)}(1-e^{-2\pi}).$$

Similarly,

$$b_n = \frac{1}{\pi}\int_0^{2\pi} e^{-x}\sin nx\,dx$$

$$= \frac{1}{\pi}\left[\frac{e^{-x}}{1+n^2}\{(-1)\sin nx - n\cos nx\}\right]_0^{2\pi}$$

$$= \frac{1}{\pi(1+n^2)}[-ne^{-2\pi} + n]$$

$$= \frac{n}{\pi(1+n^2)}(1-e^{-2\pi})$$

Substituting the values of a_0, a_n, b_n in (*i*), and remembering that the constant term is $\frac{a_0}{2}$, (and not a_0), we get that in the range $(0, 2\pi)$,

$$e^{-x} = \frac{1-e^{-2\pi}}{\pi}\left[\frac{1}{2} + \sum_{n=1}^{\infty} \frac{1}{1+n^2}(\cos nx + n \sin nx)\right] \quad ...(ii)$$

In $(0, 2\pi)$, e^{-x} is continuous. Therefore, at $x = \pi$, the value of the Fourier series equals the value of the function e^{-x}.

Hence replacing x by π in (ii),

$$e^{-\pi} = \frac{1-e^{-2\pi}}{\pi}\left[\frac{1}{2} + \sum_{n=1}^{\infty} \frac{\cos n\pi}{1+n^2}\right]$$

$$\therefore \quad \frac{1}{2} + \sum_{n=1}^{\infty} \frac{\cos n\pi}{1+n^2} = \frac{\pi e^{-\pi}}{1-e^{-2\pi}} = \frac{\pi}{e^{\pi}-e^{-\pi}} = \frac{\pi}{2\sinh \pi}$$

When $n = 1$, $\frac{\cos n\pi}{1+n^2} = -\frac{1}{2}$; further $\cos n\pi = (-1)^n$

Hence $\sum_{n=2}^{\infty} \frac{(-1)^n}{1+n^2} = \frac{\pi \text{ cose ch } \pi}{2}$

$$\therefore \quad \text{cosech } \pi = \frac{2}{\pi} \sum_{\pi=2}^{\infty} \frac{(-1)^n}{1+n^2}.$$

Example 2. *If* $f(x) = \begin{cases} -k, \text{ when } -\pi < x < 0 \\ k, \text{ when } 0 < x < \pi \end{cases}$

and $f(x + 2\pi) = f(x)$ *for all x, derive the Fourier series for* $f(x)$. *Deduce that*

$\frac{\pi}{4} = 1 - \frac{1}{3} + \frac{1}{5} - \frac{1}{7} + \cdots$ *to infinity.*

Let $f(x) = \frac{a_0}{2} + \sum_{n=1}^{\infty} (a_n \cos nx + b_n \sin nx) \quad ...(i)$

Taking $c = -\pi$ in the Euler formulas we have

$$a_0 = \frac{1}{\pi}\int_{-\pi}^{\pi} f(x)\, dx = \frac{1}{\pi}\left\{\int_{-\pi}^{0} f(x))\, dx + \int_{0}^{\pi} f(x)\, dx\right\}.$$

Now using the hypothesis for the value of $f(x)$, we get

$$a_0 = \frac{1}{\pi}\left\{\int_{-\pi}^{0} (-k)\, dx + \int_{0}^{\pi} k dx\right\} = \frac{1}{\pi}\left\{\Big(-kx\Big)_{-\pi}^{0} + \Big(kx\Big)_{0}^{\pi}\right\}$$

$$= \frac{1}{\pi} \{(0 - k\pi) + (k\pi - 0)\}$$

Thus $a_0 = 0$. Again for $n = 1, 2, 3, ...$

$$a_n = \frac{1}{\pi} \int_{-\pi}^{\pi} f(x) \cos nx\, dx$$

$$= \frac{1}{\pi} \left\{ \int_{-\pi}^{0} f(x) \cos nx\, dx + \int_{0}^{\pi} f(x) \cos nx\, dx \right\}.$$

Substituting the values supplied for $f(x)$, we have

$$a_n = \frac{1}{\pi} \left\{ \int_{-\pi}^{0} (-k) \cos nx\, dx + \int_{0}^{\pi} k \cos nx\, dx \right\}$$

$$= \frac{1}{\pi} \left\{ \left(-k \frac{\sin nx}{n} \right)_{-\pi}^{0} + \left(k \frac{\sin nx}{n} \right)_{0}^{\pi} \right\}$$

Since $\sin 0$, $\sin(-n\pi)$ and $\sin n\pi$ are all zero, we get $a_n = 0$.

$$b_n = \frac{1}{\pi} \int_{-\pi}^{\pi} f(x) \sin nx\, dx$$

$$= \frac{1}{\pi} \left\{ \int_{-\pi}^{0} f(x) \sin nx\, dx + \int_{0}^{\pi} f(x) \sin nx\, dx \right\}$$

$$= \frac{1}{\pi} \left\{ \int_{-\pi}^{0} (-k) \sin nx\, dx + \int_{0}^{\pi} k \sin nx\, dx \right\}$$

$$= \frac{1}{\pi} \left\{ \left[k \frac{\cos nx}{n} \right]_{-\pi}^{0} + \left[-k \frac{\cos nx}{n} \right]_{0}^{\pi} \right\}$$

$$= \frac{1}{\pi} \left[\left\{ \frac{k}{n} \cos 0 - \frac{k}{n} \cos(-n\pi) \right\} + \left\{ -\frac{k}{n} \cos n\pi + \frac{k}{n} \cos 0 \right\} \right]$$

But $\cos(-\alpha) = \cos \alpha$, giving $\cos(-n\pi) = \cos n\pi$; further, $\cos 0 = 1$.

Hence $b_n = \frac{k}{n\pi}[\{1 - \cos n\pi\} + \{-\cos n\pi + 1\}] = \frac{k}{n\pi}(2 - 2\cos n\pi)$

$$\therefore\ b_n = \frac{2k}{n\pi}(1 - \cos n\pi).\ \text{Now}\ \cos n\pi = \begin{cases} -1, \text{ for odd } n \\ +1, \text{ for even } n \\ (-1)^n, \text{ for any integer } n. \end{cases}$$

Hence $b_1 = \frac{4k}{\pi}$; $b_2 = 0$; $b_3 = \frac{4k}{3\pi}$; $b_4 = 0$; $b_5 = \frac{4k}{5\pi}$; $b_6 = 0$;

$$b_7 = \frac{4k}{7\pi} \ldots\ldots$$

Using the values of a_n and b_n in (i) we obtain

$$f(x) = \frac{4k}{\pi}\{\sin x + \tfrac{1}{3}\sin 3x + \tfrac{1}{5}\sin 5x + \tfrac{1}{7}\sin 7x + \cdots \text{ to } \infty\}$$

In the above equation putting $x = \pi/2$, we get

$$f\left(\frac{\pi}{2}\right) = \frac{4k}{\pi}\{1 - \tfrac{1}{3} + \tfrac{1}{5} - \tfrac{1}{7} + \cdots \text{ to } \infty\}$$

But, by hypothesis, $f\left(\frac{\pi}{2}\right) = k$.

Hence $k = \frac{4k}{\pi}\{1 - \tfrac{1}{3} + \tfrac{1}{5} - \tfrac{1}{7} + \cdots \text{ to } \infty\}$

Multiplying both the sides by $\frac{\pi}{4k}$, we have

$$\frac{\pi}{4} = 1 - \tfrac{1}{3} + \tfrac{1}{5} - \tfrac{1}{7} + \cdots \text{ to } \infty.$$

Note. Functions of the type given in this example occur as external forces acting on mechanical systems, electromotive forces in electric circuits etc.

Example 3. *Obtain the Fourier series of the periodic function defined by* $f(x) = -\pi$, if $-\pi < x < 0$ *and* $f(x) = x$, if $0 < x < \pi$. *Deduce that*

$$\frac{1}{1^2} + \frac{1}{3^2} + \frac{1}{5^2} + \cdots \text{ to } \infty = \frac{\pi^2}{8}.$$ **(*MS. 1986A*)**

$$\text{Let } f(x) = \frac{a_0}{2} + \sum_{n=1}^{\infty}(a_n \cos nx + b_n \sin nx) \qquad \ldots(i)$$

$$\text{Then } a_0 = \frac{1}{\pi}\int_{-\pi}^{\pi} f(x)\,dx = \frac{1}{\pi}\left[\int_{-\pi}^{0} f(x)\,dx + \int_{0}^{\pi} f(x)\,dx\right]$$

$$a_0 = \frac{1}{\pi}\left[\int_{-\pi}^{0}(-\pi)\,dx + \int_{0}^{\pi} x\,dx\right]$$

$$= \frac{1}{\pi}\left[-\pi^2 + \frac{\pi^2}{2}\right] = -\pi/2$$

$$a_n = \frac{1}{\pi}\int_{-\pi}^{\pi} f(x)\cos nx\, dx$$

$$= \frac{1}{\pi}\left[\int_{-\pi}^{0}(-\pi)\cos nx\, dx + \int_{0}^{\pi} x\cos nx\, dx\right]$$

$$= \frac{1}{\pi}\left[\left\{(-\pi)\left(\frac{\sin nx}{n}\right)\right\}_{-\pi}^{0} + \left\{(x)\left(\frac{\sin nx}{n}\right) - (1)\left(-\frac{\cos nx}{n^2}\right)\right\}_{0}^{\pi}\right]$$

$$= \frac{1}{\pi}\left[\frac{1}{n^2}(\cos n\pi - 1)\right]$$

$$= \frac{1}{\pi n^2}[(-1)^n - 1]$$

$$b_n = \frac{1}{\pi}\int_{-\pi}^{\pi} f(x)\sin nx\, dx$$

$$= \frac{1}{\pi}\left[\int_{-\pi}^{0}(-\pi)\sin nx\, dx + \int_{0}^{\pi} x\sin nx\, dx\right]$$

$$= \frac{1}{\pi}\left[(-\pi)\left(-\frac{\cos nx}{n}\right)_{-\pi}^{0} + \left\{(x)\left(-\frac{\cos nx}{n}\right) - (1)\left(-\frac{\sin nx}{n^2}\right)\right\}_{0}^{\pi}\right]$$

$$= \frac{1}{\pi}\left[\frac{\pi}{n}(1 - \cos n\pi) - \frac{\pi}{n}\cos n\pi\right]$$

$$= \frac{1}{n}(1 - 2\cos n\pi)$$

Substituting the values of the coefficients in (*i*),

$$f(x) = -\frac{\pi}{4} - \frac{2}{\pi}\left(\cos x + \frac{\cos 3x}{3^2} + \frac{\cos 5x}{5^2} + \ldots\right)$$

$$+ 3\sin x - \frac{\sin 2x}{2} + \frac{3\sin 3x}{3} - \frac{\sin 4x}{4} + \frac{3\sin 5x}{5} - \quad \ldots(ii)$$

Zero is a point of discontinuity of $f(x)$. But $f(0-) = -\pi$, and $f(0+) = 0$.

Hence $\dfrac{f(0-) + f(0+)}{2} = -\dfrac{\pi}{2}$.

By theorem 2 on p. 100 this is the sum of the Fourier series in (*ii*) for the value 0 of x. Hence

$$-\frac{\pi}{2} = -\frac{\pi}{4} - \frac{2}{\pi}\left(1 + \frac{1}{3^2} + \frac{1}{5^2} + \ldots\ldots\right)$$

$$\therefore \quad -\frac{\pi}{4} = -\frac{2}{\pi}\left(1 + \frac{1}{3^2} + \frac{1}{5^2} + \ldots\right)$$

Hence $\dfrac{\pi^2}{8} = 1 + \dfrac{1}{3^2} + \dfrac{1}{5^2} + \cdots$

Example 4. *What is the Fourier expansion of the periodic function, whose definition in one period is* $f(x) = \begin{cases} 0, \text{ when } -\pi < x < 0 \\ \sin x, \text{ when } 0 < x < \pi \end{cases}$?

Evaluate $\dfrac{1}{1\cdot 3} - \dfrac{1}{3\cdot 5} + \dfrac{1}{5\cdot 7} - \dfrac{1}{7\cdot 9} + \cdots$ to ∞.

Let $$f(x) = \frac{a_0}{2} + \sum_{n=1}^{\infty} (a_n \cos nx + b_n \sin nx) \qquad \ldots(i)$$

[In the great majority of the examples, we can calculate a_n and b_n for $n = 1$ to ∞. But, this is a special example, where a_1 and b_1 are to be calculated separately; but a_0 can be got from a_n.]

$$a_1 = \frac{1}{\pi}\left\{\int_{-\pi}^{0} 0 \,.\, \cos x\, dx + \int_0^{\pi} \sin x \cos x\, dx\right\}$$

$$= \frac{1}{2\pi}\left[\sin^2 x\right]_0^{\pi} = \frac{1}{\pi} \times 0 = 0.$$

For $\quad n = 0, 2, 3, 4, \ldots$

$$a_n = \frac{1}{\pi}\left\{\int_{-\pi}^{0} 0 \cdot \cos nx\, dx + \int_0^{\pi} \sin x \cos nx\, dx\right\}$$

$$= \frac{1}{2\pi}\int_0^{\pi} \{\sin (1+n)\, x + \sin (1-n)\, x\}\, dx$$

$$= \frac{1}{2\pi}\left[-\frac{\cos (1+n)\, x}{1+n} - \frac{\cos (1-n)x}{1-n}\right]_0^{\pi}, \qquad \text{for } n \neq 1$$

$$= \frac{1}{2\pi}\left[\left\{-\frac{\cos (\pi + n\pi)}{1+n} - \frac{\cos (\pi - n\pi)}{1-n}\right\} + \left\{\frac{1}{1+n} + \frac{1}{1-n}\right\}\right]$$

$$= \frac{1}{2\pi}\left[\frac{\cos n\pi}{1+n} + \frac{\cos n\pi}{1-n} + \frac{1}{1+n} + \frac{1}{1-n}\right]$$

$$= \frac{1}{2\pi}\left[\frac{1+\cos n\pi}{1+n} + \frac{1+\cos n\pi}{1-n}\right] = \frac{1+\cos n\pi}{\pi (1-n^2)}$$

i.e., $a_n = -\dfrac{1+\cos n\pi}{\pi(n-1)(n+1)}$

$$b_1 = \frac{1}{\pi}\left\{\int_{-\pi}^{0} 0 \cdot \sin x\, dx + \int_0^{\pi} \sin x \sin x\, dx\right\}$$

$$= \frac{1}{\pi}\int_0^{\pi} \sin^2 x\, dx = \frac{1}{\pi}\int_0^{\pi}\left\{\tfrac{1}{2} - \tfrac{1}{2}\cos 2x\right\} dx = \frac{1}{\pi} \times \frac{\pi}{2} = \frac{1}{2}$$

For $n = 2, 3, 4, 5....$

$$b_n = \frac{1}{\pi}\left\{\int_{-\pi}^{0} 0 . \sin nx\, dx + \int_0^{\pi} \sin x \sin nx\, dx\right\}$$

$$= \frac{1}{2\pi}\int_0^{\pi} \{\cos(n-1)x - \cos(n+1)x\}\, dx$$

$$= \frac{1}{2\pi}\left[\frac{\sin(n-1)x}{n-1} - \frac{\sin(n+1)x}{n+1}\right]_0^{\pi} = \frac{1}{2\pi} \times 0 = 0.$$

Substituting the values of the coefficients in (*i*),

$$f(x) = \frac{1}{\pi} + \frac{\sin x}{2} - \frac{2}{\pi}\left\{\frac{\cos 2x}{3} + \frac{\cos 4x}{15} + \frac{\cos 6x}{35} + \frac{\cos 8x}{63} + \cdots \text{ to } \infty\right\}$$

Putting $x = \dfrac{\pi}{2}$,

$$f\left(\frac{\pi}{2}\right) = \frac{1}{\pi} + \frac{1}{2} - \frac{2}{\pi}\left\{-\frac{1}{3} + \frac{1}{15} - \frac{1}{35} + \frac{1}{63} - \cdots \text{ to } \infty\right\}$$

But, by hypothesis $f\left(\dfrac{\pi}{2}\right) = \sin\dfrac{\pi}{2} = 1$. Hence

$$1 = \frac{1}{\pi} + \frac{1}{2} - \frac{2}{\pi}\left\{-\frac{1}{1\cdot 3} + \frac{1}{3\cdot 5} - \frac{1}{5\cdot 7} + \frac{1}{7\cdot 9} - \cdots \text{ to } \infty\right\}$$

$$\therefore\ 1 - \frac{1}{\pi} - \frac{1}{2} = \frac{2}{\pi}\left\{\frac{1}{1\cdot 3} - \frac{1}{3\cdot 5} + \frac{1}{5\cdot 7} - \frac{1}{7\cdot 9} + \cdots \text{ to } \infty\right\}$$

i.e., $$\frac{1}{2} - \frac{1}{\pi} = \frac{2}{\pi}\left\{\frac{1}{1\cdot 3} - \frac{1}{3\cdot 5} + \frac{1}{5\cdot 7} - \frac{1}{7\cdot 9} + \cdots \text{ to } \infty\right\}$$

$$\therefore\ \frac{\pi - 2}{4} = \frac{1}{1\cdot 3} - \frac{1}{3\cdot 5} + \frac{1}{5\cdot 7} - \frac{1}{7\cdot 9} + \cdots \text{ to } \infty.$$

Note. When k is a positive integer, in the Fourier expansion of any function containing either $\sin kx$ or $\cos kx$, the coefficients a_k and b_k are to be calculated separately. This is necessary in ex. 5 also.

Example 5. *Expand in Fourier series of periodicity* 2π *of* $f(x) = x \sin x$, *for* $0 < x < 2\pi$. **(MS. 1991 A, N)**

$$\text{Let } f(x) = \frac{a_0}{2} + \sum_{n=1}^{\infty} (a_n \cos nx + b_n \sin nx) \quad ...(i)$$

$$\text{Then } a_0 = \frac{1}{\pi}\int_0^{2\pi} f(x)\, dx = \frac{1}{\pi}\int_0^{2\pi} x \sin x\, dx \quad ...(ii)$$

Let us evaluate $\int_0^{2\pi} x \sin rx\, dx$, when r is an integer;

$$\int_0^{2\pi} x \sin rx\, dx = \left[(x)\left(-\frac{\cos rx}{r}\right) - (1)\left(-\frac{\sin rx}{r^2}\right)\right]_0^{2\pi}$$

$$= -\frac{2\pi}{r}\cos 2\pi r = -\frac{2\pi}{r} \quad ...(iii)$$

Using this result in (*ii*), we get $a_0 = \frac{1}{\pi}\left\{-\frac{2\pi}{1}\right\} = -2$

$$a_1 = \frac{1}{\pi}\int_0^{2\pi} f(x)\cos x\, dx = \frac{1}{\pi}\int_0^{2\pi} x \sin x \cos x\, dx = \frac{1}{2\pi}\int_0^{2\pi} x \sin 2x\, dx$$

Hence by (*iii*), $a_1 = \frac{1}{2\pi}\left\{-\frac{2\pi}{2}\right\} = -\frac{1}{2}$.

For $n = 2, 3, 4, \ldots$

$$a_n = \frac{1}{\pi}\int_0^{2\pi} x \sin x \cos nx\, dx = \frac{1}{2\pi}\int_0^{2\pi} x(2 \sin x \cos nx)\, dx$$

$$\therefore\ a_n = \frac{1}{2\pi}\int_0^{2\pi} x\{\sin (n+1)x - \sin (n-1)x\}\, dx.$$

Now using (*iii*), $a_n = \frac{1}{2\pi}\left\{-\frac{2\pi}{n+1} + \frac{2\pi}{n-1}\right\} = -\frac{1}{n+1} + \frac{1}{n-1}$

$$= \frac{-(n-1)+(n+1)}{(n+1)(n-1)} = \frac{2}{n^2-1}$$

$$b_1 = \frac{1}{2\pi}\int_0^{2\pi} f(x) \sin x\, dx = \frac{1}{\pi}\int_0^{2\pi} x \sin^2 x\, dx = \frac{1}{2\pi}\int_0^{2\pi} x(2\sin^2 x)\, dx$$

But $\cos 2x = 1 - 2\sin^2 x$, giving $2\sin^2 x = 1 - \cos 2x$.

Substituting this, $b_1 = \frac{1}{2\pi}\int_0^{2\pi} x(1 - \cos 2x)dx$ ...(*iv*)

When r is an integer,

$$\int_0^{2\pi} x\cos rx\,dx = \left[x\frac{\sin rx}{r}\right]_0^{2\pi} - \int_0^{2\pi}\frac{\sin rx}{r}dx = 0 + \left[\frac{\cos rx}{r^2}\right]_0^{2\pi} = 0 \quad ...(v)$$

Using this in (*iv*),

$$b_1 = \frac{1}{2\pi}\left\{\left[\frac{x^2}{2}\right]_0^{2\pi} - 0\right\} = \frac{1}{2\pi}\times\frac{(2\pi)^2}{2} = \frac{2\pi}{2} = \pi.$$

Lastly for $n = 2, 3, 4, \ldots$

$$b_n = \frac{1}{\pi}\int_0^{2\pi} x\sin x\sin nx\,dx = \frac{1}{2\pi}\int_0^{2\pi} x(2\sin x\sin nx)\,dx$$

$$= \frac{1}{2\pi}\int_0^{2\pi} x\{\cos(n-1)x - \cos(n+1)x\}\,dx = \frac{1}{2\pi}\{0-0\}, \text{ using } (v).$$

Thus, for $n = 2, 3, 4, \ldots, b_n = 0$.

Substituting the values of the coefficients in (*i*), we get

$$f(x) = -1 - \tfrac{1}{2}\cos x + 2\left\{\sum_{n=2}^{\infty}\frac{\cos nx}{n^2-1}\right\} + \pi\sin x.$$

Note. Putting $x = \frac{\pi}{2}$, we derive that

$$\frac{1}{1.3} - \frac{1}{3.5} + \frac{1}{5.7} - \cdots \text{ to } \infty = \frac{\pi - 2}{4}.$$

Example 6. *Derive the Fourier series of $f(x) = k$, where k is a constant, the periodicity being 2π.*

$$a_0 = \frac{1}{\pi}\int_c^{c+2\pi} f(x)dx = \frac{1}{\pi}\int_c^{c+2\pi} k\,dx = \frac{k}{\pi}\Big[x\Big]_c^{c+2\pi} = \frac{k}{\pi}\times 2\pi = 2k.$$

$$a_n = \frac{1}{\pi}\int_c^{c+2\pi} f(x)\cos nx\,dx = \frac{1}{\pi}\int_c^{c+2\pi} k\cos nx\,dx = \frac{k}{\pi}\left[\frac{\sin nx}{n}\right]_c^{c+2\pi}$$

$$= \frac{k}{\pi}\times 0 = 0.$$

$$b_n = \frac{1}{\pi}\int_c^{c+2\pi} f(x)\sin nx\,dx = \frac{1}{\pi}\int_c^{c+2\pi} k\sin nx\,dx = \frac{k}{\pi}\left[-\frac{\cos nx}{n}\right]_c^{c+2\pi}$$

$$= \frac{k}{\pi}\times 0 = 0$$

Thus the first term (constant term) in the Fourier expansion is k; *all* the other coefficients are zero each. So, no Fourier expansion, (other than what is supplied), is available for k.

Apparently, the labour involved in this example is a waste.

Note. k being any constant, the result of this example is valid for any constant number like π.

Theorem 3. *The Fourier coefficients of a sum $f_1 + f_2$ are the sums of the corresponding Fourier coefficients of f_1 and f_2.*

Theorem 4. *If c is a constant, the Fourier coefficients of cf are the corresponding Fourier coefficients of f multiplied by c.*

Example 7. *Derive the Fourier expansion for*

$$F(x) = \begin{cases} 0, \text{ when } -\pi < x < 0 \\ 2k, \text{ when } 0 < x < \pi \end{cases} \text{ and } f(x+2\pi) = f(x), \text{ for all } x.$$

$$\text{Let } f_1(x) = \begin{cases} -k, \text{ when } -\pi < x < 0 \\ +k, \text{ when } 0 < x < \pi \end{cases}$$

and $f_2(x) = k$, for all x.

Then we have developed the Fourier expansions of f_1 and f_2 in worked examples 2 and 6 respectively. But the given $F(x) = f_1(x) + f_2(x)$. Hence by theorem 3,

$$f(x) = k + \frac{4k}{\pi}\{\sin x + \tfrac{1}{3}\sin 3x + \tfrac{1}{5}\sin 5x + \tfrac{1}{7}\sin 7x + \cdots \text{ to } \infty\}.$$

Note. One can directly work out this problem as in example 2 without splitting into f_1 and f_2.

Example 8. *Derive the Fourier series for the periodic function, whose definition in one period is*

$$F(x) = \begin{cases} -\tfrac{1}{4}\pi, \text{ when } -\pi < x < 0 \\ \tfrac{1}{4}\pi + \sin x, \text{ when } 0 < x < \pi \end{cases}$$

$$\text{Let } f_1(x) = \begin{cases} -\tfrac{1}{4}\pi, \text{ when } -\pi < x < 0 \\ \tfrac{1}{4}\pi, \text{ when } 0 < x < \pi. \end{cases}$$

Then working exactly as in ex. 2,

$$f_1(x) = \sin x + \tfrac{1}{3}\sin 3x + \tfrac{1}{5}\sin 5x + \cdots \text{ to } \infty.$$

Now $F(x) = f(x) + f_1(x)$, where $f(x)$ is the one supplied in ex. 4.

Therefore using Th. 3,

$$F(x) = \frac{1}{\pi} - \frac{2}{\pi}\left\{\frac{\cos 2x}{3} + \frac{\cos 4x}{15} + \frac{\cos 6x}{35} + \frac{\cos 8x}{63} + \ldots \text{ to } \infty\right\}$$

$$+ \tfrac{3}{2}\sin x + \left\{\tfrac{1}{3}\sin 3x + \tfrac{1}{5}\sin 5x + \tfrac{1}{7}\sin 7x + \ldots \text{ to } \infty\right\}.$$

Note. Work out this example directly as in example 2.

Even and odd functions. $f(x)$ is said to be even, if $f(-x) = f(x)$; x^2, $\cos kx$, 2, $\cosh x$, $x^2 - 3\cos x + 1$ are all even functions. The graph of an even function will be symmetric about the y-axis. (Refer Fig. 1.)

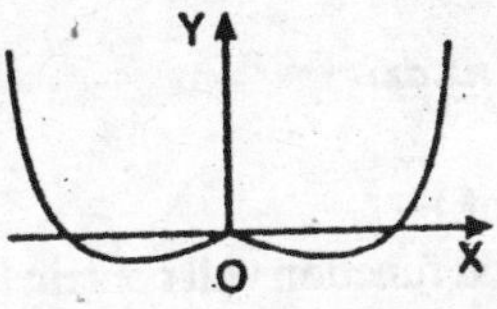

Fig. 1

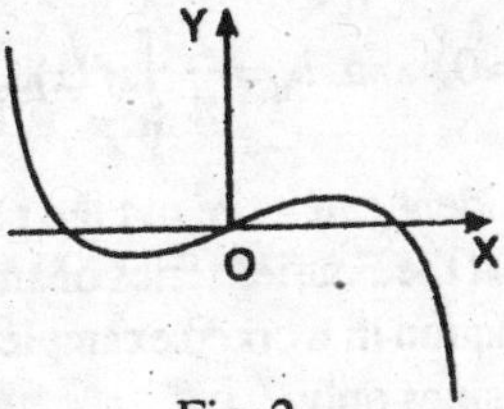

Fig. 2

The function $f(x)$ is odd, if $f(-x) = -f(x)$. For example, x, $\sin x$, $3x - 4\sin x$ are odd functions of x. The graph of an odd function will be symmetric about the origin. Refer Fig. 2.

There are functions which are neither odd nor even; $1 + x$ and e^x are such functions.

Results to Remember. (from integral calculus)

$$\int_{-a}^{a} f(x)\,dx = 2\int_{0}^{a} f(x)\,dx \text{ if } f(x) \text{ is even}$$

and $$\int_{-a}^{a} f(x)\,dx = 0 \text{ if } f(x) \text{ is odd.}$$

Theorem 5. *When an even function $f(x)$ is expanded in a Fourier series over the interval $-\pi$ to π, the series will have cosine terms only, and the coefficients of the series are given by* $a_n = \frac{2}{\pi}\int_0^{\pi} f(x)\cos nx\,dx$; *and* $b_n = 0$

Proof. We have already seen that,

$$a_n = \frac{1}{\pi} \int_{-\pi}^{\pi} f(x) \cos nx \, dx$$

$$= \frac{2}{\pi} \int_{0}^{\pi} f(x) \cos nx \, dx \text{ since the integrand is even}$$

$$b_n = \frac{1}{\pi} \int_{-\pi}^{\pi} f(x) \sin nx \, dx = 0 \because \text{ integrand is odd}$$

{$f(x)$ is even and $\sin nx$ is odd}

$$a_0 = \frac{2}{\pi} \int_{0}^{\pi} f(x) \, dx$$

Theorem 5 (a). *When an odd function $f(x)$ is developed in a Fourier series in the interval from $-\pi$ to π,*

$$a_n = 0, \text{ and } b_n = \frac{2}{\pi} \int_{0}^{\pi} f(x) \sin nx \, dx.$$

(The student can write out the proof.)

Note that the Fourier series of an odd function will contain terms in sines only. The function in worked example 2 is an odd function; and, the expansion had terms in sines only.

Example 9. *Find the Fourier series of $f(x) = x + x^2$ in $(-\pi, \pi)$ of periodicity 2π. Hence deduce $\sum \frac{1}{n^2} = \frac{\pi^2}{6}$* ***(MS. 1993 April)***

Let $f(x) = \frac{a_0}{2} + \sum_{n=1}^{\infty} (a_n \cos nx + b_n \sin nx)$ ...(1)

Then $a_0 = \frac{1}{\pi} \int_{-\pi}^{\pi} (x + x^2) dx$

$$= \frac{1}{\pi} \int_{-\pi}^{\pi} x dx + \frac{1}{\pi} \int_{-\pi}^{\pi} x^2 \, dx$$

$$= 0 + \frac{2}{\pi} \int_{0}^{\pi} x^2 \, dx$$

Since the first integral vanishes because integrand x is odd and the second integral is $\frac{2}{\pi} \int_{0}^{\pi} x^2 \, dx$ since x^2 is even.

we use the theorem, $\int_{-a}^{a} f(x)\,dx = 0$ if $f(x)$ is odd

and $\int_{-a}^{a} f(x)\,dx = 2\int_{0}^{a} f(x)\,dx$ if $f(x)$ is even.

$$\text{Hence } a_0 = \frac{2}{\pi}\left[\frac{x^3}{3}\right]_0^{\pi} = \frac{2}{3}\pi^2$$

$$a_n = \frac{1}{\pi}\int_{-\pi}^{\pi} (x + x^2)\cos nx\,dx$$

$$= \frac{1}{\pi}\int_{-\pi}^{\pi} x\cos nx\,dx + \frac{1}{\pi}\int_{-\pi}^{\pi} x^2\cos nx\,dx$$

$$= 0 + \frac{2}{\pi}\int_{0}^{\pi} x^2\cos nx\,dx, \qquad \text{since } x\cos nx \text{ is odd}$$

$$= \frac{2}{\pi}\left[(x^2)\left(\frac{\sin nx}{n}\right) - (2x)\left(-\frac{\cos nx}{n^2}\right) + (2)\left(-\frac{\sin nx}{n^3}\right)\right]_0^{\pi}$$

$$= \frac{2}{\pi}\left[\frac{2\pi}{n^2}\cos n\pi\right] = \frac{4}{n^2}(-1)^n$$

$$b_n = \frac{1}{\pi}\int_{-\pi}^{\pi} (x + x^2)\sin nx\,dx$$

$$= \frac{1}{\pi}\int_{-\pi}^{\pi} x\sin nx\,dx + \frac{1}{\pi}\int_{-\pi}^{\pi} x^2\sin nx\,dx.$$

$$= \frac{2}{\pi}\int_{0}^{\pi} x\sin nx\,dx + 0$$

$$= \frac{2}{\pi}\left[(x)\left(\frac{-\cos nx}{n}\right) - (1)\left(\frac{-\sin nx}{n^2}\right)\right]_0^{\pi}$$

$$= \frac{2}{\pi}\left[\frac{-\pi}{n}\cos n\pi\right] = -\frac{2}{n}(-1)^n$$

Substituting the values of a_0, a_n, b_n, in (1), we get

$$x + x^2 = \frac{\pi^2}{3} + \sum_{n=1}^{\infty} (-1)^n\left[\frac{4}{n^2}\cos nx - \frac{2}{n}\sin nx\right]$$

Deduction: $x = -\pi$ and $x = \pi$ are end points of the range.

Therefore the value of Fourier series at $x = \pi$ is the average of the values of $f(x)$ at $x = \pi$ and $x = -\pi$

Hence, put $x = \pi$ in Fourier series

$$\frac{\pi^2}{3} + \sum_{n=1}^{\infty} (-1)^n \frac{4}{n^2} \cos n\pi = \frac{f(\pi) + f(-\pi)}{2} = \frac{(\pi + \pi^2) + (-\pi + \pi^2)}{2}$$

$$\frac{\pi^2}{3} + 4 \sum_{n=1}^{\infty} \frac{1}{n^2} = \pi^2 \qquad \therefore \quad 4 \sum_{1}^{\infty} \frac{1}{n^2} = \frac{2}{3}\pi^2$$

$$\therefore \qquad \sum_{1}^{\infty} \frac{1}{n^2} = \frac{1}{1^2} + \frac{1}{2^2} + \frac{1}{3^2} + \cdots\text{to } \infty = \frac{\pi^2}{6}$$

Example 10. *In* $-\pi < x < \pi$, *Express sinh ax and cosh ax in Fourier series of periodicity* 2π. **(MS. 1989A; 1988 Nov.)**

Let $f(x) = \sinh ax = \dfrac{a_0}{2} + \sum_{n=1}^{\infty} (a_n \cos nx + b_n \sin nx)$

Here, $a_0 = \dfrac{1}{\pi} \int_{-\pi}^{\pi} f(x)\, dx = \dfrac{1}{\pi} \int_{-\pi}^{\pi} \sinh ax\, dx = 0$

$\because$ the integrand $f(x)$ is odd.

$$a_n = \frac{1}{\pi} \int_{-\pi}^{\pi} \sinh ax \cos nx\, dx = 0 \quad \because \text{ the integrand is odd}$$

$$b_n = \frac{1}{\pi} \int_{-\pi}^{\pi} \sinh ax \sin nx\, dx$$

$$= \frac{2}{\pi} \int_{0}^{\pi} \sinh ax \sin nx\, dx \text{ since the integrand is even.}$$

$$= \frac{2}{\pi} \int_{0}^{\pi} \frac{e^{ax} - e^{-ax}}{2} \sin nx\, dx$$

$$= \frac{1}{\pi} \int_{0}^{\pi} e^{ax} \sin nx\, dx - \frac{1}{\pi} \int_{0}^{\pi} e^{-ax} \sin nx\, dx$$

$$= \frac{1}{\pi} \cdot \left[\frac{e^{ax}}{a^2 + n^2} \{a \sin nx - n \cos nx\} - \frac{e^{-ax}}{a^2 + n^2} \{-a \sin nx - n \cos nx\} \right]_0^{\pi}$$

$$= \frac{1}{\pi} \cdot \frac{1}{a^2 + n^2} \left[-ne^{a\pi} \ (-1)^n + n + ne^{-a\pi} \ (-1)^n - n \right]$$

$$= \frac{1}{\pi} \cdot \frac{1}{a^2 + n^2} (-1)^{n-1} \cdot n(e^{a\pi} - e^{-a\pi})$$

$$= \frac{2n}{\pi} \frac{(-1)^{n-1}}{a^2 + n^2} \cdot \sinh a\pi$$

Hence $\sinh ax = \frac{2}{\pi} \sinh a\pi \sum_{n=1}^{\infty} \frac{(-1)^{n-1} n}{a^2 + n^2} \sin nx.$

(*ii*) Let $\cosh ax = \frac{a_0}{2} + \sum_{n=1}^{\infty} (a_n \cos nx + b_n \sin nx)$

$$a_0 = \frac{1}{\pi} \int_{-\pi}^{\pi} f(x)\, dx = \frac{1}{\pi} \int_{-\pi}^{\pi} \cosh ax\, dx$$

$$= \frac{2}{\pi} \int_0^{\pi} \cosh ax\, dx$$

$$= \frac{2}{\pi} \left(\frac{\sinh ax}{a} \right)_0^{\pi}$$

$$= \frac{2}{\pi a} \sinh a\pi$$

$$a_n = \frac{1}{\pi} \int_{-\pi}^{\pi} \cosh ax \cos nx\, dx$$

$$= \frac{2}{\pi} \int_0^{\pi} \cosh ax \cos nx\, dx$$

$$= \frac{2}{\pi} \int_0^{\pi} \frac{e^{ax} + e^{-ax}}{2} \cdot \cos nx\, dx$$

$$= \frac{1}{\pi} \left[\int_0^{\pi} e^{ax} \cos nx\, dx + \int_0^{\pi} e^{-ax} \cos nx\, dx \right]$$

$$= \frac{1}{\pi}\left[\frac{e^{ax}}{a^2+n^2}\{a\cos nx + n\sin nx\} + \frac{e^{-ax}}{a^2+n^2}\begin{Bmatrix}-a\cos nx \\ +n\sin nx\end{Bmatrix}\right]_0^{\pi}$$

$$= \frac{1}{\pi}\cdot\frac{1}{a^2+n^2}\left[e^{a\pi}\cdot a(-1)^n - a - e^{-a\pi}\cdot a(-1)^n + a\right]$$

$$= \frac{1}{\pi}\cdot\frac{1}{a^2+n^2}\, a(-1)^n\, [e^{a\pi} - e^{-a\pi}]$$

$$= \frac{2}{\pi}\cdot\frac{1}{a^2+n^2}\, a(-1)^n \cdot \sinh a\pi$$

$$b_n = \frac{1}{\pi}\int_{-\pi}^{\pi}\cosh ax \sin nx\, dx = 0 \quad (\because \text{ integrand is odd})$$

$$f(x) = \cosh ax = \frac{1}{\pi a}\sinh a\pi + \sum_1^{\infty}\cdot\frac{2}{\pi}\cdot\frac{1}{a^2+n^2}\, a(-1)^n \sinh a\pi \cos nx.$$

$$= \frac{2a}{\pi}\sinh a\pi\left[\frac{1}{2a^2} + \sum_1^{\infty}\frac{1}{a^2+n^2}(-1)^n\cdot\cos nx\right]$$

Example 11. *If* $f(x) = \sin x$ *in* $0 \le x \le \pi$

$= 0$ *in* $\pi \le x \le 2\pi$

find a Fourier series of periodicity 2π *and hence evaluate* $\frac{1}{1\cdot3} + \frac{1}{3\cdot5} + \frac{1}{5\cdot7} + \cdots$ *to* ∞.

Let $$f(x) = \frac{a_0}{2} + \sum_{n=1}^{\infty}(a_n\cos nx + b_n \sin nx) \quad \text{...(1)}$$

$$a_0 = \frac{1}{\pi}\int_0^{2\pi} f(x)\, dx$$

$$= \frac{1}{\pi}\left[\int_0^{\pi}\sin x\, dx + \int_{\pi}^{2\pi} 0\cdot dx\right]$$

$$= \frac{1}{\pi}(-\cos x)_0^{\pi} = \frac{2}{\pi} \quad \text{....(2)}$$

$$a_n = \frac{1}{\pi}\cdot\int_0^{\pi}\sin x \cos nx\, dx$$

$$= \frac{1}{2\pi} \int_0^{\pi} [\sin (n+1) x - \sin (n-1) x]\, dx$$

$$= \frac{1}{2\pi} \left[-\frac{\cos (n+1) x}{n+1} + \frac{\cos (n-1) x}{n-1} \right]_0^{\pi} \quad \text{if } n \neq 1$$

$$= \frac{1}{2\pi} \left[\frac{1}{n+1} \{(-1)^n + 1\} + \frac{1}{n-1} \{(-1)^{n-1} - 1\} \right]$$

$$= \frac{1}{2\pi} \frac{1}{n^2-1} [(n-1)\{1+(-1)^n\} - (n+1)\{(-1)^n + 1\}]$$

$$= \frac{1}{2\pi (n^2-1)} \{(-1)^n (-2) - 2 \}$$

$$= \frac{-1}{\pi (n^2-1)} [1 + (-1)^n]$$

$= 0$ if n is odd

$$= \frac{-2}{\pi (n^2-1)} \text{ if } n \text{ is even, } n \neq 1. \qquad \text{...(3)}$$

$$a_1 = \frac{1}{\pi} \int_0^{\pi} \sin x \cos x \, dx$$

$$= \frac{1}{2\pi} \int_0^{\pi} \sin 2x \, dx$$

$$= \frac{1}{2\pi} \left(-\frac{1}{2} \cos 2x \right)_0^{\pi}$$

$$= -\frac{1}{4\pi} (1-1) = 0$$

$$b_n = \frac{1}{\pi} \int_0^{\pi} \sin x \sin nx \, dx$$

$$= \frac{1}{2\pi} \int_0^{\pi} [\cos (n-1)x - \cos (n+1)x] \, dx$$

$$= \frac{1}{2\pi} \left[\frac{\sin (n-1) x}{n-1} - \frac{\sin (n+1) x}{n+1} \right]_0^{\pi} \quad \text{if } n \neq 1$$

$= 0$ if $n \neq 1$

$$b_1 = \frac{1}{\pi}\int_0^{\pi} \sin x \sin x \, dx$$

$$= \frac{1}{\pi}\int_0^{\pi} \frac{1-\cos 2x}{2}\, dx$$

$$= \frac{1}{2\pi}\left[x - \frac{1}{2}\sin 2x\right]_0^{\pi} = \frac{1}{2}$$

$$\therefore \quad f(x) = \frac{1}{\pi} + \frac{1}{2}\sin x - \frac{2}{\pi}\sum_{n=2,4,6}^{\infty} \frac{1}{(n^2-1)} \cos nx$$

$x = 0$ is an end point in the range

$\therefore$ The value of the Fourier series at $x = 0$ is equal to $\dfrac{f(0)+f(2\pi)}{2}$

Putting $x = 0$ in the Fourier series,

$$\frac{1}{\pi} - \frac{2}{\pi}\sum_{n=2,4,6}^{\infty} \frac{1}{n^2-1} = \frac{0+0}{2} = 0$$

$$\sum_{n=2,4,6}^{\infty} \frac{1}{(n-1)(n+1)} = \frac{1}{2}$$

i.e. $$\frac{1}{1\cdot 3} + \frac{1}{3\cdot 5} + \frac{1}{5\cdot 7} + \cdots \text{to } \infty = \frac{1}{2}$$

Example 12. *Express $f(x) = (\pi - x)^2$ as a Fourier series of periodicity 2π in $0 < x < 2\pi$ and hence deduce the sum $\sum_{n=1}^{\infty} \frac{1}{n^2}$* **(MS. 1989 April)**

Let $$f(x) = \frac{a_0}{2} + \sum_{n=1}^{\infty} (a_n \cos nx + b_n \sin nx)$$

$$a_0 = \frac{1}{\pi}\int_0^{2\pi} f(x)\, dx$$

$$= \frac{1}{\pi}\int_0^{2\pi} (\pi - x)^2\, dx$$

$$= \frac{1}{\pi}\left[\frac{-(\pi-x)^3}{3}\right]_0^{2\pi}$$

$$= -\frac{1}{3\pi}\left[-\pi^3-\pi^3\right] = \frac{2}{3}\pi^2$$

$$a_n = \frac{1}{\pi}\int_0^{2\pi}(\pi-x)^2\cos nx\,dx$$

$$= \frac{1}{\pi}\left[(\pi-x)^2\left(\frac{\sin nx}{n}\right)+2(\pi-x)\left(-\frac{\cos nx}{n^2}\right)+2\left(-\frac{\sin nx}{n^3}\right)\right]_0^{2\pi}$$

$$= \frac{1}{\pi}\left[\frac{2\pi}{n^2}+\frac{2\pi}{n^2}\right] = \frac{4}{n^2}$$

$$b_n = \frac{1}{\pi}\int_0^{2\pi}(\pi-x)^2\sin nx\,dx$$

$$= \frac{1}{\pi}\left[(\pi-x)^2\left(-\frac{\cos nx}{n}\right)+2(\pi-x)\left(-\frac{\sin nx}{n^2}\right)+2\left(\frac{\cos nx}{n^3}\right)\right]_0^{2\pi}$$

$$= \frac{1}{\pi}\left[-\frac{\pi^2}{n}+\frac{\pi^2}{n}+\frac{2}{n^3}(1-1)\right] = 0$$

$$\therefore \quad f(x) = (\pi-x)^2 = \frac{\pi^2}{3} + 4\sum_{n=1}^{\infty}\frac{1}{n^2}\cos nx$$

$x = 0$ is an end point. The value of Fourier series at $x = 0$ is $\dfrac{f(0)+f(2\pi)}{2}$

$$\therefore \quad \frac{\pi^2}{3} + 4\sum_{n=1}^{\infty}\frac{1}{n^2} = \frac{\pi^2+\pi^2}{2} = \pi^2$$

$$\therefore \quad \sum_{1}^{\infty}\frac{1}{n^2} = \frac{\pi^2}{6}.$$

Example 13. *Find the Fourier series of* $f(x)=e^x$ *in* $(-\pi,\pi)$, *of periodicity* 2π **(MS. 1991 Nov.)**

$$\text{Let}\quad f(x)=\frac{a_0}{2}+\sum_{n=1}^{\infty}(a_n\cos nx+b_n\sin nx) \qquad ...(1)$$

where

$$a_0=\frac{1}{\pi}\int_{-\pi}^{\pi}e^x\,dx$$

$$=\frac{1}{\pi}(e^x)_{-\pi}^{\pi}$$

$$=\frac{1}{\pi}(e^{\pi}-e^{-\pi})$$

$$=\frac{2}{\pi}\sinh\pi$$

$$a_n=\frac{1}{\pi}\int_{-\pi}^{\pi}e^x\cos nx\,dx$$

$$=\frac{1}{\pi}\left[\frac{e^x}{1+n^2}(\cos nx+n\sin nx)\right]_{-\pi}^{\pi}$$

$$=\frac{1}{\pi(1+n^2)}[e^{\pi}(-1)^n+e^{-\pi}(-1)^n]$$

$$=\frac{2(-1)^n}{\pi(1+n^2)}\sinh\pi$$

$$b_n=\frac{1}{\pi}\int_{-\pi}^{\pi}e^x\sin nx\,dx$$

$$=\frac{1}{\pi}\left[\frac{e^x}{1+n^2}(\sin nx-n\cos nx)\right]_{-\pi}^{\pi}$$

$$=\frac{1}{\pi(1+n^2)}[e^{\pi}(-n)(-1)^n+e^{-\pi}n(-1)^n]$$

$$=\frac{-2(-1)^n\cdot n}{\pi(1+n^2)}\sinh\pi$$

$$\therefore \qquad e^x = \frac{\sinh \pi}{\pi}\left[1 + \sum_{n=1}^{\infty} \frac{2(-1)^n}{1+n^2}(\cos nx - n \sin nx)\right]$$

Example 14. *Find the Fourier series of periodicity 2π*

$$\textit{for} \quad f(x) = \begin{cases} x \text{ when } -\pi < x < 0 \\ 0 \text{ when } 0 < x < \dfrac{\pi}{2} \\ x - \dfrac{\pi}{2} \text{ when } \dfrac{\pi}{2} < x < \pi \end{cases}$$ **(Kerala University)**

$$\text{Let} \quad f(x) = \frac{a_0}{2} + \sum_{n=1}^{\infty} (a_n \cos nx + b_n \sin nx) \qquad ...(1)$$

$$\text{where} \qquad a_0 = \frac{1}{\pi}\int_{-\pi}^{\pi} f(x)\, dx$$

$$= \frac{1}{\pi}\left[\int_{-\pi}^{0} x\, dx + \int_{0}^{\frac{\pi}{2}} 0\, dx - \int_{\frac{\pi}{2}}^{\pi}\left(x - \frac{\pi}{2}\right) dx\right]$$

$$= \frac{1}{\pi}\left[\left(\frac{x^2}{2}\right)_{-\pi}^{0} + \left\{\frac{1}{2}\left(x - \frac{\pi}{2}\right)^2\right\}_{\frac{\pi}{2}}^{\pi}\right]$$

$$= \frac{1}{\pi}\left[-\frac{\pi^2}{2} + \frac{\pi^2}{8}\right] = -\frac{3\pi}{8}$$

$$a_n = \frac{1}{\pi}\int_{-\pi}^{\pi} f(x) \cos nx\, dx$$

$$= \frac{1}{\pi}\left[\int_{-\pi}^{0} x \cos nx\, dx + \int_{0}^{\frac{\pi}{2}} 0 \cdot \cos nx\, dx + \int_{\frac{\pi}{2}}^{\pi}\left(x - \frac{\pi}{2}\right) \cos nx\, dx\right]$$

$$= \frac{1}{\pi}\left[\left\{(x)\left(\frac{\sin nx}{n}\right)-(1)\left(-\frac{\cos nx}{n^2}\right)\right\}_{-\pi}^{0}\right.$$

$$\left.+\left\{\left(x-\frac{\pi}{2}\right)\left(\frac{\sin nx}{n}\right)-(1)\left(-\frac{\cos nx}{n^2}\right)\right\}_{\frac{\pi}{2}}^{\pi}\right]$$

$$= \frac{1}{\pi}\left[\frac{1}{n^2}\{1-(-1)^n\}+\frac{1}{n^2}\left\{(-1)^n-\cos\frac{n\pi}{2}\right\}\right]$$

$$= \frac{1}{\pi}\left[\frac{1}{n^2}\left(1-\cos\frac{n\pi}{2}\right)\right]$$

$$b_n = \frac{1}{\pi}\left[\int_{-\pi}^{0} x\sin nx\,dx + \int_{\frac{\pi}{2}}^{\pi}\left(x-\frac{\pi}{2}\right)\sin nx\,dx\right]$$

$$= \frac{1}{\pi}\left[\left\{(x)\left(-\frac{\cos nx}{n}\right)-(1)\left(-\frac{\sin nx}{n^2}\right)\right\}_{-\pi}^{0}\right.$$

$$\left.+\left\{\left(x-\frac{\pi}{2}\right)\left(-\frac{\cos nx}{n}\right)-(1)\left(-\frac{\sin nx}{n^2}\right)\right\}_{\frac{\pi}{2}}^{\pi}\right]$$

$$= \frac{1}{\pi}\left[-\frac{\pi}{n}\cos n\pi - \frac{\pi}{2n}\cos n\pi - \frac{1}{n^2}\sin\frac{n\pi}{2}\right]$$

$$= -\frac{1}{\pi}\left[\frac{3\pi}{2n}(-1)^n + \frac{1}{n^2}\sin\frac{n\pi}{2}\right]$$

$$\therefore \quad f(x) = \frac{-3\pi}{16} + \frac{1}{\pi}\sum_{n=1}^{\infty}\frac{1}{n^2}\left(1-\cos\frac{n\pi}{2}\right)\cos nx - \frac{1}{\pi}\sum_{n=1}^{\infty}\left(\frac{3\pi}{2n}(-1)^n + \frac{1}{n^2}\sin\frac{n\pi}{2}\right)\sin nx$$

Example 15. *Find the Fourier series of periodicity 2π*

for $f(x) = \begin{cases} x & \text{in } (0, \pi) \\ 2\pi - x & \text{in } (\pi, 2\pi) \end{cases}$

and hence deduce $\frac{1}{1^2} + \frac{1}{3^2} + \frac{1}{5^2} = \frac{\pi^2}{8}$

Let $f(x) = \frac{a_0}{2} + \sum_{n=1}^{\infty} (a_n \cos nx + b_n \sin nx)$

$$a_0 = \frac{1}{\pi}\left[\int_0^{\pi} x\,dx + \int_{\pi}^{2\pi} (2\pi - x)\,dx\right]$$

$$= \frac{1}{\pi}\left[\left(\frac{x^2}{2}\right)_0^{\pi} - \left\{\frac{1}{2}(2\pi - x)^2\right\}_{\pi}^{2\pi}\right]$$

$$= \frac{1}{\pi}\left[\frac{\pi^2}{2} + \frac{\pi^2}{2}\right] = \pi$$

$$a_n = \frac{1}{\pi}\left[\int_0^{\pi} x\cos nx\,dx + \int_{\pi}^{2\pi} (2\pi - x)\cos nx\,dx\right]$$

$$= \frac{1}{\pi}\left[\left\{(x)\left(\frac{\sin nx}{n}\right) - (1)\left(-\frac{\cos nx}{n^2}\right)\right\}_0^{\pi}\right.$$

$$\left. + \left\{(2\pi - x)\left(\frac{\sin nx}{n}\right) - (-1)\left(-\frac{\cos nx}{n^2}\right)\right\}_{\pi}^{2\pi}\right]$$

$$= \frac{1}{\pi}\left[\frac{1}{n^2}\{(-1)^n - 1\} - \frac{1}{n^2}\{1 - (-1)^n\}\right]$$

$$= \frac{2}{\pi n^2}\,[(-1)^n - 1]$$

$$b_n = \frac{1}{\pi}\left[\int_0^{\pi} x\sin nx\,dx + \int_{\pi}^{2\pi} (2\pi - x)\sin nx\,dx\right]$$

$$= \frac{1}{\pi}\left[\left\{(x)\left(-\frac{\cos nx}{n}\right) - (1)\left(-\frac{\sin nx}{n^2}\right)\right\}_0^{\pi}\right.$$

$$\left. + \left\{(2\pi - x)\left(\frac{-\cos nx}{n}\right) - (-1)\left(-\frac{\sin nx}{n^2}\right)\right\}_{\pi}^{2\pi}\right]$$

$$= \frac{1}{\pi}\left[\frac{-\pi}{n}(-1)^n + \frac{\pi}{n}(-1)^n\right] = 0$$

Hence $f(x) = \frac{\pi}{2} + \frac{2}{\pi} \sum_{n=1}^{\infty} \frac{1}{n^2} [(-1)^n - 1] \cos nx$

$$= \frac{\pi}{2} - \frac{4}{\pi} \left[\cos x + \frac{1}{3^2} \cos 3x + \frac{1}{5^2} \cos 5x + \text{.....to } \infty \right]$$

$x = 0$ is an end point of the range.

The value of Fourier series at an end is equal to the average of $[f(0) + f(2\pi)]$

$\therefore$ At $x = 0$,

$$\frac{\pi}{2} - \frac{4}{\pi} \left(\frac{1}{1^2} + \frac{1}{3^2} + \frac{1}{5^2} + \cdots \right) = \frac{1}{2} [f(0) + f(2\pi)]$$

$$= \frac{1}{2} [0 + 0] = 0$$

$$\frac{4}{\pi} \left(\frac{1}{1^2} + \frac{1}{3^2} + \frac{1}{5^2} + \cdots \right) = \frac{\pi}{2}$$

$$\frac{1}{1^2} + \frac{1}{3^2} + \frac{1}{5^2} + \cdots = \frac{\pi^2}{8}.$$

Example 16. *In* $(-\pi, \pi)$, *find the Fourier series of periodicity* 2π *for*

$$f(x) = \begin{cases} 1 + x & \text{in } 0 < x < \pi \\ -1 + x & \text{in } -\pi < x < 0 \end{cases}$$

The function $f(x)$ *is odd evidently; for*

Let $0 < \alpha < \pi$; Than $f(\alpha) = 1 + \alpha$

$$f(-\alpha) = -1 - \alpha = -(1 + \alpha) = -f(\alpha)$$

Let $f(x) = \frac{a_0}{2} + \sum_{n=1}^{\infty} (a_n \cos nx + b_n \sin nx)$

where $a_0 = \frac{1}{\pi} \int_{-\pi}^{\pi} f(x)\, dx = 0$

$$a_n = \frac{1}{\pi} \int_{-\pi}^{\pi} f(x) \cos nx\, dx = 0 \qquad (\because \text{ integrand is o})$$

$$b_n = \frac{1}{\pi} \int_{-\pi}^{\pi} f(x) \sin nx\, dx$$

$$= \frac{2}{\pi} \int_0^{\pi} f(x) \sin nx \, dx$$

$$= \frac{2}{\pi} \int_0^{\pi} (1 + x) \sin nx \, dx$$

$$= \frac{2}{\pi} \left[(1 + x) \left(-\frac{\cos nx}{n} \right) - (1) \left(-\frac{\sin nx}{n^2} \right) \right]_0^{\pi}$$

$$= \frac{2}{\pi} \left[(1 + \pi) \left(-\frac{\cos n\pi}{n} \right) + \frac{1}{n} \right]$$

$$= \frac{2}{n\pi} [1 - (1 + \pi)(-1)^n]$$

$$\therefore \qquad f(x) = \frac{2}{\pi} \sum_{n=1}^{\infty} \frac{1}{n} [1 - (1 + \pi)(-1)^n] \sin nx.$$

Example 17. *Write down the analytic expressions of the following functions given in the graph. Hence find Fourier series of $f(x)$.*

(*a*)

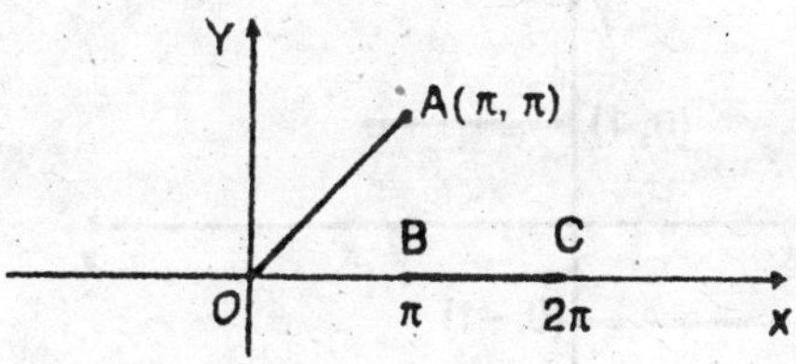

$$f(x) = x, \quad 0 < x < \pi$$
$$= 0 \quad \pi < x < 2\pi$$

(*b*) From the figure on page 128,

Slope of $BC = \dfrac{1 - 0}{0 - \pi} = \dfrac{-1}{\pi}$

equation of BC is $y = -\dfrac{1}{\pi}(x - \pi)$

Slope of AC is $\dfrac{1 - 0}{0 + \pi} = \dfrac{1}{\pi}$

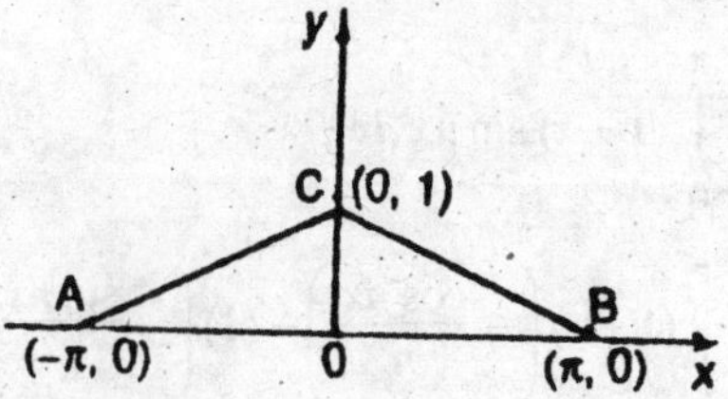

Equation of AC is $y = \frac{1}{\pi}(x + \pi)$

$$\therefore \quad f(x) = \frac{1}{\pi}(x + \pi), \qquad -\pi < x < 0$$

$$= -\frac{1}{\pi}(x - \pi), \text{ if } \quad 0 < x < \pi$$

{$f(x)$ is an even function}

(*c*)

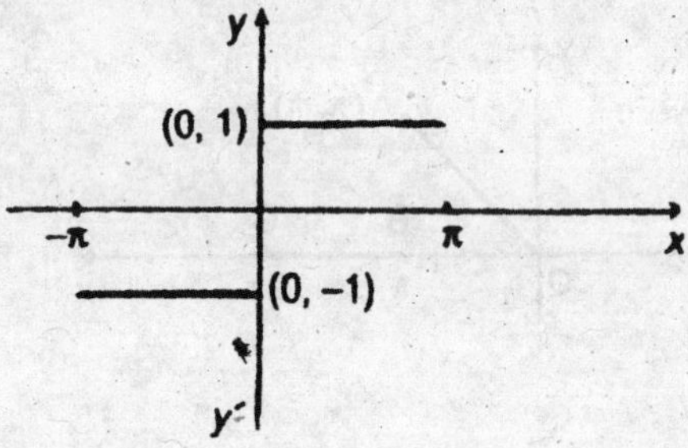

The two segments are parallel to x axis

$$\therefore \quad f(x) = -1, \quad -\pi < x < 0$$

$$= 1, \qquad 0 < x < \pi$$

(*d*) Refer to figure

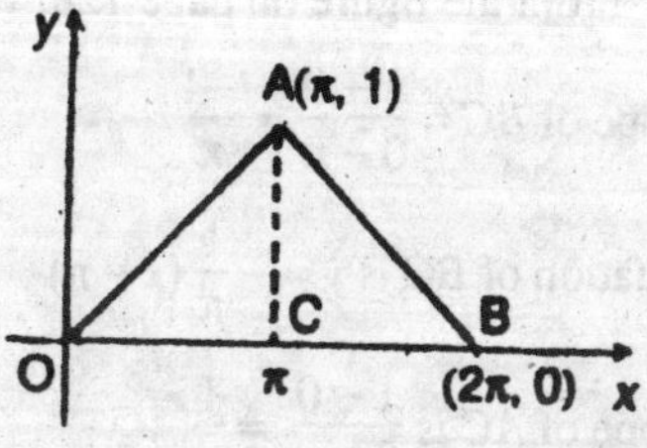

Slope of $OA = \frac{1}{\pi}$

Equation of OA is $y = \frac{1}{\pi}x$

$$\text{Slope of } AB = \frac{1-0}{\pi - 2\pi} = \frac{-1}{\pi}$$

$$\text{Equation of } AB \text{ is } y = -\frac{1}{\pi}(x - 2\pi)$$

$$\therefore \; f(x) = \frac{x}{\pi}, \quad 0 < x < \pi$$

$$= \frac{1}{\pi}(2\pi - x), \; \pi < x < 2\pi$$

Fourier series of the above functions can be got in the usual way.

Example 18. *If $f(x)$ is defined in $(-\pi, \pi)$ and if*

$f(x) = x + 1$ in $(0, \pi)$ find $f(x)$ in $(-\pi, 0)$ if

(*i*) $f(x)$ is odd (*ii*) $f(x)$ is even.

Ans: If $f(x)$ is even then,

$$f(x) = -x + 1 \quad \text{in } (-\pi, 0)$$

If $f(x)$ is odd, then

$$f(x) = x - 1 \quad \text{in } (-\pi, 0)$$

Exercises 1(c)

Expand each of the functions in 1 to 17 in a Fourier series. It is given that each of the functions has the period 2π.

1. $f(x) = e^x$, $-\pi < x < \pi$. Derive a series for $\dfrac{\pi}{\sinh \pi}$.
2. $f_1(x) = x$, $-\pi < x < \pi$. Deduce the value of $\dfrac{\pi}{4}$.
3. $F(x) = x + e^x$, $-\pi < x < \pi$.
4. $f(x) = x^2$, $-\pi < x < \pi$. Deduce the relations

$$1 + \frac{1}{2^2} + \frac{1}{3^2} + \frac{1}{4^2} + \cdots \text{ to } \infty = \frac{\pi^2}{6};$$

$$1 - \frac{1}{2^2} + \frac{1}{3^2} - \frac{1}{4^2} + \cdots \text{ to } \infty = \frac{\pi^2}{12};$$

$$1 + \frac{1}{3^2} + \frac{1}{5^2} + \frac{1}{7^2} + \cdots \text{ to } \infty = \frac{\pi^2}{8}.$$

5. $f(x) = x^3$, $-\pi < x < \pi$.
6. $f(x) = \begin{cases} -\dfrac{\pi}{4}, & \text{when } -\pi < x < 0 \\ \dfrac{\pi}{4}, & \text{when } 0 < x < \pi, \end{cases}$

and $f(-\pi) = f(0) = f(\pi) = 0$, and $f(x + 2\pi) = f(x)$ for all x.

Deduce that $\frac{\pi}{4} = 1 - \frac{1}{3} + \frac{1}{5} - \frac{1}{7} + \ldots$ to ∞.

7. $f(x) = 0$, if $-\pi \leq x < 0$; and $f(x) = \pi$, if $0 \leq x < \pi$. Show how this can be got from example 6.

8. $F(x) = -x$, if $-\pi \leq x \leq 0$; and $f(x) = \pi - x$, if $0 < x < \pi$. Show how this can be got from example 2 and example 7.

9. $f(x) = \pi$, if $-\pi < x \leq \frac{1}{2}\pi$; and $f(x) = 0$, if $\frac{1}{2}\pi < x \leq \pi$.

10. $f(x) = 1$, if $-\frac{1}{2}\pi < x < \frac{1}{2}\pi$; and $f(x) = -1$, if $\frac{1}{2}\pi < x < \frac{3}{2}\pi$.

11. $f(x) = 1$, if $0 < x < \frac{1}{2}\pi$; and $f(x) = 0$, if $\frac{1}{2}\pi < x < 2\pi$.

12. $f(x) = \begin{cases} 2, \text{ when } 0 < x < \frac{2}{3}\pi \\ 1, \text{ when } \frac{2}{3}\pi < x < \frac{4}{3}\pi \\ 0, \text{ when } \frac{4}{3}\pi < x < 2\pi \end{cases}$

13. $f(x) = 0$, when $-\pi < x < 0$ and $f(x) = x$, when $0 < x < \pi$.

14. $f(x) = x$, if $-\frac{1}{2}\pi < x < \frac{1}{2}\pi$ and $f(x) = 0$, if $\frac{1}{2}\pi < x < \frac{3}{2}\pi$.

15. $f(x) = 0$, when $-\pi < x < 0$ and $f(x) = x^2$, when $0 < x < \pi$.

16. $f(x) = \sin \frac{1}{2}x$, when $-\pi < x < \pi$.

17. $f(x) = 0$, when $-\pi < x < 0$; and $f(x) = \cos x$, when $0 \leq x \leq \pi$.

18. An alternating current of period 2π after passing through a rectifier is of the form

$$i = \begin{cases} 0, \text{ for } -\pi \leq \theta < 0, \\ I \sin \theta, \text{ for } 0 \leq \theta < \pi, \text{ where } I \text{ is the maximum current.} \end{cases}$$

Obtain the Fourier expansion for i.

19. Find the Fourier series of periodicity 2π for the following functions

(i) $f(x) = \begin{cases} 1, & 0 < x < \pi \\ -1 & \pi < x < 2\pi \end{cases}$ and $f(x + 2\pi) = f(x)$.

(ii) $f(x) = 2, \quad -\pi < x < 0$

$= 1 \quad 0 \leq x \leq \pi$ with period 2π

(iii) $f(x) = -\frac{1}{2}$ if $-\pi < x < 0$

$= \frac{1}{2}$ if $0 < x < \pi$ and $f(x + 2\pi) = f(x)$

(iv) $f(x) = x + 1, \quad 0 < x < \pi$

$= x - 1, \quad -\pi < x < 0$

(v) $f(x) = \begin{cases} 0, & -\pi < x < 0 \\ 1, & 0 < x < \pi \end{cases}$

(vi) $f(x) = (\pi - x)^2,\ 0 < x < 2\pi$ and hence deduce $1 + \frac{1}{2^2} + \frac{1}{3^2} + \cdots = \frac{\pi^2}{6}$

(MS. 1989 April)

Example 19. *Obtain the Fourier series of periodicity 2π for $f(x) = x$, in $-\pi < x < \pi$.*

$f(x)$ is an odd function of x. Hence by Th. 5(*a*), $a_n = 0$, and

$b_n = \frac{2}{\pi}\int_0^{\pi} x \sin nx\, dx$. In worked example 3, this integral was calculated as $-\frac{\pi}{n}\cos n\pi$. Hence $b_n = -2\,\frac{\cos n\pi}{n}$. Using the values of the coefficients, in the given interval,

$$f(x) = 2\left(\sin x - \frac{\sin 2x}{2} + \frac{\sin 3x}{3} - \cdots\right).$$

Note. Putting $x = \frac{\pi}{2}$, we get $1 - \frac{1}{3} + \frac{1}{5} - \frac{1}{7} + \cdots = \frac{\pi}{4}$.

Example 20. *Obtain the Fourier series of periodicity 2π for*

(*i*) $f(x) = -x$, *when* $-\pi < x \le 0$, *and* $f(x) = x$, *when* $0 < x < \pi$,

(*ii*) $f(x) = |x|$, *when* $-\pi < x < \pi$.

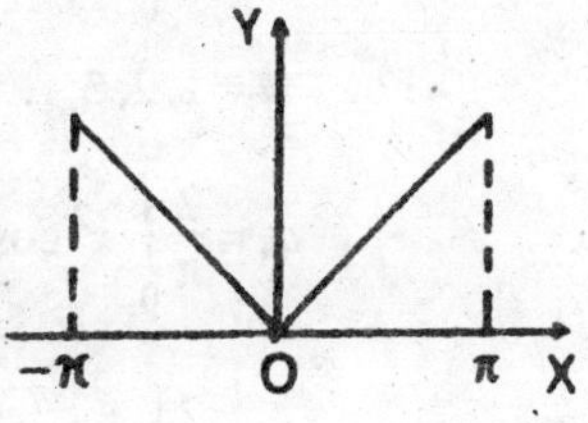

Note that both the questions are identical; they are merely different forms of wording the same question. The figure gives the graph of the function. The given function is even. Hence by Theorem 5, $b_n = 0$; and

$$a_0 = \frac{2}{\pi}\int_0^{\pi} f(x)\, dx = \frac{2}{\pi}\int_0^{\pi} x\, dx = \frac{2}{\pi}\left[\frac{x^2}{2}\right]_0^{\pi} = \frac{2}{\pi}\times\frac{\pi^2}{2} = \pi.$$

Again by Theorem 5, $a_n = \frac{2}{\pi}\int_0^{\pi} x \cos nx\, dx$. In worked example 3, this integral was calculated to be $\frac{\cos n\pi - 1}{n^2}$.

Hence $a_n = \frac{2}{\pi}\times\frac{\cos n\pi - 1}{n^2}$. Substituting the coefficients,

$$f(x) = \frac{\pi}{2} - \frac{4}{\pi}\left\{\cos x + \frac{\cos 3x}{3^2} + \frac{\cos 5x}{5^2} + \cdots\right\}$$

Note. Putting $x = 0$, we easily get $\dfrac{\pi^2}{8} = 1 + \dfrac{1}{3^2} + \dfrac{1}{5^2} + \cdots \text{to } \infty.$

Example 21. *Expand $f(x) = x^2$, when $-\pi < x < \pi$, in a Fourier series of periodicity 2π. Hence deduce that*

(*i*) $\dfrac{1}{1^2} + \dfrac{1}{2^2} + \dfrac{1}{3^2} + \cdots \text{to } \infty = \dfrac{\pi^2}{6}$

(*ii*) $\dfrac{1}{1^2} - \dfrac{1}{2^2} + \dfrac{1}{3^2} - \cdots \text{to } \infty = \dfrac{\pi^2}{12}$

(*iii*) $\dfrac{1}{1^2} + \dfrac{1}{3^2} + \dfrac{1}{5^2} + \cdots \text{to } \infty = \dfrac{\pi^2}{8}$

$f(x)$ is an even function of x in $-\pi < x < \pi$. Hence by theorem 5, $b_n = 0$ and only cosine terms will be present. Therefore,

$$f(x) = \frac{a_0}{2} + \sum_{n=1}^{\infty} a_n \cos nx \qquad \ldots.(i)$$

$$a_0 = \frac{2}{\pi}\int_0^{\pi} x^2 dx = \frac{2\pi^2}{3}$$

For $\quad n = 1, 2, 3, \ldots$

$$a_n = \frac{2}{\pi}\int_0^{\pi} x^2 \cos nx \, dx$$

$$= \frac{2}{\pi}\left[(x^2)\left(\frac{\sin nx}{n}\right) - (2x)\left(\frac{-\cos nx}{n^2}\right) + (2)\left(\frac{-\sin nx}{n^3}\right)\right]_0^{\pi}$$

$$= \frac{2}{\pi}\left[\frac{2\pi}{n^2}\cos n\pi\right] = \frac{4(-1)^n}{n^2}.$$

Substituting these values in (*i*),

$$f(x) = \frac{\pi^2}{3} + 4\sum_{n=1}^{\infty} \frac{(-1)^n}{n^2}\cos nx \qquad \ldots.(ii)$$

i.e., $x^2 = \dfrac{\pi^2}{3} - 4\left[\dfrac{\cos x}{1^2} - \dfrac{\cos 2x}{2^2} + \dfrac{\cos 3x}{3^2} - \cdots\right]$, in $-\pi < x < \pi$.

The function $f(x) = x^2$ is continuous at $x = 0$. Hence the sum of the Fourier series equals the value of the function at $x = 0$. Putting $x = 0$, in (*ii*),

$$0 = \frac{\pi^2}{3} - 4\left[\frac{1}{1^2} - \frac{1}{2^2} + \frac{1}{3^2} - \cdots\right]$$

$$\therefore \quad \frac{1}{1^2} - \frac{1}{2^2} + \frac{1}{3^2} - \cdots = \frac{\pi^2}{12} \qquad \ldots(iii)$$

$x = \pi$ is an end point. Hence the sum of the Fourier series at $x = \pi$ equals

$$\tfrac{1}{2}\{f(-\pi + 0) + f(\pi - 0)\}$$

Putting $x = \pi$ in the series of (*ii*),

$$\frac{\pi^2}{3} + 4\sum_{n=1}^{\infty} \frac{(-1)^n \cos n\pi}{n^2} = \tfrac{1}{2}\,[f(-\pi + 0) + f(\pi - 0)]$$

$$= \tfrac{1}{2}\,[\pi^2 + \pi^2] = \pi^2$$

$$\therefore \quad 4\sum_{n=1}^{\infty} \frac{1}{n^2} = \pi^2 - \frac{\pi^2}{3} = \tfrac{2}{3}\pi^2$$

i.e.,
$$\sum_{n=1}^{\infty} \frac{1}{n^2} = \frac{1}{1^2} + \frac{1}{2^2} + \frac{1}{3^2} + \cdots = \frac{\pi^2}{6} \qquad \ldots(iv)$$

Adding (*iii*) and (*iv*),

$$2\left(\frac{1}{1^2} + \frac{1}{3^2} + \frac{1}{5^2} + \cdots\right) = \frac{\pi^2}{4}$$

i.e.,
$$\frac{1}{1^2} + \frac{1}{3^2} + \frac{1}{5^2} + \cdots \text{ to } \infty = \frac{\pi^2}{8}.$$

Example 22. *Obtain the Fourier expansion of* $f(x) = |\sin x|$, *in* $-\infty < x < \infty$.

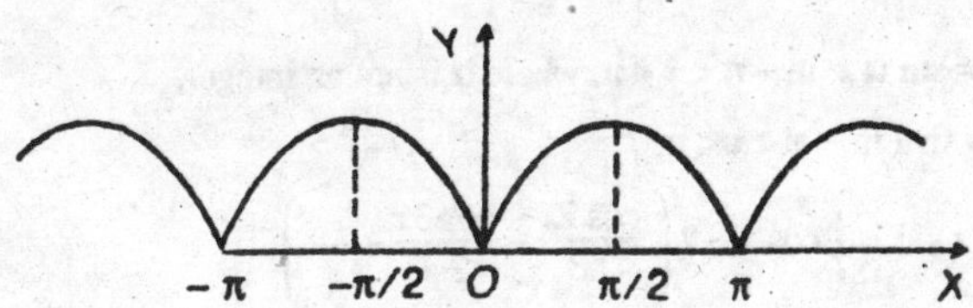

The graph of $y = |\sin x|$ is as shown above. It is periodic, with a period of π; without loss of generally, we can take a period from (0 to π). Again, the curve is evidently symmetrical about the y-axis. Hence $f(x)$ is an *even* function of x; therefore, its expansion will not contain terms in sine.

In the interval $-\pi$ to $\pi, f(x)$ can also be defined in the form $f(x) = -\sin x$, if $-\pi < x < 0$, and $f(x) = \sin x$, if $0 < x < \pi$.

Now, $a_1 = \frac{2}{\pi}\int_0^{\pi} \sin x \cos x \, dx = \frac{2}{\pi} \times 0 = 0$

For $n = 0, 2, 3, 4, ...$

$$a_n = \frac{2}{\pi}\int_0^{\pi} \sin x \cos nx \, dx = -\frac{2}{\pi} \cdot \frac{1 + \cos n\pi}{(n+1)(n-1)}$$

For the evaluation of the integral, refer worked example 4,

$$\text{Thus } f(x) = \frac{2}{\pi} - \frac{4}{\pi}\left\{\frac{\cos 2x}{3} + \frac{\cos 4x}{15} + \frac{\cos 6x}{35} + \frac{\cos 8x}{63} + ... \text{ to } \infty\right\}.$$

Example 23. *Expand $F(x) = x - x^2$, in $-\pi < x < \pi$ in a Fourier series.*

If $f(x) = x$ and $f_1(x) = x^2$, then evidently $F = f - f_1$. Hence the answer is written down using worked examples 19 and 21.

Exercises 1(d)

Obtain the Fourier expansion of the following:

1. $f(x) = \begin{cases} -x^2, & \text{if } -\pi < x < 0 \\ x^2, & \text{if } 0 < x < \pi \end{cases}$

2. $f(x) = x^3$, in $-\pi < x < \pi$.

3. $f(x) = -\cos x$, in $-\pi < x < 0$, and $f(x) = \cos x$, in $0 < x < \pi$.

4. $f(x) = \cos \alpha x$, in $-\pi < x < \pi$, where α is not an integer.

   Deduce that $\cot \pi\alpha = \frac{1}{\pi}\left\{\frac{1}{\alpha} - \sum_{n=1}^{\infty} \frac{2\alpha}{n^2 - \alpha^2}\right\}$.

5. $f(x) = \sin \alpha x$, in $-\pi < x < \pi$, where α is not an integer.

6. Show that in $-\pi < x < \pi$,

   $$x \sin x = 1 - \tfrac{1}{2}\cos x - 2\left(\frac{\cos 2x}{1 \cdot 3} - \frac{\cos 3x}{2 \cdot 4} + ...\right)$$

7. If $f(x) = 1 + \frac{2x}{\pi}$ in $-\pi \le x \le 0$

   $= 1 - \frac{2x}{\pi}$ in $0 \le x \le \pi$

   show that $f(x) = \frac{8}{\pi^2}\left(\cos x + \frac{1}{3^2}\cos 3x + \frac{1}{5^2}\cos 5x + ...\right)$

Hence show that $\sum_{n=1}^{\infty} (2n-1)^{-2} = \frac{\pi^2}{8}$.

Half-range series. In some problems, we are concerned with the interval 0 to π, (instead of the usual interval of length 2π.) Further, the conditions of the problem may require us to expand the given function in a series of sines alone or a series of consines alone.

Half-range sine series. This is achieved as follows: suppose $f(x)$ is given for $0 < x < \pi$, and we need a sine expansion for $f(x)$ in that interval. Then, we introduce a new function $F(x)$, which is defined as below.

$$F(x) = \begin{cases} f(x), \text{ when } 0 < x < \pi. \\ -f(-x), \text{ when } -\pi < x < 0. \end{cases}$$

From the definition of $F(x)$ made here, $F(x)$ is an odd function of x; and it is defined in the interval $-\pi$ to π. From theorem 5(*a*), the Fourier expansion of $F(x)$ will contain sine terms only. Since we are concerned with only the range $0 < x < \pi$, and since in this range $F(x) \equiv f(x)$, the sine series of $F(x)$ is the required series for $f(x)$ in the range $0 < x < \pi$. Further, by Theorem 5(*a*).

$$b_n = \frac{2}{\pi}\int_0^{\pi} F(x) \sin nx \, dx.$$

But as stated above, within the limits of integration, $F(x)$ coincides with $f(x)$.

Therefore, $f(x) = \sum_{1}^{\infty} b_n \sin nx$

where $$b_n = \frac{2}{\pi}\int_0^{\pi} f(x) \sin nx \, dx.$$

Since b_n depends only on $f(x)$, the actual creation of $F(x)$ is not necessary.

Half-range cosine series. On the other hand, if we need a cosine series for $f(x)$ defined in $0 < x < \pi$, define $F(x)$ by

$$F(x) = \begin{cases} f(x), \text{ when } 0 < x < \pi \\ f(-x), \text{ when } -\pi < x < 0. \end{cases}$$

Then $F(x)$ is an even function of x in the interval $(-\pi, \pi)$. By theorem 5, the Fourier series of $F(x)$ will have only the cosine terms, (including the constant). But, as before $F(x)$ coincides with $f(x)$ in the required interval $0 < x < \pi$. Hence the cosine series for $F(x)$ is the required cosine series for $f(x)$ in the given interval.

Now, $a_n = \frac{2}{\pi}\int_0^{\pi} F(x)\cos nx\, dx = \frac{2}{\pi}\int_0^{\pi} f(x)\cos nx\, dx.$

Therefore,

$$f(x) = \frac{a_0}{2} + \sum_{n=1}^{\infty} a_n \cos nx$$

where $\quad a_0 = \frac{2}{\pi}\int_0^{\pi} f(x)\, dx$

and $\quad a_n = \frac{2}{\pi}\int_0^{\pi} f(x)\cos nx\, dx.$

Thus a function $f(x)$ defined over the interval $0 < x < \pi$ is capable of expansion in two distinct types of series as above. Since these two series are valid over the half-range $0 < x < \pi$, we call them half-range series.

Since in the above discussions, we did not stipulate any condition on $f(x)$, the required type of series can be got whether $f(x)$ is odd, even or neither.

Example 24. *Expand the function $f(x) = \sin x$, $0 < x < \pi$ in Fourier cosine series.*

Let $f(x) = \frac{a_0}{2} + \sum_{n=1}^{\infty} a_n \cos nx$ where

$$a_0 = \frac{2}{\pi}\int_0^{\pi} \sin x\, dx = \frac{2}{\pi}\Big[-\cos x\Big]_0^{\pi} = \frac{2}{\pi}\left(1+1\right) = \frac{2}{\pi}\times 2 = \frac{4}{\pi}$$

$$a_1 = \frac{2}{\pi}\int_0^{\pi} \sin x \cos x\, dx = \frac{1}{\pi}\Big[\sin^2 x\Big]_0^{\pi} = \frac{1}{\pi}\times 0 = 0$$

For $n = 2, 3, 4, \ldots$ (for the working, refer worked example 4)

$$a_n = \frac{2}{\pi}\int_0^{\pi} \sin x \cos nx\, dx = -\frac{2(1+\cos n\pi)}{\pi(n-1)(n+1)}$$

Hence in the given interval,

$$f(x) = \frac{2}{\pi} - \frac{4}{\pi}\left\{\frac{\cos 2x}{1\cdot 3} + \frac{\cos 4x}{3\cdot 5} + \frac{\cos 6x}{5\cdot 7} + \ldots \text{to } \infty\right\}$$

Example 25. *Find sine half-range series for the function $f(x) = x$, in $0 < x \le \pi/2$, and $f(x) = \pi - x$ in $\frac{1}{2}\pi < x < \pi$.*

Let $f(x) = \sum_{n=1}^{\infty} b_n \sin nx$, where

$$b_n = \frac{2}{\pi}\int_0^{\pi} f(x) \sin nx \, dx = \frac{2}{\pi}\left\{\int_0^{\pi/2} x \sin nx \, dx + \int_{\pi/2}^{\pi} (\pi - x) \sin nx \, dx\right\} \quad \text{....}(i)$$

Using the method of integration by parts,

$$\int_0^{\pi/2} x \sin nx \, dx = \left[x\left(\frac{-\cos nx}{n}\right)\right]_0^{\pi/2} + \int_0^{\pi/2} \frac{\cos nx}{n} dx$$

$$= -\frac{\pi}{2n}\cos\frac{n\pi}{2} + \frac{1}{n^2}\sin\frac{n\pi}{2};$$

$$\int_{\pi/2}^{\pi} (\pi - x) \sin nx \, dx = \left[(\pi - x)\left\{-\frac{\cos nx}{n}\right\}\right]_{\pi/2}^{\pi} - \int_{\pi/2}^{\pi} \frac{\cos nx}{n} dx$$

$$= \frac{\pi}{2n}\cos\frac{n\pi}{2} + \frac{1}{n^2}\sin\frac{n\pi}{2}.$$

Using the values of the integrals in (*i*),

$$b_n = \frac{2}{\pi} \times \frac{2}{n^2}\sin\frac{n\pi}{2} = \frac{4}{\pi n^2}\sin\frac{n\pi}{2}.$$

Case 1. Let $n = 2m$; then $b_{2m} = \dfrac{4}{\pi (2m)^2}\sin m\pi = \dfrac{4}{4\pi m^2} \times 0 = 0;$

putting $m = 1, 2, 3, 4, \ldots\ldots$, we get $b_2 = b_4 = b_6 = \cdots = 0$.

Case 2. Let $n = 2r + 1$, where r is any positive integer. Then,

$$b_{2r+1} = \frac{4}{\pi (2r+1)^2}\sin\frac{(2r+1)\pi}{2} = \frac{4}{\pi (2r+1)^2}\sin\left(r\pi + \frac{\pi}{2}\right)$$

$$= \frac{4}{\pi (2r+1)^2}\cos r\pi = (-1)^r \frac{4}{\pi (2r+1)^2}.$$

Putting $r = 0, 1, 2$, we get

$$b_1 = +\frac{4}{\pi};\ b_3 = \frac{-4}{\pi \times 3^2};\ b_5 = \frac{4}{\pi \times 5^2};\ b_7 = \frac{-4}{\pi \times 7^2};\ \ldots\ldots$$

Substituting the values of the coefficients,

$$f(x) = \frac{4}{\pi}\left\{\sin x - \frac{\sin 3x}{3^2} + \frac{\sin 5x}{5^2} - \frac{\sin 7x}{7^2} + \cdots \text{ to } \infty\right\}$$

Example 26. *Find half-range Forurier cosine series and sine series.*

for $f(x) = x$ *in* $0 < x < \pi$ **(M.S. 1991 Ap.)**

(*i*) **Half-range cosine series:**

Let $f(x) = \frac{a_0}{2} + \sum_{n=1}^{\infty} a_n \cos nx$

where $$a_0 = \frac{2}{\pi} \int_0^{\pi} f(x)\, dx$$

$$= \frac{2}{\pi} \int_0^{\pi} x dx$$

$$= \frac{2}{\pi} \left(\frac{x^2}{2} \right)_0^{\pi} = \pi$$

$$a_n = \frac{2}{\pi} \int_0^{\pi} f(x) \cos nx\, dx$$

$$= \frac{2}{\pi} \int_0^{\pi} (x) \cos nx\, dx$$

$$= \frac{2}{\pi} \left[(x) \left(\frac{\sin nx}{n} \right) - (1) \left(-\frac{\cos nx}{n^2} \right) \right]_0^{\pi}$$

$$= \frac{2}{\pi} \left[\frac{1}{n^2} \{(-1)^n - 1)\} \right]$$

$$= \frac{-4}{\pi n^2} \text{ if } n \text{ is odd}$$

$$= 0 \text{ if } n \text{ is even}$$

$$\therefore \quad x = \frac{\pi}{2} - \frac{4}{\pi} \sum_{n=1,3,5\cdots} \frac{1}{n^2} \cos nx$$

(*ii*) **Half-range-sine series:**

Let $$x = \sum_{n=1}^{\infty} b_n \sin nx$$

where $b_n = \dfrac{2}{\pi}\int_0^{\pi} x \sin nx\, dx$

$$= \frac{2}{\pi}\left[(x)\left(-\frac{\cos nx}{n}\right) - (1)\left(-\frac{\sin nx}{n^2}\right)\right]_0^{\pi}$$

$$= \frac{2}{\pi}\left[-\frac{\pi}{n}(-1)^n\right]$$

$$= \frac{2}{n}(-1)^{n-1}$$

$$\therefore \qquad x = 2\sum_{n=1}^{\infty} \frac{(-1)^{n-1}}{n} \sin nx$$

Example 27. *Express $f(x) = x(\pi - x)$, $0 < x < \pi$, as a Fourier series of periodicity 2π containing (i) Sine terms only (ii) cosine terms only.*

Hence deduce, $1 - \dfrac{1}{3^3} + \dfrac{1}{5^3} - \dfrac{1}{7^3} + \cdots = \dfrac{\pi^3}{32}$

and $\dfrac{1}{1^2} - \dfrac{1}{2^2} + \dfrac{1}{3^3} - \dfrac{1}{4^2} + \cdots = \dfrac{\pi^2}{12}$

(*i*) **Sine series:**

Let $f(x) = \sum_1^{\infty} b_n \sin nx$

where $b_n = \dfrac{2}{\pi}\int_0^{\pi} x(\pi - x) \sin nx\, dx$

$$= \frac{2}{\pi}\left[\{\pi x - x^2\}\left(-\frac{\cos nx}{n}\right) - (\pi - 2x)\left(\frac{\sin nx}{n^2}\right) + (-2)\left(\frac{\cos nx}{n^3}\right)\right]_0^{\pi}$$

$$= \frac{2}{\pi}\left[-\frac{2}{n^3}\{(-1)^n - 1\}\right]$$

$$= \frac{4}{\pi n^3}\left[1 - (-1)^n\right]$$

$= 0$ if n is even

$= \dfrac{8}{\pi n^3}$ if n is odd

$$\therefore \quad f(x) = \frac{8}{\pi} \sum_{n=1}^{\infty} \frac{1}{(2n-1)^3} \sin(2n-1)x.$$

Setting $x = \pi/2$ which is a point of continuity we get first deduction.

(*ii*) **Cosine series:**

$$\text{Let } f(x) = \frac{a_0}{2} + \sum_{1}^{\infty} a_n \cos nx$$

where $$a_0 = \frac{2}{\pi} \int_0^{\pi} x(\pi - x)\, dx$$

$$= \frac{2}{\pi} \left[\frac{\pi x^2}{2} - \frac{x^3}{3} \right]_0^{\pi} = \frac{\pi^2}{3}$$

$$a_n = \frac{2}{\pi} \int_0^{\pi} (\pi x - x^2) \cos nx\, dx$$

$$= \frac{2}{\pi} \left[(\pi x - x^2)\left(\frac{\sin nx}{n}\right) - (\pi - 2x)\left(-\frac{\cos nx}{n^2}\right) + (-2)\left(-\frac{\sin nx}{n^3}\right) \right]_0^{\pi}$$

$$= \frac{2}{\pi} \left[-\frac{\pi}{n^2}(-1)^n - \frac{\pi}{n^2} \right] = -\frac{2}{n^2}[1 + (-1)^n]$$

$$= 0 \text{ for } n \text{ odd}$$

$$= -\frac{4}{n^2} \text{ for } n \text{ even}$$

$$x(\pi - x) = \frac{\pi^2}{6} - 4 \sum_{n=2,4,6\ldots}^{\infty} \frac{1}{n^2} \cos nx$$

$$x(\pi - x) = \frac{\pi^2}{6} - \sum_{1}^{\infty} \frac{1}{n^2} \cos 2nx.$$

Setting $x = \pi/2$ which is a point of continuity,

$$\frac{\pi}{2}\left(\pi - \frac{\pi}{2}\right) = \frac{\pi^2}{6} - \sum \frac{1}{n^2}(-1)^n$$

$$\sum \frac{1}{n^2}(-1)^n = \frac{\pi^2}{6} - \frac{\pi^2}{4}$$

$$\frac{1}{1^2}-\frac{1}{2^2}+\frac{1}{3^2}\,\ldots\ldots=\frac{\pi^2}{12}$$

Example 28. *Expand* $f(x)=x\sin x$ *as a cosine series in* $0<x<\pi$ *and show that*

$$1+\frac{2}{1\cdot3}-\frac{2}{3\cdot5}+\frac{2}{5\cdot7}-\cdots=\pi/2$$

Let $f(x)=\dfrac{a_0}{2}+\sum\limits_{n=1}^{\infty}a_n\cos nx$

$$a_0=\frac{2}{\pi}\int_0^{\pi}x\sin x\,dx$$

$$=\frac{2}{\pi}\Big[(x)(-\cos x)-(1)(-\sin x)\Big]_0^{\pi}$$

$$=\frac{2}{\pi}[\pi]=2$$

$$a_n=\frac{2}{\pi}\int_0^{\pi}x\sin x\cos nx\,dx$$

$$=\frac{1}{\pi}\int_0^{\pi}x[\sin(n+1)x-\sin(n-1)x]\,dx$$

$$=\frac{1}{\pi}\left[x\left\{-\frac{\cos(n+1)x}{n+1}\right\}-(1)\left\{-\frac{\sin(n+1)x}{(n+1)^2}\right\}\right.$$

$$\left.-x\left\{-\frac{\cos(n-1)x}{n-1}\right\}+(1)\left\{-\frac{\sin(n-1)x}{(n-1)^2}\right\}\right]_0^{\pi}\quad\text{if } n\neq 1$$

$$=\frac{1}{\pi}\left[\frac{-\pi(-1)^{n+1}}{n+1}+\frac{\pi}{n-1}(-1)^{n-1}\right]$$

$$=(-1)^n\left[\frac{1}{n+1}-\frac{1}{n-1}\right]=\frac{2(-1)^{n+1}}{n^2-1},\ \text{if } n\neq 1$$

$$a_1=\frac{2}{\pi}\int_0^{\pi}x\sin x\cos x\,dx$$

$$= \frac{1}{\pi} \int_0^{\pi} x \sin 2x \, dx$$

$$= \frac{1}{\pi} \left[(x)\left(-\frac{\cos 2x}{2} \right) - (1)\left(-\frac{\sin 2x}{4} \right) \right]_0^{\pi}$$

$$= \frac{1}{\pi} \left[-\frac{\pi}{2} \right] = -\frac{1}{2}$$

$$x \sin x = 1 - \frac{1}{2} \cos x - 2 \sum_{n=2}^{\infty} \frac{(-1)^n}{n^2 - 1} \cos nx$$

Deduction: Set $x = \pi/2$ which is a point of continuity.

$$\therefore \quad 1 - \frac{1}{2} \cos \frac{\pi}{2} - 2 \sum_{n=2}^{\infty} \frac{(-1)^n}{(n-1)(n+1)} \cos \frac{n\pi}{2} = \frac{\pi}{2} \sin \frac{\pi}{2}$$

$$1 - 2\left[-\frac{1}{1 \cdot 3} + \frac{1}{3 \cdot 5} - \frac{1}{5 \cdot 7} + \cdots \right] = \pi/2$$

$$1 + 2\left[\frac{1}{1 \cdot 3} - \frac{1}{3 \cdot 5} + \frac{1}{5 \cdot 7} \cdots \right] = \pi/2$$

Example 29. *If* $f(x) = \frac{\pi x}{4}, 0 < x < \pi/2$

$$= \frac{\pi}{4}(\pi - x), \pi/2 < x < \pi$$

Express $f(x)$ *in a series of cosines only (of periodicty* 2π*)* **(*M.S. 1990 A.*)**

Let $\quad f(x) = \frac{a_0}{2} + \sum_{n=1}^{\infty} a_n \cos nx$

where $\quad a_0 = \frac{2}{\pi} \int_0^{\pi} f(x) \, dx$

$$= \frac{2}{\pi} \left[\int_0^{\pi/2} \frac{\pi x}{4} dx + \frac{\pi}{4} \int_{\pi/2}^{\pi} (\pi - x) \, dx \right]$$

$$= \frac{2}{\pi} \left[\frac{\pi}{4} \left(\frac{x^2}{2} \right)_0^{\pi/2} + \frac{\pi}{4} \left(\pi x - \frac{x^2}{2} \right)_{\pi/2}^{\pi} \right]$$

$$= \frac{1}{2}\left[\frac{\pi^2}{8} + \frac{\pi^2}{2} - \frac{3}{8}\pi^2\right] = \frac{\pi^2}{8}$$

$$a_n = \frac{2}{\pi}\left[\frac{\pi}{4}\int_0^{\pi/2} x\cos nx\,dx + \frac{\pi}{4}\int_{\pi/2}^{\pi} (\pi - x)\cos nx\,dx\right]$$

$$= \frac{1}{2}\left[(x)\left(+\frac{\sin nx}{n}\right) - (1)\left(-\frac{\cos nx}{n^2}\right)\right]_0^{\pi/2}$$

$$+\frac{1}{2}\left[(\pi - x)\left(\frac{\sin nx}{n}\right) - (-1)\left(\frac{\cos nx}{n^2}\right)\right]_{\pi/2}^{\pi}$$

$$= \frac{1}{2}\left[\frac{\pi}{2n}\sin\frac{n\pi}{2} + \frac{1}{n^2}\cos\frac{n\pi}{2} - \frac{1}{n^2} - \frac{\pi}{2n}\sin\frac{n\pi}{2} - \frac{1}{n^2}\cos n\pi + \frac{1}{n^2}\cos\frac{n\pi}{2}\right]$$

$$= \frac{1}{2}\left[\frac{2}{n^2}\cos\frac{n\pi}{2} - \frac{1}{n^2} - \frac{1}{n^2}\cos n\pi\right]$$

Hence $f(x) = \dfrac{\pi^2}{16} + \dfrac{1}{2}\displaystyle\sum_{n=1}^{\infty}\left(\frac{2}{n^2}\cos\frac{n\pi}{2} - \frac{1}{n^2} - \frac{1}{n^2}\cos n\pi\right)\cos nx.$

Exercises 1(e)

1. Find the sine and cosine half range series in $0 < x < \pi$, for
   (*i*) $f(x) = x + 1$, (*ii*) $f(x) = e^x$, (*iii*) $f(x) = x^2$.

2. Find the cosine half-range series for the function

$$f(x) = \begin{cases} x, \text{ when } 0 < x < \pi/2, \\ \pi - x, \text{ when } \frac{1}{2}\pi < x < \pi. \end{cases}$$ **(*MS. 1991 A.*)**

   Deduce the sum of the series $\displaystyle\sum_{n=0}^{\infty}\frac{1}{(2n+1)^2}$

   Draw the graph of the function which the series represents in $(-2\pi, 2\pi)$.

3. Expand the function $f(x) = \sin x$, in $0 < x < \pi$ in a sine half-range series and half-range cosine series.

4. Show that $\log\left(2\sin\dfrac{x}{2}\right) = -\displaystyle\sum_{n=1}^{\infty}\frac{\cos nx}{n}$, if $0 < x < \pi$.

5. If $f(x) = kx, 0 \le x \le \pi/2,$

$$= k(\pi - x), \tfrac{1}{2}\pi \le x \le \pi,$$

Show that $f(x) = \frac{4k}{\pi}\left(\frac{\sin x}{1^2} - \frac{\sin 3x}{3^2} + \frac{\sin 5x}{5^2} - \ldots\right)$

6. If c is a constant, then, in $0 < x < \pi$, show that

$$c = \frac{4c}{\pi}\left(\sin x + \frac{\sin 3x}{3} + \frac{1}{5}\sin 5x + \ldots\right)$$

7. Show that in $(0, \pi)$

$$x^2 = \frac{2}{\pi}\left[\left(\frac{\pi^2}{1} - \frac{4}{1^3}\right)\sin x - \frac{\pi^2}{2}\sin 2x + \left(\frac{\pi^2}{3} - \frac{4}{3^3}\right)\sin 3x - \ldots\right]$$

Deduce the sum of the series $\frac{1}{1^3} - \frac{1}{3^3} + \frac{1}{5^3} - \frac{1}{7^3} + \ldots$

8. Obtain cosine series for

$$f(x) = \begin{cases} \cos x \text{ in } 0 < x < \pi/2 \\ 0 \text{ in } \pi/2 < x < \pi \end{cases}$$ *(M.S. 1990 Ap.)*

$$\left[\textbf{Ans :} \ a_0 = \frac{2}{\pi}, a_n = \frac{-2\cos n\pi/2}{\pi(n^2 - 1)}, a_1 = 1/2\right]$$

9. Find half-range sine series for $f(x)$ in $(0, \pi)$ where $f(x) = \sin kx$ for k, not an integer.

$$\left[\textbf{Ans.} \ b_n = \frac{2n}{\pi}\frac{(-1)^{n+1}\sin k\pi}{\pi(n^2 - k^2)}\right]$$

10. Find half-range cosine series for $f(x) = x^2$ in $0 \le x \le \pi$ and hence deduce the sum $1 - \frac{1}{2^2} + \frac{1}{3^2} - \frac{1}{4^2} + \ldots$ *(M.S. 1989 Nov.)*

$$\left[\textbf{Ans.} \ f(x) = \frac{\pi^2}{3} - 4\left(\cos x - \frac{\cos 2x}{2^2} + \frac{\cos 3x}{3^2} - \ldots\right)\right]$$

11. Find the half-range sine series for $f(x) = x^3$ in $(0, \pi)$.

$$\left[\textbf{Ans.} \ x^3 = 2\sum_{1}^{\infty}(-1)^n\left[\frac{6}{n^3} - \frac{\pi^2}{n}\right]\sin nx\right]$$

12. Find Fourier cosine series in $(0, \pi)$ for

$$f(x) = x \text{ in } 0 < x < \pi/2$$
$$= \pi - x \text{ in } \pi/2 < x < \pi$$

$$\left[\textbf{Ans.} \ f(x) = \frac{\pi}{4} - \frac{2}{\pi}\left(\frac{\cos 2x}{1^2} + \frac{\cos 6x}{3^2} + \frac{\cos 10x}{5^2} + \ldots\right)\right]$$

13. Obtain cosine series for

$$f(x) = \cos x, \ 0 < x < \pi/2$$
$$= 0, \ \pi/2 < x < \pi$$

14. Obtain sine series for $f(x) = \pi x - x^2$ in $(0, \pi)$

15. In $(0, \pi)$, prove

$$\sin x = \frac{4}{\pi}\left(\frac{1}{2} - \frac{\cos 2x}{1 \cdot 3} - \frac{\cos 4x}{3 \cdot 5} - \frac{\cos 6x}{5 \cdot 7} - \dots\right)$$

and $$\cos x = \frac{4}{\pi}\left(\frac{2 \sin 2x}{2^2 - 1} + \frac{4 \sin 4x}{4^2 - 1} + \frac{6 \sin 6x}{6^2 - 1} + \dots\right)$$

16. Find half range sine series and cosine series for

(*i*) $f(x) = x - x^2$ in $0 < x < \pi$

(*ii*) $f(x) = x$ in $0 < x < \pi/2$

$= 0$ in $\frac{\pi}{2} < x < \pi$

(*iii*) $f(x) = x \sin x$ in $0 < x < \pi$.

Change of interval. In very many engineering applications of Fourier series, we require an expansion of a given function over an interval of length other than π or 2π. It is therefore a matter of importance to determine how the foregoing theory can be applied to the representation of periodic functions of arbitrary period. The problem is not a difficult one, for all that is needed is a proportional change of scale.

Suppose we have to obtain the Fourier series of a function $f(x)$ defined for the interval $-L < x < L$, where L is *any* positive number. Introduce a new variable z given by $z = \frac{\pi}{L} x \cdots$ (*i*). When $x = -L$, $z = -\pi$; and when $x = L$, $z = \pi$. Thus, when x changes from $-L$ to L, z changes from $-\pi$ to π. From (*i*), $x = \frac{Lz}{\pi}$; hence in terms of the variable z, $f(x)$ becomes $f\left(\frac{Lz}{\pi}\right)$. For the sake of clarity, let us call $f\left(\frac{Lz}{\pi}\right)$ by $F(z)$. Now, $F(z)$ is a function of z; since $f(x)$ is defined in the interval $(-L, L)$ and since $F(z)$ is derived from $f(z)$ using (*i*), $F(z)$ is defined in the interval $(-\pi, \pi)$. This is precisely the situation that we have mastered. Expanding $F(z)$ into a Fourier series, we get

$$F(z) = \frac{a_0}{2} + \sum_{n=1}^{\infty} (a_n \cos nz + b_n \sin nz),$$

where, $$a_n = \frac{1}{\pi}\int_{-\pi}^{\pi} F(z) \cos nz \, dz, \text{ and } b_n = \frac{1}{\pi}\int_{-\pi}^{\pi} F(z) \sin nz \, dz.$$

We introduced the notation $F(z)$ only for the sake of familiarity. It was nothing but $f\left(\frac{Lz}{\pi}\right)$. Making that change throughout, we get

$$f\left(\frac{Lz}{\pi}\right)=\frac{a_0}{2}+\sum_{n=1}^{\infty}(a_n\cos nz+b_n\sin nz),$$

where $\quad a_n=\frac{1}{\pi}\int_{-\pi}^{\pi}f\left(\frac{Lz}{\pi}\right)\cos nz\,dz$, and $b_n=\frac{1}{\pi}\int_{-\pi}^{\pi}f\left(\frac{Lz}{\pi}\right)\sin nz\,dz.$

Now that we have managed to get an expansion in z, let us go back to x; this is done with the help of (*i*). From (*i*), $dz=\frac{\pi}{L}dx$. In the above result, changing every z in favour of x, we have

$$f(x)=\frac{a_0}{2}+\sum_{n=1}^{\infty}\left\{a_n\cos\frac{n\pi x}{L}+b_n\sin\frac{n\pi x}{L}\right\},$$

where $\quad a_n=\frac{1}{\pi}\int_{-L}^{L}f(x)\left\{\cos\frac{n\pi x}{L}\right\}\frac{\pi}{L}dx$, and $b_n=\frac{1}{\pi}\int_{-L}^{L}f(x)\left\{\sin\frac{n\pi x}{L}\right\}\frac{\pi}{L}dx.$

i.e., $\quad a_n=\frac{1}{L}\int_{-L}^{L}f(x)\cos\frac{n\pi x}{L}dx$, and $b_n=\frac{1}{L}\int_{-L}^{L}f(x)\sin\frac{n\pi x}{L}dx.$

Note 1. The Fourier expansion of an even function $f(x)$ defined in the interval $-L$ to L will contain cosine terms only (i.e., $b_n=0$, and $a_n=\frac{2}{L}\int_{0}^{L}f(x)\cos\frac{n\pi x}{L}dx$).

Note 2. The Fourier expansion of an odd function $f(x)$ defined in the interval $(-L, L)$ will contain sine terms only (*i.e.*, $a_n=0$), and $b_n=\frac{2}{L}\int_{0}^{L}f(x)\sin\frac{n\pi x}{L}dx.$

Note 3. *Any* function $f(x)$ defined in the half-range 0 to L admits of

(*i*) a cosine expansion, where $a_n=\frac{2}{L}\int_{0}^{L}f(x)\cos\frac{n\pi x}{L}dx,$

(*ii*) a sine expansion, where $b_n=\frac{2}{L}\int_{0}^{L}f(x)\sin\frac{n\pi x}{L}dx.$

Note 4. If n is any integer, then

$$\cos\frac{n\pi(x+2L)}{L}=\cos\left(\frac{n\pi x}{L}+2n\pi\right)=\cos\frac{n\pi x}{L}$$

and similarly for sines. Hence each term of the expansion for $f(x)$ has period $2L$, and therefore the sum also has period $2L$. For this reason, the sum cannot represent an arbitrary function on $(-\infty, \infty)$; it represents periodic functions only.

Note 5. It is presumed that the function $f(x)$ introduced in the main discussion above or the notes, satisfies Dirichlet's conditions within the interval of its definition.

Note 6. If the function $f(x)$ is defined in the interval c to $c + 2L$, then the Fourier expansion is given by

$$f(x) = \frac{a_0}{2} + \sum_{n=1}^{\infty} \left(a_n \cos \frac{n\pi x}{L} + b_n \sin \frac{n\pi x}{L} \right),$$

where $a_n = \frac{1}{L} \int_c^{c+2L} f(x) \cos \frac{n\pi x}{L} dx$, and $b_n = \frac{1}{L} \int_c^{c+2L} f(x) \sin \frac{n\pi x}{L} dx$.

Example 30. *Obtain the Fourier series expansion of $f(x)$, given that*

$$f(x) = \begin{cases} 1 \text{ for } 0 < x < 1 \\ 2 \text{ for } 1 < x < 3 \end{cases}$$

and $f(x) = \frac{3}{2}$ when $x = 0, 1, 3$ and $f(x + 3) = f(x)$ for all x.

$f(x)$ is defined in the interval 0 to 3. Further, we are informed that $f(x)$ has the period 3. Thus $2L = 3$, giving $L = \frac{3}{2}$. Hence

$$f(x) = \frac{a_0}{2} + \sum_{n=1}^{\infty} \left(a_n \cos \frac{n\pi x}{\frac{3}{2}} + b_n \sin \frac{n\pi x}{\frac{3}{2}} \right)$$

$$\text{i.e., } f(x) = \frac{a_0}{2} + \sum_{n=1}^{\infty} \left(a_n \cos \frac{2n\pi x}{3} + b_n \sin \frac{2n\pi x}{3} \right) \qquad \text{...(}i\text{)}$$

where, $a_0 = \dfrac{1}{\frac{3}{2}} \displaystyle\int_0^3 f(x)\, dx = \frac{2}{3} \left\{ \int_0^1 1\, dx + \int_1^3 2\, dx \right\} = \frac{2}{3} \left\{ [x]_0^1 + [2x]_1^3 \right\}$

$$= \frac{2}{3} \{ (1 - 0) + (6 - 2) \} = \frac{2}{3} \times 5 = \frac{10}{3}$$

For $n = 1, 2, 3, \cdots$,

$$a_n = \frac{1}{\frac{3}{2}} \int_0^3 f(x) \cos \frac{n\pi x}{\frac{3}{2}} dx = \frac{2}{3} \left\{ \int_0^1 1 \cos \frac{2n\pi x}{3} dx + \int_1^3 2 \cos \frac{2n\pi x}{3} dx \right\}.$$

The following adjustment about the integrals is available.

$$a_n = \frac{2}{3}\left\{ \int_0^1 \cos \frac{2n\pi x}{3}\, dx + 2 \int_1^3 \cos \frac{2n\pi x}{3}\, dx \right\}$$

$$= \frac{2}{3}\left\{ \left[\frac{3}{2n\pi} \sin \frac{2n\pi x}{3} \right]_0^1 + \left[\frac{6}{2n\pi} \sin \frac{2n\pi x}{3} \right]_1^3 \right\}$$

$$= -\frac{1}{n\pi} \sin \frac{2n\pi}{3}$$

When n is a multiple of 3, that is when $n = 3m$, where m is an integer, we get

$$a_n = -\frac{1}{3m\pi} \sin 2m\pi = -\frac{1}{3m\pi} \times 0 = 0.$$

Thus $a_3 = a_6 = a_9 = \cdots = 0.$

Now $a_1 = -\frac{1}{\pi} \sin \frac{2\pi}{3} = -\frac{1}{\pi} \times \frac{\sqrt{3}}{2} = -\frac{\sqrt{3}}{2\pi}$

$$a_2 = -\frac{1}{2\pi} \sin \frac{4\pi}{3} = -\frac{1}{2\pi}\left(-\frac{\sqrt{3}}{2} \right) = \frac{\sqrt{3}}{2\pi}\left(\frac{1}{2} \right).$$

Similarly, $a_4 = -\left(\frac{\sqrt{3}}{2\pi} \right)\frac{1}{4};\ a_5 = \frac{\sqrt{3}}{25}\left(\frac{1}{5} \right); \cdots$

Again, $b_n = \frac{2}{3}\left\{ \int_0^1 \sin \frac{2n\pi x}{3}\, dx + 2 \int_1^3 \sin \frac{2n\pi x}{3}\, dx \right\}$

$$= \frac{2}{3}\left\{ \left[-\frac{3}{2n\pi} \cos \frac{2n\pi x}{3} \right]_0^1 + \left[-\frac{6}{2n\pi} \cos \frac{2n\pi x}{3} \right]_1^3 \right\}$$

$$= -\frac{1}{n\pi}\left(1 - \cos \frac{2n\pi}{3} \right)$$

When n is a multiple of 3, that is when $n = 3m$, where m is an integer,

$$b_n = -\frac{1}{3m\pi}(1 - \cos 2m\pi) = -\frac{1}{3m\pi}(1 - 1) = 0.$$

Also, $b_1 = -\frac{1}{\pi}\left(1 - \cos \frac{2\pi}{3} \right) = -\frac{1}{\pi}\left(1 + \frac{1}{2} \right) = -\frac{3}{2\pi}$

$$b_2 = -\frac{1}{2\pi}\left(1 - \cos\frac{4\pi}{3}\right) = -\frac{1}{2\pi}\left(1 + \frac{1}{2}\right) = -\frac{3}{2\pi}\left(\frac{1}{2}\right)$$

$$b_4 = -\frac{3}{2\pi}\left(\frac{1}{4}\right);\ b_5 = -\frac{3}{2\pi}\left(\frac{1}{5}\right);\ \ldots$$

Substituting the coefficients in (*i*), we have

$$f(x) = \frac{5}{3} - \frac{\sqrt{3}}{2\pi}\left\{\cos\frac{2\pi x}{3} - \frac{1}{2}\cos\frac{4\pi x}{3} + \frac{1}{4}\cos\frac{8\pi x}{3} - \frac{1}{5}\cos\frac{10\pi x}{3} + \cdots\right\}$$

$$- \frac{3}{2\pi}\left\{\sin\frac{2\pi x}{3} + \frac{1}{2}\sin\frac{4\pi x}{3} + \frac{1}{4}\sin\frac{8\pi x}{3} + \frac{1}{5}\sin\frac{10\pi x}{3} + \cdots\right\}$$

Note : Since the period of the function is 3, $f(0-) = f(3-) = 2$, by hypothesis. Again by hypothesis, $f(0+) = 1$. Hence $\frac{f(0-)+f(0+)}{2} = \frac{3}{2}$. According to Dirichlet's theorem, the sum of the series for $x = 0$ will be $\frac{3}{2}$; the hypothesis, defines $f(0) = \frac{3}{2}$. Hence the derived series represents the $f(x)$ at the discontinuity 0 also. This is true of the other discontinuities $x = 1$ and 3.

Example 31. *Find the Fourier series of the function*

$$f(t) = \begin{cases} 0, \text{ when } -2 < t < -1 \\ k, \text{ when } -1 < t < 1 \\ 0, \text{ when } 1 < t < 2 \end{cases} \quad \text{and } f(t+4) = f(t).$$

[Note that the name of the variable need not be necessarily x; here, it is t,]

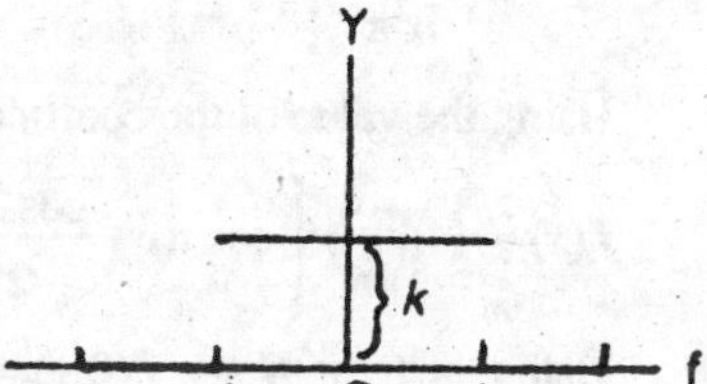

The period is 4; the graph is symmetrical about the y-axis in the interval -2 to 2. That is because $f(x)$ is even. Hence there will be no sine terms in its expansion. Since $L = 2$,

$$a_0 = \frac{2}{2}\int_0^2 f(t)\,dt = \int_0^1 k\,dt + \int_1^2 0\,dt = [kt]_0^1 = k$$

$$a_n = \frac{2}{2}\int_0^2 f(t)\cos\frac{n\pi t}{2}\,dt = \int_0^1 k\cos\frac{n\pi t}{2}\,dt = \frac{2k}{n\pi}\left[\sin\frac{n\pi t}{2}\right]_0^1$$

i.e., $a_n = \frac{2k}{n\pi}\sin\frac{n\pi}{2}$. Hence when n is even, $a_n = 0$.

Further, when $n = 1, 5, 9, \ldots$, $a_n = \frac{2k}{n\pi}$ and for $n = 3, 7, 11, \ldots$,

$$a_n = -\frac{2k}{n\pi}.$$

$$\therefore\ f(t)=\frac{k}{2}+\frac{2k}{\pi}\left\{\cos\frac{\pi t}{2}-\frac{1}{3}\cos\frac{3\pi t}{2}+\frac{1}{5}\cos\frac{5\pi t}{2}-\cdots\right\}$$

Example 32. *Find the half-range cosine series for the function $f(x)=(x-1)^2$ in the interval $0<x<1$. Hence show that*

$$\pi^2=6\left\{1+\frac{1}{2^2}+\frac{1}{3^2}+\cdots\right\}.$$

Here $L=1$

$$a_0=\frac{2}{1}\int_0^1 (x-1)^2\,dx=2\left[\frac{(x-1)^3}{3}\right]_0^1=2\left\{0-\left(-\frac{1}{3}\right)\right\}=\frac{2}{3}$$

$$a_n=\frac{2}{1}\int_0^1 (x-1)^2\cos\frac{n\pi x}{1}\,dx=2\int_0^1 (x-1)^2\cos n\pi x\,dx$$

$$=2\left[(x-1)^2\left(\frac{\sin n\pi x}{n\pi}\right)-2(x-1)\left(-\frac{\cos n\pi x}{n^2\pi^2}\right)+(2)\left(-\frac{\sin n\pi x}{n^3\pi^3}\right)\right]_0^1$$

$$=2\left[\frac{2}{n^2\pi^2}\right]=\frac{4}{n^2\pi^2}.$$

Using the value of the coefficients,

$$f(x)=\frac{1}{3}+\frac{4}{\pi^2}\left\{\cos\pi x+\frac{\cos 2\pi x}{2^2}+\frac{\cos 3\pi x}{3^2}+\frac{\cos 4\pi x}{4^2}+\cdots\right\}$$

Using the hypothesis, $f(0+)=1$; since we have developed the half range *cosine* series, $F(x)$ in the interval $(-1, 1)$ is an even function. Hence $F(0-)=F(0+)=f(0+)=1$.

$\therefore\ \dfrac{F(0-)+F(0+)}{2}=\dfrac{1+1}{2}=1.$ Though the function is *not defined* at $x=0$, by Dirichlet's theorem, the value of the series at $x=0$, will be 1. Hence

$$1=\frac{1}{3}+\frac{4}{\pi^2}\left\{1+\frac{1}{2^2}+\frac{1}{3^2}+\cdots\right\}$$

Example 33. *If* $f(x)=\begin{cases}\sin x \text{ for } 0\le x\le \pi/4\\ \cos x \text{ for } \frac{1}{4}\pi\le x\le \pi/2\end{cases}$

express $f(x)$ in a series of sines.

In this question, the half range is 0 to $\pi/2$; hence $L=\pi/2$.

Now, $b_n = \dfrac{2}{\frac{\pi}{2}} \displaystyle\int_0^{\pi/2} f(x) \sin \dfrac{n\pi x}{\frac{\pi}{2}}\, dx = \dfrac{4}{\pi} \int_0^{\pi/2} f(x) \sin 2nx\, dx$

$$= \frac{4}{\pi} \left\{ \int_0^{\pi/4} \sin x \sin 2nx\, dx + \int_{\pi/4}^{\pi/2} \cos x \sin 2nx\, dx \right\}$$

$$= \frac{2}{\pi} \left\{ \int_0^{\pi/4} [\cos (2n-1)x - \cos (2n+1)x]\, dx + \int_{\pi/4}^{\pi/2} [\sin (2n+1)x + \sin (2n-1)x]\, dx \right\}$$

$$= \frac{2}{\pi} \left\{ \left[\frac{\sin (2n-1)x}{2n-1} - \frac{\sin (2n+1)x}{2n+1} \right]_0^{\pi/4} - \left[\frac{\cos (2n+1)x}{2n+1} + \frac{\cos (2n-1)x}{2n-1} \right]_{\pi/4}^{\pi/2} \right\}$$

$$= \frac{2}{\pi} \left\{ \frac{\sin\left(\frac{n\pi}{2} - \frac{\pi}{4}\right)}{2n-1} - \frac{\sin\left(\frac{n\pi}{2} + \frac{\pi}{4}\right)}{2n+1} - \frac{\cos\left(n\pi + \frac{\pi}{2}\right)}{2n+1} - \frac{\cos\left(n\pi - \frac{\pi}{2}\right)}{2n-1} + \frac{\cos\left(\frac{n\pi}{2} + \frac{\pi}{4}\right)}{2n+1} + \frac{\cos\left(\frac{n\pi}{2} - \frac{\pi}{4}\right)}{2n-1} \right\}$$

Expanding expressions like $\sin\left(\dfrac{n\pi}{2} - \dfrac{\pi}{4}\right)$, and using the value

$\sin \dfrac{\pi}{4} = \dfrac{1}{\sqrt{2}} = \cos \dfrac{\pi}{4}$, we get

$$b_n = \frac{2}{\pi} \left[\frac{1}{2n-1} \left\{ \left(\frac{1}{\sqrt{2}} \sin \frac{n\pi}{2} - \frac{1}{\sqrt{2}} \cos \frac{n\pi}{2} \right) + \left(\frac{1}{\sqrt{2}} \cos \frac{n\pi}{2} + \frac{1}{\sqrt{2}} \sin \frac{n\pi}{2} \right) \right\} - \frac{1}{2n+1} \left\{ \left(\frac{1}{\sqrt{2}} \sin \frac{n\pi}{2} + \frac{1}{\sqrt{2}} \cos \frac{n\pi}{2} \right) - \left(\frac{1}{\sqrt{2}} \cos \frac{n\pi}{2} - \frac{1}{\sqrt{2}} \sin \frac{n\pi}{2} \right) \right\} \right],$$

$$\text{since } \cos\left(n\pi + \frac{\pi}{2}\right) = 0 = \cos\left(n\pi - \frac{\pi}{2}\right)$$

$$= \frac{2}{\pi}\left\{\frac{\sqrt{2}}{2n-1}\sin\frac{n\pi}{2} - \frac{\sqrt{2}}{2n+1}\sin\frac{n\pi}{2}\right\}$$

$$= \frac{2\sqrt{2}}{\pi}\left(\sin\frac{n\pi}{2}\right)\frac{(2n+1)-(2n-1)}{(2n-1)(2n+1)} = \frac{4\sqrt{2}}{\pi\,(2n-1)(2n+1)}\sin\frac{n\pi}{2}$$

It is evident that when n is even, $b_n = 0$; further for $n = 1, 5, 9, \cdots$

$$b_n = \frac{4\sqrt{2}}{\pi\,(2n-1)\,(2n+1)} \text{ and for } n = 3, 7, 11, \ldots,$$

$$b_n = -\frac{4\sqrt{2}}{\pi\,(2n-1)(2n+1)}$$

$$\therefore\ f(x) = \frac{4\sqrt{2}}{\pi}\left\{\frac{\sin 2x}{1\cdot 3} - \frac{\sin 6x}{5\cdot 7} + \frac{\sin 10x}{9\cdot 11} - \frac{\sin 14x}{13\cdot 15} + \cdots\right\}$$

Exercises 1(f)

Develop the Fourier expansion of (1 to 11):

1. (*a*) $f(x) = 0$, in $-2 < x < 0$ and $f(x) = 1$, in $0 < x < 2$.

   (*b*) $f(x) = 0$, in $-2 < x < 0$ and $f(x) = k$, in $0 < x < 2$, k being $\neq 0$.

2. $f(x) = -1$, in $(-2, 0)$ and $f(x) = 1$, in $(0, 2)$.

3. $f(x) = 1$, in $-1 < x < 1$ and $f(x) = 0$, in $1 < x < 3$.

4. $f(x) = x$, in $-1 < x < 1$.

5. $f(x) = |x|$, in $-1 < x < 1$.

6. $f(x) = \frac{1}{4} - x$, in $0 < x < \frac{1}{2}$, and $f(x) = x - \frac{3}{4}$, in $\frac{1}{2} < x < 1$.

7. $f(x) = \begin{cases} 0, \text{ in } -2 < x < -1 \\ 1 + x, \text{ in } -1 < x < 0 \\ 1 - x, \text{ in } 0 < x < 1 \\ 0, \text{ in } 1 < x < 2. \end{cases}$

8. $f(x) = x^2$, in $-1 < x < 1$.

9. $f(x) = \cos x$, when $-\frac{\pi}{2} < x < \frac{\pi}{2}$.

10. $f(x) = 1 + \sin x$, in $-1 < x < 1$.

11. $f(x) = \begin{cases} 0, \text{ in } -3 < x < -1 \\ 1 + \cos \pi x, \text{ in } -1 < x < 1 \\ 0, \text{ in } 1 < x < 3. \end{cases}$

12. A sinusoidal voltage $E \sin \omega t$ is passed through a half wave rectifier which clips the negative portion of the wave. The resulting periodic function is given by.

$$u(t) = \begin{cases} 0, \text{ when } -\dfrac{\pi}{\omega} < t < 0 \\[2ex] E \sin \omega t, \text{ when } 0 < t < \pi/\omega. \end{cases}$$

Develop $u(t)$, in a Fourier series.

13. Find the Fourier series of $f(x)$ of period 4 given by

$$f(x) = \begin{cases} 1, \text{ when } -2 < x < 0 \\ e^{-x}, \text{ when } 0 < x < 2 \end{cases}$$

14. Show that the series $\frac{l}{\pi} \sum_{n=1}^{\infty} \frac{1}{n} \sin \frac{2n\pi x}{l}$ represents $\frac{1}{2} l - x$, when $0 < x < l$.

15. Find the expansion in the series of cosines, if $f(x) = 1$, $0 < x < \pi$, and $f(x) = 0$, $\pi < x < 2\pi$.

16. Find the half-range expansions of the function $f(t) = t - t^2$, $0 < t < 1$,

17. Find the half-range expansions of the function

$$f(t) = \begin{cases} \frac{2kt}{l}, \text{ when } 0 < t < \frac{l}{2} \\ \frac{2k(l-t)}{l}, \text{ when } \frac{l}{2} < t < l. \end{cases}$$

Expand the following (18 to 24) :

18. $f(t) = \begin{cases} t^2, 0 < t < 1 \\ 2 - t, 1 < t < 2 \end{cases}$ in a cosine series.

19. $f(x) = x^2 - 2$, $0 < x < 2$, in a sine series.

Deduce that $\pi^2 = 32\left(1 - \frac{1}{3^3} + \frac{1}{5^3} - \frac{1}{7^3} + \cdots\right)$

20. $f(x) = (x-1)^2$, $0 < x < 1$, in a sine series.

21. $\cos t$ in a sine series for $0 < t < 1$.

22. $\cos \frac{1}{2} x$ in a sine series in $(0, \pi)$.

23. $\sin \frac{\pi x}{l}$ in a cosine series for $0 < x < l$.

24. e^x in a sine series for $0 < x < 1$.

25. If $f(x) = lx - x^2$ in $(0, l)$, show that the half range sine series for

$f(x)$ is $\frac{8l^2}{\pi^3} \sum_{n=1}^{\infty} \frac{1}{(2n+1)^3} \sin \frac{(2n+1)\pi x}{l}$.

26. For the function defined by the graph OAB in the figure below, show that the half range Fourier sine series of periodicity $2l$ is

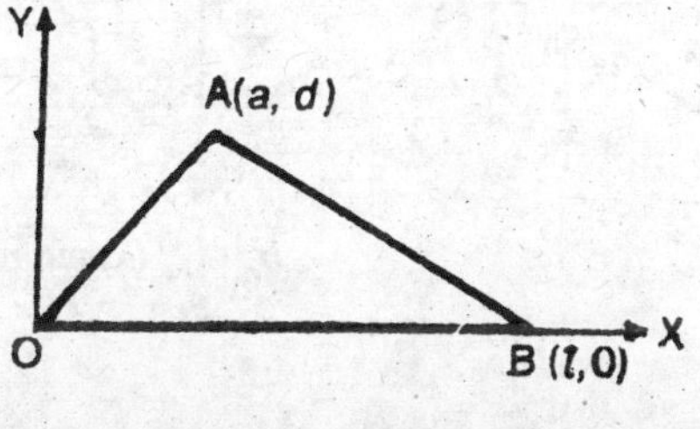

$$\frac{2dl^2}{a\pi^2 (l-a)} \sum_{n=1}^{\infty} \frac{l}{n^2} \sin \frac{n\pi a}{l} \sin \frac{n\pi x}{l}$$

Root-Mean square value of a function. PARSEVAL'S THEOREM.

In a number of physical applications, we have occasion to deal with the concept of root-mean-value (r.m.s), or effective value of a function.

The root-mean-square value of a function $y=f(x)$ over a given interval (a, b) is defined as

$$\bar{y} = \sqrt{\left\{ \frac{\int_a^b y^2\, dx}{b-a} \right\}} \quad ...(i)$$

If the interval is taken as $(c, c+2\pi)$, then

$$\bar{y}^2 = \frac{1}{2\pi} \int_c^{c+2\pi} y^2\, dx$$

Suppose that $y=f(x)$ is expressed as a Fourier-series of periodicity 2π in $(c, c+2\pi)$. then,

$$y = f(x) = \frac{a_0}{2} + \sum_{n=1}^{\infty} (a_n \cos nx + b_n \sin nx) \quad ...(ii)$$

$$\left.\begin{aligned} \text{where} \quad a_0 &= \frac{1}{\pi} \int_c^{c+2\pi} f(x)\, dx \\ a_n &= \frac{1}{\pi} \int_c^{c+2\pi} f(x) \cos nx\, dx \\ \text{and} \quad b_n &= \frac{1}{\pi} \int_c^{c+2\pi} f(x) \sin nx\, dx. \end{aligned}\right\} \quad ...(iii)$$

Multiply *(ii)* by $f(x)$ and integrate term by term with respect to x over the given range. Thus,

$$\int_c^{c+2\pi} [f(x)]^2\, dx = \frac{a_0}{2} \int_c^{c+2\pi} f(x)\, dx + \sum_{n=1}^{\infty} \left[a_n \int_c^{c+2\pi} f(x) \cos nx\, dx + b_n \int_c^{c+2\pi} f(x) \sin nx\, dx \right]$$

$$= \frac{a_0}{2}(\pi a_0) + \sum_{n=1}^{\infty} [a_n(\pi a_n) + b_n(\pi b_n)] \text{ using } (iii)$$

$$\int_c^{c+2\pi} [f(x)]^2\, dx = 2\pi\left[\frac{a_0^2}{4} + \frac{1}{2}\sum_{n=1}^{\infty}(a_n^2 + b_n^2)\right]$$

$$= (\text{Range})\left[\frac{a_0^2}{4} + \frac{1}{2}\sum_{n=1}^{\infty}(a_n^2 + b_n^2)\right]$$

$$\bar{y}^2 = \frac{a_0^2}{4} + \frac{1}{2}\sum_{n=1}^{\infty}(a_n^2 + b_n^2).$$

Ex. 34. *Find the Fourier series of periodicity 2π for $f(x) = x^2$, in $-\pi < x < \pi$. Hence show that*

$$\frac{1}{1^4} + \frac{1}{2^4} + \frac{1}{3^4} + \cdots + \text{to } \infty = \frac{\pi^4}{90}.$$

In example 21, we have proved

$f(x) = \dfrac{\pi^2}{3} + 4\displaystyle\sum_{n=1}^{\infty} \frac{(-1)^n}{n^2} \cos nx$, which is the first part of this problem. The coefficients a_0, a_n, b_n were seen to be

$$a_0 = \frac{2\pi^2}{3},\ a_n = \frac{4(-1)^n}{n^2},\ b_n = 0.$$

Hence using the root-mean-square value in series,

$$2\pi\left[\frac{a_0^2}{4} + \frac{1}{2}\sum_{n=1}^{\infty}(a_n^2 + b_n^2)\right] = \int_{-\pi}^{\pi} [f(x)^2]\, dx = \int_{-\pi}^{\pi} x^4 dx$$

$$2\pi\left[\frac{\pi^4}{9} + \frac{1}{2}\sum_{n=1}^{\infty}\frac{16}{n^4}\right] = \frac{2}{5}\pi^5$$

$$8\sum_{n=1}^{\infty}\frac{1}{n^4} = \frac{\pi^4}{5} - \frac{\pi^4}{9} = \frac{4\pi^4}{45}$$

$$\therefore\ \sum_{n=1}^{\infty}\frac{1}{n^4} = \frac{\pi^4}{90}.$$

Example 35. *The instantaneous current i at time t of an alternating-current wave is given by*

$i = I_1 \sin(\omega t + \alpha_1) + I_3 \sin(3\omega t + \alpha_3) + I_5 \sin(5\omega t + \alpha_5) + \ldots$ *Find the effective value of the current.*

Expanding expressions like $\sin(\omega t + \alpha_1)$ using the trigonometric addition formula we get

$$i = I_1 (\sin \omega t \cos \alpha_1 + \cos \omega t \sin \alpha_1) + I_3 (\sin 3\omega t \cos \alpha_3 + \cos 3\omega t \sin \alpha_3) + I_5 (\sin 5\omega t \cos \alpha_5 + \cos 5\omega t \sin \alpha_5) + \cdots$$

Let $A_n = I_n \sin \alpha_n$ and $B_n = I_n \cos \alpha_n$ ...(i)

Then, $i = A_1 \cos \omega t + A_3 \cos 3\omega t + A_5 \cos 5\omega t + \ldots$

$+ B_1 \sin \omega t + B_3 \sin 3\omega t + B_5 \sin 5\omega t + \ldots$

The R.H.S. is the Fourier expansion for i. If i denotes the required (r.m.s.) value of i, we have

$$i^2 = \frac{1}{2}\{(A_1^2 + B_1^2) + (A_3^2 + B_3^2) + (A_5^2 + B_5^2) + \ldots\}$$

But from (*i*), $A_n^2 + B_n^2 = I_n^2 (\sin^2 \alpha_n + \cos^2 \alpha_n) = I_n^2 \times 1 = I_n^2$.

Hence $i^2 = \frac{1}{2}(I_1^2 + I_3^2 + I_5^2 + \ldots)$.

Example 36. *If* $f(x) = \frac{x}{l},\ 0 < x < l$

$$= \frac{2l - x}{l},\ l < x < 2l$$

express $f(x)$ as a Fourier series of periodicity $2l$.

$$\text{Let } f(x) = \frac{a_0}{2} + \sum_{n=1}^{\infty}\left(a_n \cos\frac{n\pi x}{l} + b_n \sin\frac{n\pi x}{l}\right)$$

where $$a_0 = \frac{1}{l}\int_0^{2l} f(x)\, dx$$

$$= \frac{1}{l}\left[\int_0^{l} \frac{x}{l}\, dx + \int_l^{2l} \frac{2l - x}{l}\, dx\right]$$

$$= \frac{1}{l^2}\left[\left(\frac{x^2}{2}\right)_0^l + \left(2lx - \frac{x^2}{2}\right)_l^{2l}\right]$$

$$= \frac{1}{l^2}\left[\frac{l^2}{2} + 2l^2 - \frac{3}{2}l^2\right]$$

$$= 1$$

$$a_n = \frac{1}{l}\left[\int_0^l \frac{x}{l}\cos\frac{n\pi x}{l}\,dx + \int_l^{2l} \frac{(2l-x)}{l}\cos\frac{n\pi x}{l}\,dx\right]$$

$$= \frac{1}{l^2}\left[(x)\left(\frac{\sin\frac{n\pi x}{l}}{\frac{n\pi}{l}}\right) - (1)\left(-\frac{\cos\frac{n\pi x}{l}}{\frac{n^2\pi^2}{l^2}}\right)\right]_0^l$$

$$+ \frac{1}{l^2}\left[(2l-x)\left(\frac{\sin\frac{n\pi x}{l}}{\frac{n\pi}{l}}\right) - (-1)\left(-\frac{\cos\frac{n\pi x}{l}}{\frac{n^2\pi^2}{l^2}}\right)\right]_l^{2l}$$

$$= \frac{1}{l^2}\left[\frac{l^2}{n^2\pi^2}\{(-1)^n - 1\} - \frac{l^2}{n^2\pi^2}\{1 - (-1)^n\}\right]$$

$$= \frac{2}{n^2\pi^2}\{(-1)^n - 1\}$$

$= 0$ if n is even

$= -\dfrac{4}{n^2\pi^2}$ if n is odd

$$b_n = \frac{1}{l}\left[\int_0^l \frac{x}{l}\sin\frac{n\pi x}{l}\,dx + \int_l^{2l} \frac{2l-x}{l}\,.\,\sin\frac{n\pi x}{l}\,dx\right]$$

$$= \frac{1}{l^2}\left[(x)\left(-\frac{\cos\frac{n\pi x}{l}}{\frac{n\pi}{l}}\right) - (1)\left(-\frac{\sin\frac{n\pi x}{l}}{\frac{n^2\pi^2}{l^2}}\right)\right]_0^l$$

$$+ \frac{1}{l^2}\left[(2l-x)\left(-\frac{\cos\frac{n\pi x}{l}}{\frac{n\pi}{l}}\right) - (-1)\left(-\frac{\sin\frac{n\pi x}{l}}{\frac{n^2\pi^2}{l^2}}\right)\right]_l^{2l}$$

$$= \frac{1}{l^2}\left[-\frac{l^2}{n\pi}\cos n\pi + \frac{l^2}{n\pi}\cos n\pi\right] = 0$$

$$\therefore\ f(x) = \frac{1}{2} - \frac{4}{\pi^2} \sum_{n=1,3,5...}^{\infty} \frac{1}{n^2} \cos\frac{n\pi x}{l}$$

Example 37. *Find the Fourier series of periodicity 3 for $f(x) = 2x - x^2$, in $0 < x < 3$*

Here the range and the period are same (equal to 3)

It is a full range series.

$$\therefore\ 2l = 3\ ;\ l = \frac{3}{2}$$

$$\text{Let } f(x) = \frac{a_0}{2} + \sum_{n=1}^{\infty}\left(a_n \cos\frac{2n\pi x}{3} + b_n \sin\frac{2n\pi x}{3}\right)$$

where $$a_0 = \frac{1}{\frac{3}{2}}\int_0^3 (2x - x^2)\,dx = \frac{2}{3}\left[x^2 - \frac{x^3}{3}\right]_0^3 = 0$$

$$a_n = \frac{1}{\frac{3}{2}}\int_0^3 (2x - x^2)\cos\frac{2n\pi x}{3}\,dx$$

$$= \frac{2}{3}\left[(2x - x^2)\left(\frac{\sin\frac{2n\pi x}{3}}{\frac{2n\pi}{3}}\right) - (2 - 2x)\left(-\frac{\cos\frac{2n\pi x}{3}}{\frac{4n^2\pi^2}{9}}\right) + (-2)\left(-\frac{\sin\frac{2n\pi x}{3}}{\frac{8n^3\pi^3}{27}}\right)\right]_0^3$$

$$= \frac{2}{3}\left[\frac{-9}{n^2\pi^2} - \frac{9}{2n^2\pi^2}\right]$$

$$= -\frac{9}{n^2\pi^2}$$

$$b_n = \frac{1}{\frac{3}{2}}\int_0^3 (2x - x^2)\sin\frac{2n\pi x}{3}\,dx$$

$$= \frac{2}{3}\left[(2x - x^2)\left(-\frac{\cos \frac{2n\pi x}{3}}{\frac{2n\pi}{3}} \right) - (2 - 2x)\left(-\frac{\sin \frac{2n\pi x}{3}}{\frac{4n^2\pi^2}{9}} \right) + (-2)\left(\frac{\cos \frac{2n\pi x}{3}}{\frac{8n^3\pi^3}{27}} \right) \right]_0^3$$

$$= \frac{3}{n\pi}$$

$$\therefore \quad f(x) = -\frac{9}{\pi^2} \sum_{n=1}^{\infty} \frac{1}{n^2} \cos\left(\frac{2n\pi x}{3}\right) + \frac{3}{\pi} \sum_{n=1}^{\infty} \frac{1}{n} \sin\left(\frac{2n\pi x}{3}\right)$$

Example 38. *$f(x)$ is defined in (–2, 2) as follows. Express $f(x)$ in a Fourier series of periodicity 4.* ***(MS. 1987 Nov.)***

$$f(x) = \begin{cases} 0 & -2 < x < -1 \\ 1 + x & -1 < x < 0 \\ 1 - x & 0 < x < 1 \\ 0 & 1 < x < 2 \end{cases}$$

Sol. Here $2l = 4 \quad \therefore \ l = 2$

$$\text{Hence } f(x) = \frac{a_0}{2} + \sum_{n=1}^{\infty}\left(a_n \cos \frac{n\pi x}{2} + b_n \sin \frac{n\pi x}{2} \right)$$

$$\text{where} \quad a_0 = \frac{1}{2}\int_{-2}^{2} f(x)\, dx$$

$$= \frac{1}{2}\left[\int_{-2}^{-1} a dx + \int_{-1}^{0} (1 + x)dx + \int_{0}^{1} (1 - x)\, dx + \int_{1}^{2} 0 \,.\, dx \right]$$

$$= \frac{1}{2}\left[\frac{(1 + x)^2}{2} \right]_{-1}^{0} + \frac{1}{2}\left[\frac{(1 - x)^2}{-2} \right]_{0}^{1}$$

$$= \frac{1}{4} + \frac{1}{4} = \frac{1}{2}$$

$$a_n = \frac{1}{2}\int_{-2}^{2} f(x) \cos \frac{n\pi x}{2}\, dx$$

$$= \frac{1}{2}\left[\int_{-1}^{0} (1+x)\cos\frac{n\pi x}{2}\,dx + \int_{0}^{1} (1-x)\cos\frac{n\pi x}{2}\,dx\right],$$

since the other integrals vanish

$$= \frac{1}{2}\left[(1+x)\left(\frac{\sin\frac{n\pi x}{2}}{\frac{n\pi}{2}}\right) - (1)\left(-\frac{\cos\frac{n\pi x}{2}}{\frac{n^2\pi^2}{4}}\right)\right]_{-1}^{0}$$

$$+ \frac{1}{2}\left[(1-x)\left(\frac{\sin\frac{n\pi x}{2}}{\frac{n\pi}{2}}\right) - (-1)\left(-\frac{\cos\frac{n\pi x}{2}}{\frac{n^2\pi^2}{4}}\right)\right]_{0}^{1}$$

$$= \frac{1}{2}\left[\frac{4}{n^2\pi^2} - \frac{4}{n^2\pi^2}\cos\frac{n\pi}{2} - \frac{4}{n^2\pi^2}\cos\frac{n\pi}{2} + \frac{4}{n^2\pi^2}\right]$$

$$= \frac{4}{n^2\pi^2}\left[1 - \cos\frac{n\pi}{2}\right]$$

$$b_n = \frac{1}{2}\int_{-2}^{2} f(x)\,.\,\sin\frac{n\pi x}{2}\,dx$$

$$= \frac{1}{2}\left[\int_{-1}^{0} (1+x)\sin\frac{n\pi x}{2}\,dx + \int_{0}^{1} (1-x)\sin\frac{n\pi x}{2}\,dx\right],$$

the other integrals being zero.

$$= \frac{1}{2}\left[(1+x)\left(-\frac{\cos\frac{n\pi x}{2}}{\frac{n\pi}{2}}\right) - (1)\left(-\frac{\sin\frac{n\pi x}{2}}{\frac{n^2\pi^2}{4}}\right)\right]_{-1}^{0}$$

$$+ \frac{1}{2}\left[(1-x)\left(-\frac{\cos\frac{n\pi x}{2}}{\frac{n\pi}{2}}\right) - (-1)\left(-\frac{\sin\frac{n\pi x}{2}}{\frac{n^2\pi^2}{4}}\right)\right]_{0}^{1}$$

$$= \frac{1}{2}\left[-\frac{2}{n\pi} + \frac{4}{n^2\pi^2}\sin\frac{n\pi}{2} - \frac{4}{n^2\pi^2}\sin\frac{n\pi}{2} + \frac{2}{n\pi}\right]$$

$$= 0.$$

Therefore, $f(x) = \frac{1}{4} + \frac{4}{\pi^2} \sum_{n=1}^{\infty} \frac{1}{n^2}\left(1 - \cos\frac{n\pi}{2}\right)\cos\frac{n\pi x}{2}$

$$= \frac{1}{4} + \frac{8}{\pi^2} \sum_{n=1}^{\infty} \frac{1}{n^2} \sin^2\frac{n\pi}{4} \cos\frac{n\pi x}{2}.$$

Example 39. *Find Fourier series of periodicity 2 for $f(x)$*

given $f(x) = \begin{cases} 0 \text{ in } -1 < x < 0 \\ 1 \text{ in } \ 0 < x < 1 \end{cases}$ **(*MS. 1986 Ap.*)**

Here the period and the range are equal.

Hence, $2l = 2;\ l = 1$

Let $f(x) = \frac{a_0}{2} + \sum_{n=1}^{\infty} (a_n \cos n\pi x + b_n \sin n\pi x)$

where $a_0 = \frac{1}{1}\int_{-1}^{1} f(x)dx$

$$= \int_0^1 1\, dx$$

$$= (x)_0^1 = 1$$

$$a_n = \frac{1}{1}\int_{-1}^{1} f(x) \cos n\pi x\, dx$$

$$= \int_0^1 \cos n\pi x\, dx$$

$$= \left(\frac{\sin n\pi x}{n\pi}\right)_0^1$$

$$= 0$$

$$b_n = \frac{1}{1}\int_{-1}^{1} f(x) \sin n\pi x\, dx$$

$$= \int_0^1 \sin n\pi x\, dx$$

$$= -\left(\frac{\cos n\pi x}{n\pi}\right)_0^1$$

$$= -\frac{1}{n\pi}((-1)^n - 1)$$

$= 0$ if n is even

$= \dfrac{2}{n\pi}$ if n is odd

$$\therefore \quad f(x) = \frac{1}{2} + \frac{2}{\pi} \sum_{n=1,3,5...}^{\infty} \frac{1}{n} \sin n\pi x.$$

$$= \frac{1}{2} + \frac{2}{\pi} \sum_{n=1}^{\infty} \frac{1}{(2n-1)} \sin(2n-1)\pi x.$$

Example 40. *Find Fourier series of periodicity 2 for*

$$f(x) = \begin{cases} x & in -1 < x \leq 0 \\ x+2 & in \quad 0 < x \leq 1 \end{cases}$$

and hence deduce the sum of $1 - \frac{1}{3} + \frac{1}{5} - \frac{1}{7} + \ldots$ *to* ∞.

The periodicity and range are equal to 2.

Hence $2l = 2$; $l = 1$

$$\text{Let} \quad f(x) = \frac{a_0}{2} + \sum_{n=1}^{\infty}\left(a_n \cos\frac{n\pi x}{1} + b_n \sin\frac{n\pi x}{1}\right)$$

where $\quad a_0 = \dfrac{1}{1}\displaystyle\int_{-1}^{1} f(x)dx$

$$= \int_{-1}^{0} x\,dx + \int_{0}^{1} (x+2)\,dx$$

$$= \left(\frac{x^2}{2}\right)_{-1}^{0} + \left[\frac{(x+2)^2}{2}\right]_0^1$$

$$= \frac{-1}{2} + \frac{9}{2} - \frac{4}{2}$$

$$= 2$$

$$a_n = \int_{-1}^{0} x \cos n\pi x\,dn + \int_{0}^{1} (x+2) \cos n\pi x\,dx$$

$$= \left[(x)\left(\frac{\sin n\pi x}{n\pi}\right) - (1)\left(-\frac{\cos n\pi x}{n^2\pi^2}\right) \right]_{-1}^{0}$$

$$+ \left[(x+2)\left(\frac{\sin n\pi x}{n\pi}\right) - (1)\left(-\frac{\cos n\pi x}{n^2\pi^2}\right) \right]_{0}^{1}$$

$$= \frac{1}{n^2\pi^2}[1-(-1)^n] + \frac{1}{n^2\pi^2}[(-1)^n - 1]$$

$$= 0$$

$$b_n = \int_{-1}^{0} x \sin n\pi x \, dx + \int_{0}^{1} (x+2) \sin n\pi x \, dx$$

$$= \left[(x)\left(-\frac{\cos n\pi x}{n\pi}\right) - \left(-\frac{\sin n\pi x}{n^2\pi^2}\right) \right]_{-1}^{0}$$

$$+ \left[(x+2)\left(-\frac{\cos n\pi x}{n\pi}\right) - \left(-\frac{\sin n\pi x}{n^2\pi^2}\right) \right]_{0}^{1}$$

$$= \left[-\frac{1}{n\pi}(-1)^n - \frac{3}{n\pi}(-1)^n + \frac{2}{n\pi} \right]$$

$$\therefore \quad f(x) = 1 + \frac{2}{\pi}\sum_{n=1}^{\infty}\frac{1}{n}\{1 - 2(-1)^n\} \sin n\pi x$$

$$= 1 + \frac{2}{\pi}\left[\frac{3}{1}\sin \pi x - \frac{1}{2}\sin 2\pi x + \frac{3}{3}\sin 3\pi x - \cdots\right]$$

Putting $x = \frac{1}{2}$ which is a point of continuity, the value of the Fourier series at $x = \frac{1}{2}$ is $f\left(\frac{1}{2}\right)$

$$\therefore \quad 1 + \frac{2}{\pi}\left[3\cdot\frac{1}{1} - 3\cdot\frac{1}{3} + 3\cdot\frac{1}{5} - \cdots\right] = \frac{1}{2} + 2$$

$$\frac{6}{\pi}\left(1 - \frac{1}{3} + \frac{1}{5} - \cdots\right) = \frac{3}{2}$$

$$\therefore \quad 1 - \frac{1}{3} + \frac{1}{5} - \frac{1}{7} + \cdots = \frac{\pi}{4}$$

Example 41. *Express $f(x) = x$ in half range cosine series and sine series of periodicity $2l$ in the range $0 < x < l$ and deduce the value of $\frac{1}{1^4} + \frac{1}{3^4} + \frac{1}{5^4} + \cdots$ to ∞.*

Case 1. *Cosine series* :

Let $f(x) = \frac{a_0}{2} + \sum_{n=1}^{\infty} a_n \cos \frac{n\pi x}{l}$

where $a_0 = \frac{2}{l}\int_0^l f(x)\, dx$

$= \frac{2}{l}\int_0^l x\, dx$

$= \frac{2}{l}\left(\frac{x^2}{2}\right)_0^l$

$= l$

$a_n = \frac{2}{l}\int_0^l x \cos \frac{n\pi x}{l}\, dx$

$$= \frac{2}{l}\left[(x)\left(\frac{\sin \frac{n\pi x}{l}}{\frac{n\pi}{l}}\right) - (1)\left(-\frac{\cos\left(\frac{n\pi x}{l}\right)}{\frac{n^2\pi^2}{l^2}}\right)\right]_0^l$$

$$= \frac{2}{l}\left[\frac{l^2}{n^2\pi^2}\{(-1)^n - 1\}\right]$$

$= \frac{-4l}{n^2\pi^2}$ if n is odd

$= 0$ if n is even

Hence $x = \frac{l}{2} - \frac{4l}{\pi^2}\sum_{n=1}^{\infty}\frac{1}{(2n-1)^2}\cos\frac{(2n-1)\pi x}{l}$ ...(1)

Case 2. *Sine series* :

Let $f(x) = \sum_{n=1}^{\infty} b_n \sin \frac{n\pi x}{l}$

where $b_n = \frac{2}{l}\int_0^l x \sin \frac{n\pi x}{l}\, dx$

$$= \frac{2}{l}\left[(x)\left(-\frac{\cos\frac{n\pi x}{l}}{\frac{n\pi}{l}}\right) - (1)\left(-\frac{\sin\frac{n\pi x}{l}}{\frac{n^2\pi^2}{l^2}}\right)\right]_0^l$$

$$= \frac{2}{l}\left[-\frac{l^2}{n\pi}(-1)^n\right]$$

$$= \frac{2l}{n\pi}(-1)^{n+1}$$

$$\therefore \quad x = \frac{2l}{\pi}\sum_{n=1}^{\infty}\frac{(-1)^{n+1}}{n}\sin\frac{n\pi x}{l} \qquad \text{...(2)}$$

Deduction: By Parsevel's theorem,

$$(\text{Range})\left[\frac{a_0^{\,2}}{4} + \frac{1}{2}\Sigma a_n^{\,2}\right] = \int_0^l [f(x)]^2\, dx \quad \text{(for cosine series)}$$

$$l\left[\frac{l^2}{4} + \frac{1}{2}\sum_{1,3,5,\dots}\frac{16l^2}{n^4\pi^4}\right] = \int_0^l x^2 dx = \frac{l^3}{3}$$

$$8\sum_{1,3,5,\dots}^{\infty}\frac{1}{n^4} = \frac{\pi^4}{12}$$

$$\frac{1}{1^4} + \frac{1}{3^4} + \frac{1}{5^4} + \dots = \frac{\pi^4}{96}$$

Again

$$(\text{Range})\left[\frac{1}{2}\sum b_n^{\,2}\right] = \int_0^l [f(x)]^2\, dx \quad \text{(for sine series)}$$

$$l\left[\frac{1}{2}\sum_1^{\infty}\frac{4l^2}{n^2\pi^2}\right] = \int_0^l x^2\, dx = \frac{l^3}{3}$$

$$\sum_1^{\infty}\frac{1}{n^2} = \frac{\pi^2}{6}.$$

Example 42. *Expand $f(x) = x - x^2$ as a Fourier series in $-1 < x < 1$ and using this series find the R.M.S. value of $f(x)$ in the interval.* **(BR. 1995 Ap.)**

Here $2l = 2$; $l = 1$

Let $f(x) = \frac{a_0}{2} + \sum_{n=1}^{\infty} (a_n \cos n\pi x + b_n \sin n\pi x)$

where $a_0 = \frac{1}{1}\int_{-1}^{1} f(x)\, dx$

$$= \int_{-1}^{1} (x - x^2)\, dx$$

$$= \int_{-1}^{1} x dx - \int_{-1}^{1} x^2 dx$$

$$= 0 - 2\int_{0}^{1} x^2 dx$$

$$= -2\left(\frac{x^3}{3}\right)_0^1$$

$$= -\frac{2}{3}$$

$$a_n = \frac{1}{1}\int_{-1}^{1} (x - x^2) \cos n\pi x\, dx$$

$$= \int_{-1}^{1} x \cos n\pi x\, dx - \int_{-1}^{1} x^2 \cos n\pi x\, dx$$

$$= -2\int_{0}^{1} x^2 \cos n\pi x\, dx \quad (\because \text{ first integral vanishes})$$

$$= -2\left[(x^2)\left(\frac{\sin n\pi x}{n\pi}\right) - (2x)\left(-\frac{\cos n\pi x}{n^2\pi^2}\right) + (2)\left(-\frac{\sin n\pi x}{n^3\pi^3}\right)\right]_0^1$$

$$= -2\left[\frac{2}{n^2\pi^2}(-1)^n\right] = \frac{4(-1)^{n+1}}{n^2\pi^2}.$$

$$b_n = \int_{-1}^{1} (x - x^2) \sin n\pi x\, dx$$

$$= \int_{-1}^{1} x \sin n\pi x \, dx, \quad \text{(the other integral vanishes)}$$

$$= 2\int_{0}^{1} x \sin n\pi x \, dx$$

$$= 2\left[(x)\left(-\frac{\cos n\pi x}{n\pi}\right) - \left(-\frac{\sin n\pi x}{n^2\pi^2}\right)\right]_0^1$$

$$= -\frac{2}{n\pi}(-1)^n$$

$$\therefore \; x - x^2 = -\frac{1}{3} + \sum_{n=1}^{\infty}\left[\frac{4}{n^2\pi^2}(-1)^{n+1}\cos n\pi x + \frac{2}{n\pi}(-1)^{n+1}\sin n\pi x\right]$$

for $-1 < x < 1$

$$\text{R.M.S. value of } f(x) = \sqrt{\frac{\int_{-1}^{1}(x - x^2)^2 \, dx}{2}}$$

$$= \sqrt{\frac{1}{2}\int_{-1}^{1}(x^2 + x^4 - 2x^3)\, dx}$$

$$= \sqrt{\frac{1}{2} \times 2\int_{0}^{1}(x^2 + x^4)\, dx}$$

$$= \sqrt{\left(\frac{x^3}{3} + \frac{x^5}{5}\right)_0^1}$$

$$= \sqrt{\frac{8}{15}}$$

Example 43. *Find Fourier cosine series of* $f(x) = \begin{cases} x^2, & 0 < x < l \\ 2 - x, & l < x < 2 \end{cases}$

Here half-range l = 2.

$$\text{Let} \quad f(x) = \frac{a_0}{2} + \sum_{n=1}^{\infty} a_n \cos\frac{n\pi x}{2}$$

where $a_0 = \frac{2}{2}\int_0^2 f(x)\, dx$

$$= \int_0^1 x^2\, dx + \int_1^2 (2-x)\, dx$$

$$= \left(\frac{x^3}{3}\right)_0^1 + \left[\frac{(2-x)^2}{-2}\right]_1^2$$

$$= \frac{1}{3} + \frac{1}{2} = \frac{5}{6}$$

$$a_n = \int_0^1 x^2 \cos\frac{n\pi x}{2}\, dx + \int_1^2 (2-x)\cos\frac{n\pi x}{2}\, dx$$

$$= \left[(x)^2\left(\frac{\sin\frac{n\pi x}{2}}{\frac{n\pi}{2}}\right) - (2x)\left(-\frac{\cos\frac{n\pi x}{2}}{\frac{n^2\pi^2}{4}}\right) + (2)\left(-\frac{\sin\frac{n\pi x}{2}}{\frac{n^3\pi^3}{8}}\right)\right]_0^1$$

$$+\left[(2-x)\left(\frac{\sin\frac{n\pi x}{2}}{\frac{n\pi}{2}}\right) - (-1)\left(-\frac{\cos\frac{n\pi x}{2}}{\frac{n^2\pi^2}{4}}\right)\right]_1^2$$

$$= \frac{2}{n\pi}\sin\frac{n\pi}{2} + \frac{8}{n^2\pi^2}\cos\frac{n\pi}{2} - \frac{16}{n^3\pi^3}\sin\frac{n\pi}{2}$$

$$+\frac{4}{n^2\pi^2}\cos\frac{n\pi}{2} - \frac{4}{n^2\pi^2}\cos n\pi - \frac{2}{n\pi}\sin\frac{n\pi}{2}$$

$$= \frac{12}{n^2\pi^2}\cos\frac{n\pi}{2} - \frac{16}{n^3\pi^3}\sin\frac{n\pi}{2} - \frac{4}{n^2\pi^2}\cos n\pi$$

$$\therefore\quad f(x) = \frac{5}{12} + \sum_{n=1}^{\infty}\left(\frac{12}{n^2\pi^2}\cos\frac{n\pi}{2} - \frac{16}{n^3\pi^3}\sin\frac{n\pi}{2} - \frac{4}{n^2\pi^2}\cos n\pi\right)\cos\frac{n\pi x}{2}.$$

Example 44. *Expand $f(x) = (x-1)^2$, $0 < x < 1$ in a Fourier series of sines only.*

Here $l = 1$; we want half range sine series.

Let $f(x) = \sum_1^{\infty} b_n \sin n\pi x$

where $b_n = \frac{2}{1}\int_0^1 (x-1)^2 \sin n\pi x dx$

$$= 2\left[(x-1)^2\left(-\frac{\cos n\pi x}{n\pi}\right) - 2(x-1)\left(-\frac{\sin n\pi x}{n^2\pi^2}\right) + 2\left(\frac{\cos n\pi x}{n^3\pi^3}\right)\right]_0^1$$

$$= 2\left[\frac{2(-1)^n}{n^3\pi^3} + \frac{1}{n\pi} - \frac{2}{n^3\pi^3}\right]$$

$$= \frac{2}{\pi^3}\left[\frac{2(-1)^n - 2}{n^3} + \frac{\pi^2}{n}\right]$$

$$\therefore \quad f(x) = \frac{2}{\pi^3}\sum_{n=1}^{\infty}\left(\frac{2(-1)^n - 2}{n^3} + \frac{\pi^2}{n}\right)\sin n\pi x.$$

Example 45. *Find the Fourier series for $f(x) = |\cos x|$ in $(-\pi, \pi)$ of periodicity* 2π. **(*MS. 1992 Ap.*)**

$$f(x) = |\cos x| \text{ is even since } f(-x) = f(x).$$

Hence $f(x)$ will contain only cosine terms.

Let $f(x) = \frac{a_0}{2} + \sum_{n=1}^{\infty} a_n \cos nx$

$$a_0 = \frac{1}{\pi}\int_{-\pi}^{\pi} |\cos x|\, dx$$

$$= \frac{2}{\pi}\int_0^{\pi} |\cos x|\, dx$$

$$= \frac{2}{\pi}\left[\int_0^{\pi/2}\cos x\, dx + \int_{\pi/2}^{\pi}(-\cos x)\, dx\right]$$

(since in $(0, \pi/2)$, $\cos x$ is positive and in $(\pi/2, \pi)$ $\cos x$ is negative.

$$= \frac{2}{\pi}\left[(\sin x)_0^{\pi/2} - (\sin x)_{\pi/2}^{\pi}\right] = \frac{4}{\pi}$$

$$a_n = \frac{1}{\pi}\int_{-\pi}^{\pi} |\cos x| \cos nx\, dx$$

$$= \frac{2}{\pi}\int_0^{\pi} |\cos x| \cos nx\, dx \quad \text{since integrand is even}$$

$$= \frac{2}{\pi}\left[\int_0^{\pi/2} \cos x \cos nx\, dx + \int_{\pi/2}^{\pi} (-\cos x \cos nx)\, dx\right]$$

$$= \frac{1}{\pi}\left[\int_0^{\pi/2} [\cos (n+1)\, x + \cos (n-1)\, x]\, dx - \int_{\pi/2}^{\pi} [\cos (n+1)x + \cos (n-1)\, x]\, dx\right]$$

$$= \frac{1}{\pi}\left[\left\{\frac{\sin (n+1)x}{n+1} + \frac{\sin (n-1)x}{n-1}\right\}_0^{\pi/2} - \left\{\frac{\sin (n+1)x}{n+1} + \frac{\sin (n-1)x}{n-1}\right\}_{\pi/2}^{\pi}\right]$$

$$= \frac{1}{\pi}\left[\frac{\sin (n+1)\pi/2}{n+1} + \frac{\sin (n-1)\pi/2}{n-1} + \frac{\sin (n+1)\pi/2}{n+1} + \frac{\sin (n-1)\pi/2}{n-1}\right] \text{ if } n \neq 1$$

$$= \frac{2}{\pi}\left[\frac{1}{n+1}\left\{\sin \frac{n\pi}{2} \cos \frac{\pi}{2} + \cos \frac{n\pi}{2} \sin \frac{\pi}{2}\right\} + \frac{1}{n-1} \times \left\{\sin \frac{n\pi}{2} \cos \frac{\pi}{2} - \cos \frac{n\pi}{2} \sin \frac{\pi}{2}\right\}\right] \text{ if } n \neq 1$$

$$= \frac{2}{\pi}\left[\frac{1}{n+1} - \frac{1}{n-1}\right] \cos \frac{n\pi}{2}$$

$$= -\frac{4}{\pi (n^2 - 1)} \cos \frac{n\pi}{2} \text{ if } n \neq 1$$

$$a_1 = \frac{1}{\pi}\int_{-\pi}^{\pi} |\cos x| \cos x\, dx$$

$$= \frac{2}{\pi}\left[\int_0^{\pi} |\cos x| \cos x\right]$$

$$= \frac{2}{\pi}\left[\int_0^{\pi/2} \cos^2 x\, dx - \int_{\pi/2}^{\pi} \cos^2 x\, dx\right]$$

$$= \frac{2}{\pi}\left[\frac{1}{2}\cdot\frac{\pi}{2} - \int_{\pi/2}^{\pi} \frac{1+\cos 2x}{2}\, dx\right]$$

$$= \frac{2}{\pi}\left[\frac{\pi}{4} - \frac{1}{2}\left(x + \frac{\sin 2x}{2}\right)_{\pi/2}^{\pi}\right]$$

$$= \frac{2}{\pi}\left[\frac{\pi}{4} - \frac{\pi}{4}\right]$$

$$= 0.$$

$$\therefore \quad |\cos x| = \frac{2}{\pi} - \frac{4}{\pi}\sum_{n=2}^{\infty} \frac{1}{n^2-1}\cos\frac{n\pi}{2}\cos nx$$

Example 46. *Find the Fourier series of $f(x) = |\sin x|$ in $(-\pi, \pi)$ of periodicity* **(MS. 1991 Ap.) (BR Ap. 1997)**

$f(x) = |\sin x|$ is an even function of x though $\sin x$ is odd.

$\therefore$ $f(x)$ will contain only cosine terms in its Fourier series.

$$\text{Let} \quad f(x) = \frac{a_0}{2} + \sum_{1}^{\infty} a_n \cos nx$$

$$a_0 = \frac{1}{\pi}\int_{-\pi}^{\pi} |\sin x|\, dx$$

$$= \frac{2}{\pi}\int_0^{\pi} |\sin x|\, dx \qquad (\because \text{ even integrand})$$

$$= \frac{2}{\pi}\int_0^{\pi} \sin x\, dx$$

$$= \frac{2}{\pi}(-\cos x)_0^{\pi}$$

$$= \frac{2}{\pi}[1+1] = \frac{4}{\pi}$$

$$a_n = \frac{1}{\pi}\int_{-\pi}^{\pi} |\sin x| \cos nx\, dx$$

$$= \frac{2}{\pi}\int_0^{\pi} \sin x \cos nx \, dx$$

$$= \frac{1}{\pi}\int_0^{\pi} [\sin (n+1)x - \sin (n-1)x] \, dx$$

$$= \frac{1}{\pi}\left[-\frac{\cos (n+1)x}{n+1} + \frac{\cos (n-1)x}{n-1}\right]_0^{\pi} \text{ if } n \neq 1$$

$$= \frac{1}{\pi}\left[\frac{-1}{n+1}\left\{(-1)^{n+1} - 1\right\} + \frac{1}{n-1}\left\{(-1)^{n-1} - 1\right\}\right]$$

$$= \frac{1}{\pi}\left[\frac{1}{n+1}\left\{1 + (-1)^n\right\} - \frac{1}{n-1}\left\{1 + (-1)^n\right\}\right]$$

$$= \frac{1}{\pi}\left(\frac{-2}{n^2-1}\right)\left\{1 + (-1)^n\right\}$$

$$= 0 \text{ if } n \text{ is odd}$$

$$= -\frac{4}{\pi(n^2-1)} \text{ if } n \text{ is even and } n \neq 1$$

$$a_1 = \frac{2}{\pi}\int_0^{\pi} \sin x \cos x \, dx = \frac{1}{\pi}\left(-\frac{\cos 2x}{2}\right)_0^{\pi} = 0$$

$$\therefore \quad f(x) = \frac{2}{\pi} - \frac{4}{\pi}\sum_{n=1}^{\infty} \frac{1}{(4n^2-1)} \cos 2nx.$$

Exercises 1(g)

1. If $f(x) = \frac{a_0}{2} + \sum_{n=1}^{\infty}\left(a_n \cos \frac{n\pi x}{l} + b_n \sin \frac{n\pi x}{l}\right)$

   in $0 < x < 2l$, prove $\int_0^{2l} [f(x)]^2 \, dx = 2l\left[\frac{a_0^2}{4} + \frac{1}{2} \; \Sigma(a_n^2 + b_n^2)\right]$.

2. Show that whenever

   $f(x) = \sum_{n=1}^{\infty} b_n \sin \frac{n\pi x}{L}$ in the interval $0 < x < L$, the coefficients are given by

   $b_n = \frac{2}{L}\int_0^{L} f(x) \sin \frac{n\pi x}{L} \, dx$ and then show that $\int_0^{L} [f(x)]^2 dx = \frac{1}{2}L \sum_{n=1}^{\infty} b_n^2$.

3. Check that for $0 < x < L$,

$$1 = \frac{4}{\pi}\left(\sin\frac{\pi x}{L} + \frac{1}{3}\sin\frac{3\pi x}{L} + \frac{1}{5}\sin\frac{5\pi x}{L} + \cdots\right)$$

and deduce that

$$(i)\ 1 - \frac{1}{3} + \frac{1}{5} - \frac{1}{7} + \cdots = \frac{\pi}{4}$$

$$\text{and } (ii)\ \frac{1}{1^2} + \frac{1}{3^2} + \frac{1}{5^2} + \ldots = \frac{\pi^2}{8}.$$

4. If $f(x) = \dfrac{a_0}{2} + \sum_{n=1}^{\infty} a_0 \cos\dfrac{n\pi x}{l}$, in $0 < x < l$ show that

$$\int_0^l [f(x)]^2 dx = \frac{1}{2}l\left(\frac{1}{2}a_0^2 + a_1^2 + a_2^2 + \ldots\right)$$

5. The impressed voltage e of an alternating-current wave is given by

$$e = E_1 \sin(\omega t + \beta_1) + E_3 \sin(3\omega t + \beta_3) + E_5 \sin(5\omega t + \beta_5)$$

$+ \ldots$. Show that the effective voltage $\bar{e}$ is given by

$$\bar{e}^2 = \frac{1}{2}(E_1^2 + E_3^2 + E_5^2 + \cdots).$$

6. Prove $\sin\dfrac{\pi x}{l} = \dfrac{2}{\pi} - \dfrac{4}{\pi}\left[\dfrac{\cos\frac{2\pi x}{l}}{1\cdot 3} + \dfrac{\cos\frac{4\pi x}{l}}{3\cdot 5} + \dfrac{\cos\frac{6\pi x}{l}}{5\cdot 7} + \cdots\right]$, in $0 < x < l$

7. Prove $e^x = \pi \sum_{1}^{\infty} \dfrac{n[1 - e(-1)^n]}{1 + n^2\pi^2} \sin n\pi x$, in $0 < x < 1$.

8. Prove $x = -\dfrac{4}{\pi} \sum_{1}^{\infty} \dfrac{(-1)^n}{n} \sin\dfrac{n\pi x}{2}$, in $0 < x < 2$.

9. If $f(x) = x$, in $0 < x < \dfrac{l}{2}$

$= \dfrac{2a}{l}(l - x)$, in $\dfrac{l}{2} < x < l$, prove

$$f(x) = \frac{8a}{\pi^2}\left[\sin\frac{\pi x}{l} - \frac{1}{3^2}\sin\frac{3\pi x}{l} + \frac{1}{5^2}\sin\frac{5\pi x}{l} - \cdots\right]$$

10. If $f(x) = \dfrac{2kx}{l}$, $0 < x < \dfrac{l}{2}$

$= \dfrac{2k}{l}(l - x)$, $\dfrac{l}{2} < x < l$ prove

$$f(x) = \frac{k}{2} - \frac{16k}{\pi^2}\left(\frac{1}{2^2}\cos\frac{2\pi x}{l} + \frac{1}{6^2}\cos\frac{6\pi x}{l} + \cdots\right)$$

11. If $f(x) = \sin x, 0 < x < \pi/4$

$= \cos x, \pi/4 < x < \pi/2$

expand $f(x)$ as a sine series.

12. If $f(x) = \frac{\pi}{3}$, in $0 < x < \frac{\pi}{3}$

$= 0$, in $\frac{\pi}{3} < x < \frac{2\pi}{3}$

$= -\frac{\pi}{3}$, in $\frac{2\pi}{3} < x < \pi$

prove $f(x) = \sin 2x + \frac{1}{2}\sin 4x + \frac{1}{4}\sin 8x + \dots$

13. In $0 < x < 4$, find cosine series and sine series for $f(x) = 3x - 2$.

14. Find half range sine series and cosine series of period 4 for $f(x)$ where

$f(x) = 2x$, in $0 < x < 1$

$= 4 - 2x$, in $1 < x < 2$.

15. Find half range sine series for

$f(x) = \frac{2x}{l}$, in $0 < x < \frac{l}{2}$

$= \frac{2}{l}(l - x)$, in $\frac{l}{2} < x < l$.

16. If $f(x)) = c - x$, in $0 < x < c$, prove

$$f(x) = \frac{c}{2} + \frac{2c}{\pi^2}\left[\frac{\cos\frac{\pi x}{c}}{1^2} + \frac{\cos\frac{3\pi x}{c}}{3^2} + \frac{\cos\frac{5\pi x}{c}}{5^2} + \dots\right]$$

17. If $f(x) = \begin{cases} \pi x, & \text{in } 0 < x < 1 \\ 0 & \text{at } x = 1 \\ \pi(x-2), & \text{in } 1 < x < 2 \end{cases}$

Prove $f(x) = 2\left[\sin \pi x - \frac{1}{2}\sin 2\pi x + \frac{1}{3}\sin 3\pi x - \dots\right]$

18. If $f(x) = \frac{1}{4} - x$, in $0 < x < \frac{1}{2}$

$= x - \frac{3}{4}$, in $\frac{1}{2} < x < 1$

Prove $f(x) = \left(\frac{1}{\pi} - \frac{4}{\pi^2}\right)\sin \pi x + \left(\frac{1}{3\pi} - \frac{4}{3^2\pi^2}\right)\sin 3\pi x + \dots$

19. If $f(x) = -1$, in $-2 \le x < -1$

$= x$, in $-1 < x < 1$

$= 1$, in $1 < x \le 2$

find Fourier series of $f(x)$, in $(-2, 2)$ of periodicity 4.

Rectangular rule for integration. In engineering applications, we come across definite integrals, whose integrands are given by a table of numerical values which are obtained from a physical experiment. In such a situation, we resort to numerical integration. The simplest formula for numerical integration is the rectangular rule.

Suppose we require $\int_a^b f(x)\,dx$. Let us draw the graph $y = f(x)$, and the ordinates $x = a, x = b$. Then we know that the required integral is the area enclosed by the x-axis and the curves drawn above. Now divide the length AB into q equal parts. Let these point of division be $(a =) x_0, x_1, x_2, ..., x_q (= b)$. Then the length of each part is $h = \frac{b-a}{q}$. Let $x_1^{\,1}\; x_2^{\,1}, x_3^{\,1}, ..., x_q^{\,1}$ be the mid-points of these n intervals. The n rectangles of the above figure have areas given by $hf(x_1^{\,1}),\; hf(x_2^{\,1}), ...\; hf(x_q^{\,1})$. Since the sum of the areas of the

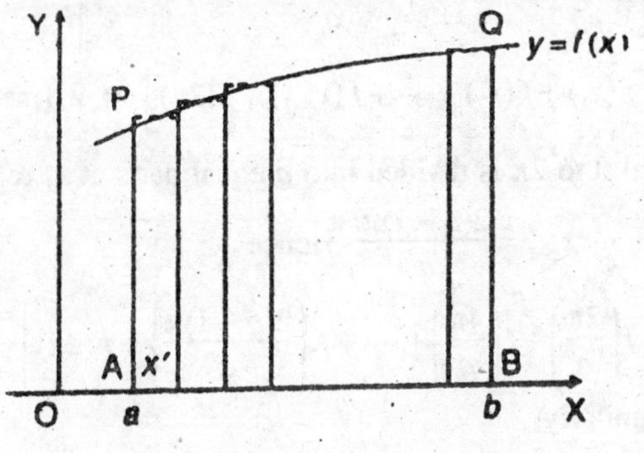

rectangles is nearly equal to the area representing the definite integral, we have

$$\int_a^b f(x)dx$$

$$\approx h[f(x_1^{\,1}) + f(x_2^{\,1}) + \cdots + f(x_q^{\,1})].$$ Substituting for h, we get that

$$\int_a^b f(x)\,dx \approx (l - a)\frac{f(x_1^{\,1}) + f(x_2^{\,1}) + \cdots + f(x_q^{\,1})}{q}.$$

This is the rectangular rule. Since $\frac{f(x_1^{\,1}) + f(x_2^{\,1}) + \cdots + f(x_q^{\,1})}{q}$ is the mean of the mid-values (of the function), we have

$$\int_a^b f(x)\,dx \approx \text{interval} \times \text{mean of the mid-values.}$$

Note. An approximation to the above approximate formula is got by changing the mid-values into the end values. Thus, we get

$$\int_a^b f(x)\,dx \approx (b-a)\frac{f(x_1)+f(x_2)+\cdots+f(x_{q-1})+f(b)}{q}.$$

Harmonic analysis. When a function $f(x)$ is given by its numerical values at q equally spaced points, the coefficients in the Fourier series representing $f(x)$ can be obtained by numerical integration—for example, by the rectangular rule as given in the above note.

By theorem, $a_0 = \dfrac{1}{\pi}\displaystyle\int_0^{2\pi} f(x)\,dx.$

Hence remembering that in the present case $b - a = 2\pi$, and using the result in the above note, we get

$$a_0 = \frac{1}{\pi}\left[2\pi\,.\,\frac{f(x_1)+f(x_2)+\cdots+f(x_{q-1})+f(2\pi)}{q}\right].$$

i.e., $$a_0 = \frac{2}{q}\left[f(x_1)+f(x_2)+\cdots+f(x_{q-1})+f(2\pi)\right] = 2\ [\text{mean value of } f(x)]$$

Since the interval 0 to 2π is divided into q equal parts at $x_1, x_2, \ldots$ we have $x_1 = \dfrac{2\pi}{q}, x_2 = \dfrac{4\pi}{q}, \ldots, x_{q-1} = \dfrac{2(q-1)\pi}{q}$. Hence

$$a_0 = \frac{2}{q}\left[f\left(\frac{2\pi}{q}\right)+f\left(\frac{4\pi}{q}\right)+\cdots+f\left\{\frac{2(q-1)\pi}{q}\right\}+f(2\pi)\right]$$

Working similarly,

$$a_n = \frac{2}{q}\left[f\left(\frac{2\pi}{q}\right)\cos\frac{2n\pi}{q}+f\left(\frac{4\pi}{q}\right)\cos\frac{4n\pi}{q}+\cdots+f\left\{\frac{2(q-1)\pi}{q}\right\}\right.$$
$$\left.\cos\frac{2(q-1)n\pi}{q}+f(2\pi)\cos 2n\pi\right] = 2\ [\text{mean value of } f(x)\cos nx]$$

and $$b_n = \frac{2}{q}\left[f\left(\frac{2\pi}{q}\right)\sin\frac{2n\pi}{q}+f\left(\frac{4\pi}{q}\right)\sin\frac{4n\pi}{q}+\cdots\right.$$
$$\left.+f\left\{\frac{2(q-1)\pi}{q}\right\}\sin\frac{2(q-1)n\pi}{q}+f(2\pi)\sin 2n\pi\right].$$
$$= 2\ [\text{mean value of } f(x)\sin nx]$$

The process of finding the Fourier series for a function given by numerical values is known as harmonic analysis.

Note 1. Since the values of sine and cosine are repeated in the four quadrants, it is advantageous to have q as a multiple of 4.

Note 2. In a Fourier expansion, the term $(a_1 \cos x + b_1 \sin x)$ is called the *fundamental* or *first harmonic*, the term $(a_2 \cos 2x + b_2 \sin 2x)$, the *second* *harmonic*, and so on. The harmonics of successive harmonics decrease rapidly; hence a few earlier harmonics will in general give a good approximation of the given function.

Example 1. *Compute the first three harmonics of the Fourier series for $f(x)$ from the following data:*

x	30°	60°	90°	120°	150°	180°	210°	240°	270°	300°	330°	360°
$f(x)$	2·34	3·01	3·68	4·15	3·69	2·20	0·83	0·51	0·88	1·09	1·19	1·64

Let the Fourier series upto the third harmonic representing $f(x)$ in $(0, 2\pi)$ be

$$f(x) = \tfrac{1}{2} a_0 + a_1 \cos x + a_2 \cos 2x + a_3 \cos 3x$$

$$+ b_1 \sin x + b_2 \sin 2x + b_3 \sin 3x \qquad \text{...(i)}$$

To evaluate the coefficients, the following table is constructed.

x	$f(x)$	$\cos x$	$\sin x$	$\cos 2x$	$\sin 2x$	$\cos 3x$	$\sin 3x$
30°	2·34	0·866	0·5	0·5	0·866	0	1
60°	3·01	0·5	·866	–0·5	0·866	–1	0
90°	3·68	0	1	–1	0	0	–1
120°	4·15	–0·5	0·866	– 0·5	–0·866	1	0
150°	3·69	–0·866	0·5	0·5	–0·866	0	1
180°	2·20	–1	0	1	0	–1	0
210°	0·83	–0·866	–0·5	0·5	0·866	0	–1
240°	0·51	–0·5	–0·866	–0·5	0·866	1	0
270°	0·88	0	–1	–1	0	0	1
300°	1·09	0·5	–0·866	–0·5	–0·866	–1	0
330°	1·19	0·866	–0·5	0·5	–0·866	0	–1
360°	1·64	1	0	1	0	1	0

$$a_0 = \tfrac{2}{12} \Sigma f(x)$$

$$= \tfrac{1}{6} \times 25{\cdot}21 = 4{\cdot}202$$

$$a_1 = \tfrac{2}{12} \times \Sigma f(x) \cos x$$

$$= \tfrac{1}{6} [2{\cdot}34 \times 0{\cdot}866 + 3{\cdot}01 \times 0{\cdot}5 - 4{\cdot}15 \times 0{\cdot}5 - 3{\cdot}69 \times 0{\cdot}86 - 2{\cdot}20$$

$$\times 1 - 0{\cdot}83 \times 0{\cdot}866 - 0{\cdot}51 \times 0{\cdot}5 + 1{\cdot}09 \times 0{\cdot}5$$

$$+ 1{\cdot}19 \times 0{\cdot}866 + 1{\cdot}64 \times 1]$$

$$= \tfrac{1}{6} [(2{\cdot}34 - 3{\cdot}69 - 0{\cdot}83 + 1{\cdot}19) \times 0{\cdot}866 + (3{\cdot}01 - 4{\cdot}15 - 0{\cdot}51$$

$$+ 1{\cdot}09) \times 0{\cdot}5 + (- 2{\cdot}20 + 1{\cdot}64)]$$

$$= -0.280$$

$$b_1 = \frac{2}{12} \times \Sigma f(x) \sin x$$

$$= \frac{1}{6}[(2.34 + 3.69 - 0.83 - 1.19) \times 0.5$$

$$+ (3.01 + 4.15 - 0.51 - 1.09) \times 0.866 + (3.68 - 0.88)]$$

$$= 1.618$$

$$a_2 = \frac{2}{12} \times \Sigma f(x) \cos 2x$$

$$= \frac{1}{6}[(2.34 - 3.01 - 4.15 + 3.69 + 0.83 - 0.51 - 1.09 + 1.19) \times 0.5$$

$$+ (-3.68 + 2.20 - 0.88 + 1.64)]$$

$$= -0.178.$$

$$b_2 = \frac{2}{12} \times \Sigma f(x) \sin 2x$$

$$= \frac{1}{6}[(2.34 + 3.01 - 4.15 - 3.69 + 0.83 + 0.51 - 1.09 - 1.19) \times 0.866]$$

$$= -0.495$$

$$a_3 = \frac{2}{12} \times \Sigma f(x) \cos 3x$$

$$= \frac{1}{6}[-3.01 + 4.15 - 2.20 + 0.51 - 1.09 + 1.64]$$

$$= 0$$

$$b_3 = \frac{2}{12} \times \Sigma f(x) \sin 3x$$

$$= \frac{1}{6}[2.34 - 3.68 + 3.69 - 0.83 + 0.88 - 1.19]$$

$$= 0.202.$$

$$f(x) = \frac{1}{2}a_0 + a_1 \cos x + a_2 \cos 2x + a_3 \cos 3x + \cdots$$

$$+ b_1 \sin x + b_2 \sin 2x + b_3 \sin 3x + \cdots$$

$$= 2.101 - 0.280 \cos x - 0.178 \cos 2x + 0(\cos 3x) + \cdots$$

$$+ 1.618 \sin x - 0.495 \sin 2x + 0.202 \sin 3x + \cdots$$

Example 2. *Compute the first three harmonics of the Fourier series of $f(x)$ given by the following table.* *(MS. 1991 Ap.)*

x	0	$\pi/3$	$2\pi/3$	π	$4\pi/3$	$5\pi/3$	2π
$f(x)$	1·0	1·4	1·9	1·7	1·5	1·2	1·0

We will form the table for the convenience of work.

We exclude the last point $x = 2\pi$.

x	$f(x)$	$\cos x$	$\sin x$	$\cos 2x$	$\sin 2x$	$\cos 3x$	$\sin 3x$
0	1·0	1	0	1	0	1	0
$\frac{\pi}{3}$	1·4	0·5	0·866	–0·5	0·866	–1	0
$\frac{2\pi}{3}$	1·9	–0·5	0·866	–0·5	–0·866	1	0
π	1·7	–1	0	1	0	–1	0
$\frac{4\pi}{3}$	1·5	–0·5	–0·866	–0·5	0·866	1	0
$\frac{5\pi}{3}$	1·2	0·5	–0·866	–0·5	–0·866	–1	0

$$a_0 = \frac{2}{6}\Sigma f(x) = \frac{1}{3}(1{\cdot}0 + 1{\cdot}4 + 1{\cdot}9 + 1{\cdot}7 + 1{\cdot}5 + 1{\cdot}2) = 2{\cdot}9$$

$$a_1 = \frac{2}{6}\Sigma f(x)\cos x = \frac{1}{6}(1 + 0{\cdot}7 - 0{\cdot}95 - 1{\cdot}7 - 0{\cdot}75 + 0{\cdot}6)$$

$$= -\,0{\cdot}37$$

$$a_2 = \frac{2}{6}\Sigma f(x)\cos 2x = -\,0{\cdot}1$$

$$a_3 = \frac{2}{6}\Sigma f(x)\cos 3x = 0{\cdot}03$$

$$b_1 = \frac{2}{6}\Sigma f(x)\sin x = 0{\cdot}17$$

$$b_2 = \frac{2}{6}\Sigma f(x)\sin 2x = -\,0{\cdot}06$$

$$b_3 = \frac{2}{6}\Sigma f(x)\sin 3x = 0$$

$$\therefore\ f(x) = 1{\cdot}45 - 0{\cdot}33\cos x - 0{\cdot}1\cos 2x + 0{\cdot}03\cos 3x + \ldots$$
$$+ 0{\cdot}17\sin x - 0{\cdot}06\sin 2x$$
$$+ 0\sin 3x + \ldots$$

Example 3. *Compute the first three harmonics of Fourier series of $f(x)$ given by the following table*:

x	0°	30°	60°	90°	120°	150°	180°	210°	240°	270°	300°	330°
$y = f(x)$	6·824	7·976	8·026	7·204	5·676	3·674	1·764	0·552	0·262	0·904	2·492	4·736

We will form the table to calculate a's and b's of the Fourier coefficients.

x	$y = f(x)$	$\cos x$	$y \cos x$	$\cos 2x$	$y \cos 2x$	$\cos 3x$	$y \cos 3x$	$\sin x$	$y \sin x$	$\sin 2x$	$y \sin 2x$	$\sin 3x$	$y \sin 3x$
0°	6·824	1·000	6·824	1	6·824	1	6·824	0	0	0	0	0	0
30°	7·976	0·866	6·907	0·5	3·988	0	0	·5	3·988	0·866	6·907	1	7·976
60°	8·026	0·5	4·013	−0·5	−4·013	−1	−8·026	0·866	6·950	0·866	6·950	0	0
90°	7·204	0	0	−1	−7·204	0	0	1	7·204	0	0	−1	−7·24
120°	5·676	−0·5	−2·838	−0·5	−2·838	1	5·676	0·866	4·916	−·866	−4·918	0	0
150°	3·674	−0·866	−3·182	0·5	1·837	0	0	0·5	1·187	−·866	−3·182	1	3·674
180°	1·764	−1	−1·764	1	1·764	−1	−1·764	0	0	0	0	0	0
210°	0·552	−0·866	−0·478	0·5	0·276	0	0	−·5	−0·276	0·866	0·478	−1	−0·552
240°	0·262	−0·5	0·131	−0·5	−0·131	1	0·262	−·866	−0·227	0·866	0·227	0	0
270°	0·904	0	0	−1	−·904	0	0	−1	−0·904	0	0	1	0·904
300°	2·492	0·5	1·246	−0·5	−1·246	−1	−2·492	−·866	−2·158	−·866	−2·158	0	0
330°	4·736	0·866	4·102	0·5	2·368	0	0	−0·5	−2·368	−·866	−4·102	−1	−4·736
	50·090		14·699		0·721		0·48		18·962		0·204		0·062

$$a_0 = \frac{2}{12} \Sigma f(x) = 8{\cdot}348 \qquad b_1 = \frac{2}{12} \Sigma y \sin x = 3{\cdot}160$$

$$a_1 = \frac{2}{12} \Sigma f(x) \cos x = 2{\cdot}450 \qquad b_2 = \frac{2}{12} \times 0{\cdot}204 = 0{\cdot}034$$

$$a_2 = \frac{2}{12} (0{\cdot}721) = 0{\cdot}120 \qquad b_3 = \frac{2}{12} \times 0{\cdot}062 = 0{\cdot}010$$

$$a_3 = \frac{2}{12} (0{\cdot}48) = 0{\cdot}080$$

$$\therefore \quad f(x) = 4{\cdot}174 + 2{\cdot}450 \cos x + 0{\cdot}120 \cos 2x$$
$$+ 0{\cdot}080 \cos 3x + 3{\cdot}160 \sin x + 0{\cdot}034 \sin 2x$$
$$+ 0{\cdot}010 \sin 3x.$$

Example 4. *The values of x and the corresponding values of $f(x)$ over a period T are given below. Show that $f(x) = 0{\cdot}75 + 0{\cdot}37 \cos \theta + 1{\cdot}004 \sin \theta$* where $\theta = \dfrac{2\pi x}{T}$.

x	0	$T/6$	$T/3$	$T/2$	$2T/3$	$5T/6$	T
$f(x)$	1·98	1·30	1·05	1·30	−0·88	−0·25	1·98

We omit the last value since $f(x)$ at $x = 0$ is known. $\theta = \dfrac{2\pi x}{T}$

When x varies from 0 to T, θ varies from 0 to 2π with an increase of $\dfrac{2\pi}{6}$

Let $f(x) = F(\theta) = \dfrac{a_0}{2} + a_1 \cos \theta + b_1 \sin \theta$

$$a_0 = \frac{2}{6} \Sigma f(x) = \frac{4{\cdot}6}{3} = 1{\cdot}5$$

θ	y	cos θ	sin θ	y cos θ	y sin θ
0	1·98	1·0	0	1·98	0
π/3	1·30	0·5	0·866	0·65	1·1258
2π/3	1·05	−0·5	0·866	−0·525	0·9093
π	1·30	−1	0	−1·3	0
4π/3	−0·88	−0·5	−0·866	0·44	0·762
5π/3	−0·25	0·5	−·866	−0·125	0·2165
Σ	4·6			1·12	3·013

$$a_1 = 2\frac{(1\cdot 12)}{6} = 0\cdot 37$$

$$b_1 = \frac{2}{6}(3\cdot 013) = 1\cdot 005 \quad \therefore \quad f(x) = 0\cdot 75 + 0\cdot 37 \cos\theta + 1\cdot 005 \sin\theta$$

Example 5. *Find the first three harmonics of Fourier series of $y = f(x)$ from the following data.*

x	0°	30°	60°	90°	120°	150°	180°	210°	240°	270°	300°	330°
y	298	356	373	337	254	155	80	51	60	93	147	221

Let $\quad y = \dfrac{a_0}{2} + \sum_{n=1}^{\infty}\left(a_n \cos nx + b_n \sin nx\right)$

Here $a_0 = \dfrac{2}{12}\Sigma y = \dfrac{1}{6}(2425) \approx 404$

$$a_1 = \frac{2}{12}\Sigma y \cos x = 107\cdot 048 \approx 107$$

$$a_2 = \frac{2}{12}\Sigma y \cos 2x \approx -13$$

$$a_3 = \frac{2}{12}\Sigma y \cos 3x \approx 2\cdot 0$$

$$b_1 = \frac{2}{12}\Sigma y \sin x \approx 121$$

$$b_2 = \frac{2}{12}\Sigma y \sin 2x \approx 9$$

$$b_3 = \frac{2}{12}\Sigma y \sin 3x \approx -1$$

The table can be formulated by the reader in the usual way.

$\therefore \; y \approx 202 + 107 \cos x - 13 \cos 2x + 2 \cos 3x + 121 \sin x + 9 \sin 2x - \sin 3x.$

Exercises 1 (h)

1. The following values of x and y are taken from a periodic curve which has a period 2π. Express y as a Fourier series up to the third harmonic.

x	0°	30°	60°	90°	120°	150°	180°	210°	240°	270°	300°	330°
y	0·50	0·60	0·83	1	0·80	0·42	0·00	– 0·34	–0·50	–0·20	0·67	0·70

[**Note.** In this question, take the approximation to the rectangular rule in the form:

$$\left[\int_a^b f(x)\,dx = (b-a)\times\frac{f(a)+f(x_1)+f(x_2)+\cdots+f(x_{q-1})}{q}\right]$$

2. The table of values of the function $y = f(x)$ is given below:

x	0	$\frac{\pi}{3}$	$\frac{2\pi}{3}$	π	$\frac{4\pi}{3}$	$\frac{5\pi}{3}$	2π
y	1·0	1·4	1·9	1·7	1·5	1·2	1·0

Find a Fourier series upto the third harmonic to represent $f(x)$ in terms of x.

[**Note.** For the application of the formula, *one* of the end points 0 or 2π can be ignored.]

3. Analyse the current i given by the table below into *its* constituent harmonics as far as the third one:

θ (deg.)	0	30	60	90	120	150	180	210	240	270	300	330
i (amp.)	0	24·0	32·5	27·5	18·2	13·0	0	–24	–33·5	–27·5	–18·2	–13

[**Note.** The table shows that $f(180° + x) = -f(x)$; hence, the even harmonics will be zero.]

4. The displacement y of a part of a mechanism are tabulated with the corresponding angular movement of the crank. Express y as a Fourier series neglecting the harmonics above the third:

$x°$	0	30	60	90	120	180	210	240	270	300	330	150
y	1·8	1·1	0·30	0·16	0·50	2·16	1·25	1·30	1·52	1·76	2·00	1·3

5. The turning moment T lb-ft. on the crank-shaft of a steam engine is given for a series of values of the crank-angle θ degrees:

θ	0	30	60	90	120	150	180
T	0	5224	8097	7850	5499	2626	0

Obtain the first four terms in a series of sines to represent T; also calculate T when 0 – 75°.

6. The following table gives the variations of a periodic current over a period:–

t sec.	0	T/6	T/3	T/2	2T/3	5T/6	T
A amps.	1·98	1·30	1·05	1·30	–0·88	0·25	1·98

By numerical analysis, show that there is a direct current part of 0·75 amp. in the variable current; obtain the amplitude of the first harmonic.

7. Using six ordinates analyse harmonically the following data to two harmonics. ***(MS. 1988 Nov.)***

x	0	$\pi/3$	$2\pi/3$	π	$4\pi/3$	$5\pi/3$	π
y	10	12	15	20	17	11	10

8. Complete the first two harmonics of the Fourier series of $f(x)$ from the data given below. ***(MS 1989 Nov.)***

x	30°	60°	90°	120°	150°	180°	210°	240°	270°	300°	330°	360°
y	2·34	3·01	3·68	4·15	3·69	2·2	0·83	0·51	0·88	1·09	1·19	1·64

9. From the following table find the first two harmonics of the Fourier series of $f(x)$.

x	0°	30°	60°	90°	120°	150°	180°	210°	240°	270°	300°	330°
y	137	164	265	325	156	–54	–137	–164	–265	–325	–156	54

10. Determine the first two harmonics of the Fourier series for $f(x)$ from the table below.

x	30°	60°	90°	120°	150°	180°	210°	240°	270°	300°	330°	360°
y	3·5	6·09	7·82	8·58	8·43	7·73	6·98	6·19	6·04	5·55	5·01	5·35

Fourier series in complex form. The series for $f(x)$ defined in the interval c to $c + 2\pi$, and satisfying Dirichlet's conditions can be given in the form

$$f(x) = \sum_{n=-\infty}^{\infty} c_n e^{inx} \qquad \text{...(i)}$$

where
$$c_n = \frac{1}{2\pi} \int_c^{c+2\pi} f(x)\, e^{-inx}\, dx.$$

Proof. Multiply both the sides of (i) by e^{-imx}, where m is an integer; then, integrate between the limits c and $c + 2\pi$. Thus we have

$$\int_c^{c+2\pi} f(x) e^{-imx}\, dx = \sum_{n=-\infty}^{\infty} c_n \int_c^{c+2\pi} e^{i(n-m)x}\, dx \qquad \text{...(ii)},$$

on the assumption that term by term integration is permissible.

When $n = m$, $\displaystyle\int_c^{c+2\pi} e^{i(n-m)x}\, dx = \int_c^{c+2\pi} e^0\, dx = \int_c^{c+2\pi} dx = \Big[x\Big]_c^{c+2\pi} = 2\pi$ ...(iii)

For all other values of n, that is if $n \neq m$,

$$\int_c^{c+2\pi} e^{i(n-m)x}\, dx = \left[\frac{e^{i(n-m)x}}{i(n-m)}\right]_c^{c+2\pi} = \frac{e^{i(n-m)(c+2\pi)} - e^{i(n-m)c}}{i(n-m)}$$

$$= \frac{e^{i(n-m)c}}{i(n-m)}\left[e^{2i(n-m)\pi}\right] \qquad ...(iv)$$

Since n and m are integers, $n - m$ of an integer; let us denote it by p. Then,

$$e^{2i(n-m)\pi} - 1 = e^{2ip\pi} - 1 = \cos 2p\pi + i \sin 2p\pi - 1 = 1 + i \times 0 - 1 = 0.$$

Using this in (*iv*), we get that when $n \neq m$,

$$\int_c^{c+2\pi} e^{i(n-m)x}\, dx = 0 \qquad ...(v)$$

Using the results (*iii*) and (*v*) in (*ii*), we get

$$\int_c^{c+2\pi} f(x)\, e^{imx}\, dx = c_m \times 2\pi.$$

Hence $c_m = \dfrac{1}{2\pi} \cdot \displaystyle\int_c^{c+2\pi} f(x) e^{-imx}\, dx$; or $c_n = \dfrac{1}{2\pi} \displaystyle\int_c^{c+2\pi} f(x)\, e^{-inx}\, dx.$

This proves the statement (*i*).

Note 1. When the R.H.S. of (*i*) is written out in full, (*i*) becomes

$$f(x) = c_0 + c_1 e^{ix} + c_2 e^{2ix} + c_3 e^{3ix} + \cdots + c_n e^{inx} + \cdots \text{ to } \infty$$

$$+ c_{-1}\, e^{-ix} + c_{-2}\, e^{-2ix} + c_{-3} e^{-3ix} + \cdots + c_{-n} e^{-inx} + \cdots \text{ to } \infty.$$

Thus, the index n runs through all positive and negative values including 0.

The term independent of the exponential factor is c_0 and *not* $\frac{1}{2}c_0$: remember that in the trigonometric form of the series, the independent term was given by $\frac{1}{2}a_0$. Putting $n = 0$, $c_0 = \dfrac{1}{2\pi} \displaystyle\int_c^{c+2\pi} f(x)\, dx.$

Note 2. In very many cases, the lower limit c will be either 0 or $-\pi$. Thus, for example when $f(x)$ satisfies Dirichlet's conditions in the interval $-\pi$ to π,

$$f(x) = \sum_{n=-\infty}^{\infty} c_n e^{inx}, \text{ where } c_n = \frac{1}{2\pi} \int_{-\pi}^{\pi} f(x)\, e^{-inx}\, dx.$$

Note 3. We can show that if $f(x)$ is defined in $(-L, L)$, then

$$f(x) = \sum_{n=-\infty}^{\infty} c_n\, e^{in\pi x/L}, \text{ where } c_n = \frac{1}{2L} \int_{-L}^{L} f(x)\, e^{-in\pi x/L}\, dx.$$

Theorem. Derive the trigonometric form of Fourier series from the complex form.

We have

$$f(x) = c_0 + c_1 e^{ix} + c_2 e^{2ix} + \cdots + c_n e^{inx} + \cdots \text{ to } \infty$$

$$+ c_{-1} e^{-ix} + c_{-2} e^{-2ix} + \cdots + c_{-n} e^{-inx} + \cdots \text{ to } \infty.$$

This can be written in the form

$$f(x) = c_0 + \sum_{n=1}^{\infty} (c_n e^{inx} + c_{-n} e^{-inx})$$

$$= c_0 + \sum_{n=1}^{\infty} \{c_n (\cos nx + i \sin nx) + c_{-n} (\cos nx - i \sin nx)\}$$

i.e. $$f(x) = c_0 + \sum_{n=1}^{\infty} [(\cos nx)(c_n + c_{-n}) + (\sin nx)\{i (c_n - c_{-n})\}] \qquad ...(i)$$

Let us denote c_0 by $\frac{1}{2} a_0$, $c_n + c_{-n}$ by a_n and $i (c_n - c_{-n})$ by b_n.

Then $a_0 = 2c_0 = 2 \times \dfrac{1}{2\pi} \displaystyle\int_c^{c+2\pi} f(x)\, dx = \dfrac{1}{\pi} \int_c^{c+2\pi} f(x)\, dx$

$$a_n = c_n + c_{-n} = \frac{1}{2\pi} \left\{ \int_c^{c+2\pi} f(x)\, e^{-inx}\, dx + \int_c^{c+2\pi} f(x)\, e^{inx}\, dx \right\}$$

$$= \frac{1}{2\pi} \int_c^{c+2\pi} f(x)\, (e^{-inx} + e^{inx})\, dx = \frac{1}{2\pi} \int_c^{c+2\pi} \{f(x)\}\, 2 \cos nx\, dx$$

i.e., $$a_n = \frac{1}{\pi} \int_c^{c+2\pi} f(x) \cos nx\, dx.$$

Again, $b_n = i (c_n - c_{-n}) = i \left[\dfrac{1}{2\pi} \displaystyle\int_c^{c+2\pi} f(x)\, (e^{-inx} - e^{inx})\, dx \right.$

$$= \frac{i}{2\pi} \int_c^{c+2\pi} \{f(x)\}\, (-2i \sin nx)\, dx$$

$$= \frac{-i^2}{\pi} \int_c^{c+2\pi} f(x) \sin nx\, dx.$$

Thus $$b_n = \frac{1}{\pi} \int_c^{c+2\pi} f(x) \sin nx\, dx.$$

Changing the coefficients c_n and c_{-n} in (*i*) in favour of a_n and b_n, we get,

$$f(x) = \frac{a_0}{2} + \sum_{n=1}^{\infty} (a_n \cos nx + b_n \sin nx),$$

where we have calculated that

$$a_n = \frac{1}{\pi} \int_c^{c+2\pi} f(x) \cos nx \, dx, \text{ and } b_n = \frac{1}{\pi} \int_c^{c+2\pi} f(x) \sin nx \, dx.$$

Example 1. *Derive the complex form of the Fourier series for $f(x) = e^{ax}, -\pi < x < \pi$, given that a is a real constant. Deduce that (i) when α is a constant other than an integer,*

$$\cos \alpha x = \frac{\sin \pi \alpha}{\pi} \sum_{n=-\infty}^{\infty} (-1)^n \frac{\alpha}{\alpha^2 - n^2} e^{inx}, -\pi < x < \pi,$$

$$(ii) \sum_{n=-\infty}^{\infty} \frac{(-1)^n}{n^2 + a^2} = \frac{\pi}{a \sinh a\pi}.$$

By theorem, $f(x) = \sum_{n=-\infty}^{\infty} c_n e^{inx}$ ...(*i*),

where $c_n = \frac{1}{2\pi} \int_{-\pi}^{\pi} e^{ax} e^{-inx} dx = \frac{1}{2\pi} \int_{-\infty}^{\infty} e^{(a-in)x} dx$

$$= \frac{1}{2\pi} \left[\frac{e^{(a-in)x}}{a - in} \right]_{-\pi}^{\pi} = \frac{1}{2\pi (a - in)} \left\{ e^{(a-in)\pi} - e^{-(a-in)\pi} \right\}$$

i.e., $c_n = \frac{1}{2\pi (a - in)} \left\{ e^{a\pi} . e^{-in\pi} - e^{-a\pi} . e^{in\pi} \right\}$

But $e^{\pm in\pi} = \cos (\pm n\pi) + i \sin (\pm n\pi) = (-1)^n + i \times 0 = (-1)^n$.

Hence $c_n = \frac{1}{2\pi (a - in)} (-1)^n (e^{a\pi} - e^{-a\pi})$

$$= (-1)^n \frac{1}{\pi (a - in)} \times \frac{e^{a\pi} - e^{-a\pi}}{2}$$

$$= (-1)^n \frac{a + in}{\pi (a - in)(a + in)} \sinh a\pi$$

i.e. $c_n = \frac{\sinh a\pi}{\pi} \left\{ (-1)^n \frac{a + in}{a^2 + n^2} \right\}.$

Substituting the value of c_n in (*i*) we get

$$f(x) = \sum_{n=-\infty}^{\infty} \frac{\sinh a\pi}{\pi} (-1)^n \frac{a+in}{a^2+n^2} e^{inx}$$

Since $f(x) = e^{ax}, -\pi < x < \pi$, we get

$$e^{ax} = \frac{\sinh a\pi}{\pi} \sum_{n=-\infty}^{\infty} (-1)^n \frac{a+in}{a^2+n^2} e^{inx}, \; in -\pi < x < \pi \qquad ...(ii)$$

(*i*) Now $\cos \alpha x = \frac{1}{2}(e^{i\alpha x} + e^{-i\alpha x})$ ...(*iii*)

The Fourier series for $e^{i\alpha x}$ is got from (*ii*) by putting $a = i\alpha$

Thus

$$e^{i\alpha x} = \frac{\sinh i\alpha\pi}{\pi} \sum_{n=-\infty}^{\infty} (-1)^n \frac{i\alpha + in}{(i\alpha)^2 + n^2} e^{inx}$$

$$= \frac{\sinh i\alpha\pi}{\pi} \sum_{n=-\infty}^{\infty} (-1)^n \frac{i(\alpha+n)}{-\alpha^2+n^2} e^{inx}$$

i.e.,

$$e^{i\alpha x} = \frac{-i \sinh i\alpha\pi}{\pi} \sum_{n=-\infty}^{\infty} (-1)^n \frac{\alpha+n}{\alpha^2-n^2} e^{inx}$$

But we know that $\sin i\theta = i \sinh \theta$; putting $\theta = i\alpha\pi$, $\sin i^2 \alpha\pi = i \sin h\, i\alpha\pi$ sinh $i\alpha\pi$ *i.e*, $-\sin \alpha\pi = i \sinh\; i\alpha\pi$ giving $\sin \alpha\pi = -i \sinh i\alpha\pi$. Hence

$$e^{i\alpha x} = \frac{\sin \alpha\pi}{\pi} \sum_{n=-\infty}^{\infty} (-1)^n \frac{\alpha+n}{\alpha^2-n^2} e^{inx}$$

Changing α into $-\alpha$, we have

$$e^{-i\alpha x} = \frac{\sin \alpha\pi}{\pi} \sum_{n=-\infty}^{\infty} (-1)^n \frac{\alpha-n}{\alpha^2-n^2} e^{inx}$$

Adding, $e^{i\alpha x} + e^{-i\alpha x} = \dfrac{2 \sin \alpha\pi}{\pi} \displaystyle\sum_{n=-\infty}^{\infty} (-1)^n \frac{\alpha}{\alpha^2-n^2} e^{inx}$

Hence $\cos \alpha x = \frac{1}{2}(e^{i\alpha x} + e^{-i\alpha x})$

$$= \frac{\sin \alpha\pi}{\pi} \sum_{n=-\infty}^{\infty} (-1)^n \frac{\alpha}{\alpha^2-n^2} e^{inx}$$

Note. Subtracting the value of $e^{-i\alpha x}$ from that of $e^{i\alpha x}$, we get

$$\sin \alpha x = \frac{\sin \alpha\pi}{\pi i} \sum_{n=-\infty}^{\infty} (-1)^n \frac{n}{\alpha^2 - n^2} e^{inx}$$

(*ii*) In (*ii*), putting $x = 0$, we get

$$1 = \frac{\sinh a\pi}{\pi} \sum_{n=-\infty}^{\infty} (-1)^n \frac{a + in}{a^2 + n^2}$$

Equating the real parts we have

$$1 = \frac{\sinh a\pi}{\pi} \sum_{n=-\infty}^{\infty} (-1)^n \frac{a}{a^2 + n^2}$$

i.e.,

$$1 = \frac{a \sinh a\pi}{\pi} \sum_{n=-\infty}^{\infty} \frac{(-1)^n}{n^2 + a^2}$$

$$\therefore \quad \frac{\pi}{a \sinh a\pi} = \sum_{n=-\infty}^{\infty} \frac{(-1)^n}{n^2 + a^2}$$

Note. If the range is $(c, c + 2l)$ the complex form is

$$f(x) = \sum_{n=-\infty}^{\infty} c_n e^{\frac{i n\pi x}{l}} \qquad ...(1)$$

where

$$c_n = \frac{1}{2l} \int_c^{c+2l} f(x)\, e^{\frac{-in\pi x}{l}}\, dx \qquad ...(2)$$

Example 2. *Find the complex form of the Fourier series of* $f(x) = e^{-x}$ *in* $-1 < x < 1$ **(*M.S. 1997*)**

Here $2l = 2$; $l = 1$

Let $f(x) = \sum_{n=-\infty}^{\infty} c_n\, e^{in\pi x}$

where
$$c_n = \frac{1}{2} \int_{-1}^{1} e^{-x}\, e^{-in\pi x}\, dx$$

$$= \frac{1}{2} \int_{-1}^{1} e^{-(1 + in\pi)x}\, dx$$

$$= \frac{1}{2} \left[\frac{e^{-(1 + in\pi)x}}{-(1 + in\pi)} \right]_{-1}^{1}$$

$$= -\frac{1}{2(1+in\pi)}\left[e^{-(1+in\pi)} - e^{(1+in\pi)}\right]$$

$$= \frac{-(1-in\pi)}{2(1+n^2\pi^2)}\left[e^{-1}(\cos n\pi - i\sin n\pi) - e(\cos n\pi + i\sin n\pi)\right]$$

$$= \frac{(1-in\pi)}{2(1+n^2\pi^2)}\cos n\pi\,(e - e^{-1})$$

$$= \frac{(1-in\pi)}{1+n^2\pi^2}(-1)^n \sinh 1$$

$$f(x) = \sum_{n=-\infty}^{\infty} \frac{(1-in\pi)}{1+n^2\pi^2}(-1)^n \sinh 1\; e^{in\pi x}$$

Exercises 1(i)

1. Starting with the complex form of the Fourier series for e^{ax} obtained in the worked example 1, derive the real Fourier series for e^{ax}.
2. Using the worked example 1, deduce the real Fourier series for (*i*) $\cos \alpha x$, (*ii*) $\sin \alpha x$.
3. Find the complex form of the Fourier series of the function whose definition in one period is $f(x) = e^{-x}, -1 \le x \le 1$.

   Deduce the real Fourier series for e^{-x}, in the given interval.
4. Find the complex form of the Fourier series of the periodic function whose definition in one period is given by $f(x) = \sin x$, in $-\pi < x < \pi$.
5. Show the complex form of the Fourier series of $f(x) = e^x$, when $-\pi < x < \pi$ and $f(x + 2\pi) = f(x)$ is

$$f(x) = \frac{\text{Sin}\,h\,\pi}{\pi}\sum_{n=-\infty}^{\infty}(-1)^n \frac{1+in}{1+n^2} e^{inx}.$$

6. Find the complex form of the Fourier series of the periodic function $f(x) = \text{Sin}\, x, 0 < x < \pi$
7. Find the complex form of the Fourier series of e^{ax}, in $-l < x < l$.
8. Find the complex Fourier series for $f(x)$

   where

   (*i*) $f(x) = \sin ax$, a is not an integer in $(-\pi, \pi)$

   (*ii*) $f(x) = e^x$, in $-\pi < x < \pi$

   (*iii*) $f(x) = e^{-x}$, in $-1 < x < 1$

ANSWERS

(FOURIER SERIES)

Exercises 1(a) Page 93

1. (*i*) π, (*ii*) $\frac{2}{3}\pi$, (*iii*) $2\pi/n$, (*iv*) 1, (*v*) k.

Exercises 1(b) Page 95

1. 9; 9; 9. **2.** $\frac{1}{2}; \frac{1}{2}; \frac{1}{2}$. **3.** $+\infty; -\infty$. **4.** 1, 0, does not exist.

5. 1, 0; $f(x)$ is not continuous at $x = 0$; jump at 0 is 1; $f(0) = 1$.

6. 1, 2; 1.

Exercises 1(c) Page 129

1. $f(x) = \dfrac{e^{2\pi} - 1}{\pi e^{\pi}} \left\{ \frac{1}{2} + \sum_{n=1}^{\infty} \dfrac{(-1)^n}{n^2 + 1} (\cos nx - n \sin nx) \right\}$.

2. $f_1(x) = 2\left(\sin x - \frac{1}{2} \sin 2x + \frac{1}{3} \sin 3x - \cdots\right)$ $\dfrac{\pi}{4} = 1 - \frac{1}{3} + \frac{1}{5} - \frac{1}{7} + \ldots$

3. $F(x) = f(x) + f_1(x)$, where $f(x)$ and $f_1(x)$ are given above.

4. $f(x) = \dfrac{\pi^2}{3} + 4 \sum_{n=1}^{\infty} \dfrac{(-1)^n}{n^2} \cos nx$.

5. $f(x) = 2\left\{ \left(\dfrac{\pi^2}{1} - \dfrac{6}{1^3} \right) \sin x - \left(\dfrac{\pi^2}{2} - \dfrac{6}{2^3} \right) \sin 2x + \left(\dfrac{\pi^2}{3} - \dfrac{6}{3^3} \right) \sin 3x - \cdots \right\}$.

6. $f(x) = \sin x + \frac{1}{3} \sin 3x + \frac{1}{5} \sin 5x + \cdots$ to ∞.

7. $f(x) = \frac{1}{2}\pi + 2\left(\sin x + \frac{1}{3} \sin 3x + \frac{1}{5} \sin 5x + \cdots \text{ to } \infty\right)$.

8. $f(x) = \frac{1}{2}\pi + 2\left(\frac{1}{2} \sin 2x + \frac{1}{4} \sin 4x + \frac{1}{6} \sin 6x + \cdots \text{ to } \infty\right)$.

9. $f(x) = \dfrac{3\pi}{4} + \sum_{n=1}^{\infty} \dfrac{(-1)^{n-1} \cos (2n-1)x - \sin (2n-1)x + \sin 2(2n-1)x}{2n-1}$

10. $f(x) = \dfrac{4}{\pi}\left(\cos x - \dfrac{\cos 3x}{3} + \dfrac{\cos 5x}{5} - \cdots \text{ to } \infty \right)$.

11. $a_0 = \frac{1}{2}$; $a_n = \begin{cases} 0, n = 2, 4, 6, \cdots \\ 1/n\pi, n = 1, 5, 9, \cdots \\ -1/n\pi, n = 3, 7, 11, \cdots \end{cases}$ $b_n = \begin{cases} 1/n\pi, n = 1, 3, 5, \cdots \\ 2/n\pi, n = 2, 6, 10, \cdots \\ 0, n = 4, 8, 12 \cdots \end{cases}$

12. $a_0 = 2$; $a_n = 0$, when $n \neq 0$; $b_n = \begin{cases} 3/n\pi, n = 1, 2, 4, 5, 7, 8, \cdots \\ 0, n = 3, 6, 9, \cdots \end{cases}$

13. $a_0 = \dfrac{\pi}{2}$; $a_n = \dfrac{-2}{n^2\pi}$ for odd n, and = 0 for even n; $b_n = \pm\dfrac{1}{n}$, as n is odd or even.

14. $f(x) = \dfrac{2}{\pi}\sin x + \dfrac{1}{2}\sin 2x - \dfrac{2}{9\pi}\sin 3x - \dfrac{1}{4}\sin 4x + \dfrac{2}{25\pi}\sin 5x + \cdots$ to ∞.

15. $a_0 = \dfrac{\pi^2}{3}$; $a_n = \dfrac{(-1)^n 2}{n^2}$ for $n \neq 0$; $b_n = \dfrac{\pi}{n} - \dfrac{4}{\pi n^3}$ for odd n, and $= -\dfrac{\pi}{n}$ for even n.

16. $a_n = 0$; $b_n = \dfrac{(-1)^{n+1}\, 8n}{\pi\,(4n^2 - 1)}$.

17. $\dfrac{\cos x}{2} + \dfrac{4}{\pi}\sum_{n=1}^{\infty}\dfrac{n\sin 2nx}{4n^2 - 1}$

18. $i = \dfrac{2I}{\pi}\left\{1 - \dfrac{\cos 2\theta}{1\cdot 3} - \dfrac{\cos 4\theta}{3\cdot 5} - \dfrac{\cos 6\theta}{5\cdot 7} - \cdots\right\}$

19. (*i*) $f(x) = \dfrac{4}{\pi}\sum\dfrac{1}{n}\sin nx$ for n odd.

(*ii*) $f(x) = \dfrac{3}{2} - \dfrac{2}{\pi}\left[\sin x + \dfrac{1}{3}\sin 3x + \dfrac{1}{5}\sin 5x + \cdots\right]$

(*iii*) $f(x) = \dfrac{2}{\pi}\left[\sin x + \dfrac{1}{3}\sin 3x + \dfrac{1}{5}\sin 5x + \cdots\right]$

(*v*) $f(x) = \dfrac{1}{2} + \dfrac{2}{\pi}\left(\sin x + \dfrac{1}{3}\sin 3x + \dfrac{1}{5}\sin 3x + \cdots\right)$

(*vi*) $f(x) = \dfrac{\pi^2}{3} + 4\left(\dfrac{\cos x}{1^2} + \dfrac{\cos 2x}{2^2} + \dfrac{\cos 3x}{3^2} + \cdots\right)$

Exercises 1(d) Page 134

1. $\dfrac{2}{\pi}\left[\left(\dfrac{\pi^2}{1} - \dfrac{4}{1^2}\right)\sin x - \dfrac{\pi^2}{2}\sin 2x + \left(\dfrac{\pi^2}{3} - \dfrac{4}{3^2}\right)\sin 3x - \dfrac{\pi^2}{4}\sin 4x + \cdots\right]$

2. $2\left[\left(\dfrac{\pi^2}{1} - \dfrac{6}{1^3}\right)\sin x - \left(\dfrac{\pi^2}{3} - \dfrac{6}{2^3}\right)\sin 2x + \left(\dfrac{\pi^2}{3} - \dfrac{6}{3^3}\right)\sin 3x - \cdots\right]$

3. $-\dfrac{1}{\pi}\sin x - \dfrac{8}{\pi}\sum_{x=1}^{\infty}\dfrac{r}{4r^2 - 1}\sin 2r\pi.$

4. $\dfrac{\sin \pi\alpha}{\pi\alpha} + \sum_{r=1}^{\infty}(-1)^n\dfrac{2\alpha\sin \pi\alpha}{\pi\,(\alpha^2 - n^2)}\cos nx.$

5. $\dfrac{2\sin \pi\alpha}{\pi}\left\{\dfrac{\sin x}{1^2 - \alpha^2} - \dfrac{2\sin 2x}{2^2 - \alpha^2} + \dfrac{3\sin 3x}{3^2 - \alpha^2}\cdots\right\}$

Exercises 1(e) Page 143

1. (*i*) $\dfrac{2}{\pi}\left\{(\pi + 2)\sin x - \dfrac{\pi\sin 2x}{2} + \dfrac{(\pi + 2)\sin 3x}{3} - \dfrac{\pi\sin 4x}{4} + \cdots\right\}$;

$$\frac{\pi}{2}+1-\frac{4}{\pi}\left\{\cos x+\frac{\cos 3x}{3^2}+\frac{\cos 5x}{5^2}+\cdots\right\}$$

(ii) $$\frac{2}{\pi}\sum_{n=1}^{\infty}\frac{n\{1-(-1)^n e^{\pi}\}}{n^2+1}\sin nx;$$

$$\frac{e^{\pi}-1}{\pi}-\frac{2}{\pi}\sum_{n=1}^{\infty}\frac{1-(-1)^n e^{\pi}}{n^2+1}\cos nx$$

(iii) The same as exercises 1 (d), ex. 1;

$$\frac{\pi^2}{3}-4\left\{\cos x-\frac{\cos 2x}{2^2}+\frac{\cos 3x}{3^2}-\frac{\cos 4x}{4^2}+\cdots\right\}.$$

2. $$\frac{\pi}{4}-\frac{2}{\pi}\left\{\cos 2x+\frac{\cos 6x}{3^2}+\frac{\cos 10x}{5^2}+\cdots+\frac{\cos 2(2r+1)x}{(2r+1)^2}+\cdots\right\}.$$

3. The sine series for $\sin x$ is $\sin x$. This answer is consistent with the theorem. If two Fourier series in sines alone (or cosines alone) converge to the same sum for all values of x, then the corresponding coefficients are equal.

Exercises 1(f) Page 152

1. (a) The same as in 1 (b), with $k = 1$.

(b) $$\frac{k}{2}+\frac{2k}{\pi}\left\{\sin\frac{\pi x}{2}+\frac{1}{3}\sin\frac{3\pi x}{2}+\frac{1}{5}\sin\frac{5\pi x}{2}+\cdots\right\}.$$

2. $$\frac{4}{\pi}\sum_{0}^{\infty}\frac{1}{2n+1}\sin\frac{(2n+1)\pi x}{2}.$$ 3. $$\frac{1}{2}+\frac{2}{\pi}\left\{\cos\frac{\pi x}{2}-\frac{1}{3}\cos\frac{3\pi x}{2}+\frac{1}{5}\cos\frac{5\pi x}{2}-\cdots\right\}.$$

4. $$\frac{2}{\pi}\left\{\sin \pi x-\frac{\sin 2\pi x}{2}+\frac{\sin 3\pi x}{3}-\cdots\right\}.$$ 5. $$\frac{1}{2}-\sum_{n=1}^{\infty}\frac{4\cos(2n-1)\pi x}{\pi^2(2n-1)^2}.$$

6. $$4\sum_{n=1}^{\infty}\frac{\cos 2n\pi x}{4n^2\pi^2}.$$ 7. $$b_n=0;\ a_0=\frac{1}{2};\ a_n=\begin{cases}\dfrac{4}{n^2\pi^2}, & n=1,3,5,\cdots\\ 0, & n=4,8,12,\cdots\\ -\dfrac{8}{n^2\pi^2}, & n=2,6,10,\cdots\end{cases}$$

8. $$\frac{1}{3}-\frac{4}{\pi^2}\left\{\cos\pi x-\frac{\cos 2\pi x}{4}+\frac{\cos 3\pi x}{6}-\frac{\cos 4\pi x}{8}+\cdots\right\}.$$

9. $$\frac{2}{\pi}-\frac{4}{\pi}\left\{\frac{\cos 2x}{1\cdot 3}+\frac{\cos 4x}{3\cdot 5}+\frac{\cos 6x}{5\cdot 6}-\frac{\cos 4\pi x}{7\cdot 9}+\cdots\right\}.$$

10. $1 + (2\pi \sin 1) \sum_{n=1}^{\infty} \frac{(-1)^{n+1} n \sin n\pi x}{n^2\pi^2 - 1}$

11. $a_0 = \frac{2}{3}$ and $a_3 = \frac{1}{3}$. For $n = 1, 2, 4, 5, 6, 7, \cdots$

$$a_n = \frac{18}{(n^2 - 9)n\pi} \sin \frac{n\pi}{3};\ b_n = 0, \text{ for all } n.$$

12. $\frac{E}{\pi} + \frac{E}{2} \sin \omega t + \frac{2E}{\pi} \left\{ \frac{\cos 2\omega t}{1\cdot 3} + \frac{\cos 4\omega t}{3\cdot 5} + \frac{\cos 6\omega t}{5\cdot 7} - \cdots \right\}$

13. $a_0 = \frac{1}{2}(3 - e^{-2});\ a_n = \frac{2k_n}{l_n}$ where $k_n = 1 - (-1)^n e^{-2},\ l_n = 4 + n^2\pi^2$.

For even n, $b_n = \frac{n\pi k_n}{l_n}$ and for odd n, $b_n = \frac{n\pi k_n}{l_n} - \frac{2}{n\pi}$.

15. $\frac{1}{2} - \sum_{n=1}^{\infty} \frac{(-1)^n}{2n-1} \cos \frac{(2n-1)x}{2}$.

16. $\frac{1}{6} - \frac{4}{\pi^2} \left\{ \frac{\cos 2\pi t}{4} + \frac{\cos 4\pi t}{16} + \frac{\cos 6\pi t}{9} + \frac{\cos 8\pi t}{64} + \cdots \right\};$

$$\frac{8}{\pi^2} \left\{ \frac{\sin \pi t}{1} + \frac{\sin 3\pi t}{27} + \frac{\sin 5\pi t}{125} + \cdots \right\}$$

17. $\frac{k}{2} - \frac{16k}{\pi^2} \left\{ \frac{1}{2^2} \cos \frac{2\pi t}{l} + \frac{1}{6^2} \cos \frac{6\pi t}{l} + \cdots \right\};$

$$\frac{8k}{\pi^2} \left\{ \sin \frac{\pi t}{l} - \frac{1}{3^2} \sin \frac{3\pi t}{l} + \frac{1}{5^2} \sin \frac{5\pi t}{l} - \cdots \right\}.$$

18. $a_0 = \frac{5}{6};\ a_n = \begin{cases} \frac{4}{n^2\pi^2} - \frac{16}{n^3\pi^3},\ n = 1, 5, 9, \cdots \\ -\frac{16}{n^2\pi^2},\ n = 2, 6, 10, \cdots \\ \frac{4}{n^2\pi^2} + \frac{16}{n^3\pi^3},\ n = 3, 7, 11, \cdots \\ \frac{8}{n^2\pi^2},\ n = 4, 8, 12, \cdots \end{cases}$

19. $-\frac{32}{\pi^3} \left\{ \sin \frac{\pi x}{2} + \frac{\pi^2}{8} \sin \frac{2\pi x}{2} + \frac{1}{3^3} \sin \frac{3\pi x}{2} + \frac{\pi^2}{16} \sin \frac{4\pi x}{2} + \cdots \right\}$

20. $\frac{2}{\pi^3} \left\{ \left(\frac{\pi^2}{1} - \frac{4}{1^3} \right) \sin \pi x + \frac{\pi^2}{2} \sin 2\pi x + \left(\frac{\pi^2}{3} - \frac{4}{3^3} \right) \sin 3\pi x + \frac{\pi^2}{4} \sin 4\pi x + \cdots \right\}$

21. For even n, $b_n = 0$ and for odd n, $b_n = \frac{4n}{(n^2 - 4)\pi}$.

22. $\frac{8}{\pi}\left\{\frac{\sin x}{1\cdot 3}+\frac{2\sin 2x}{3\cdot 5}+\frac{3\sin 3x}{5\cdot 7}+\frac{4\sin 4x}{7\cdot 9}+\dots\right\}$.

23. $\frac{2}{\pi}-\frac{4}{\pi}\left\{\frac{1}{1\cdot 3}\cos\frac{2\pi x}{l}+\frac{1}{3\cdot 5}\cos\frac{4\pi x}{l}+\frac{1}{5\cdot 7}\cos\frac{6\pi x}{l}\dots\right\}$.

24. $\sum_{n=1}^{\infty}\frac{n\pi}{1+n^2\pi^2}(1-e\cos n\pi)\sin n\pi x$.

Exercises 1(h) Page 182

1. $y = 0{\cdot}373 + 0{\cdot}359\cos x - 0{\cdot}085\cos 2x - 0{\cdot}117\cos 3x + \cdots$
$+ 0{\cdot}466\sin x - 0{\cdot}289\sin 2x - 0{\cdot}090\sin 3x + \cdots$

2. $1{\cdot}45 - 0{\cdot}367\cos x - 0{\cdot}1\cos 2x + 0{\cdot}033\cos 3x$
$+ 0{\cdot}173\sin x - 0{\cdot}067\sin 2x + 0\,(\sin 3x) + \cdots$

3. $5{\cdot}73\cos\theta + 30{\cdot}26\sin\theta - 5{\cdot}10\cos 3\theta + 3{\cdot}17\sin 3\theta$

4. $1{\cdot}26 + 0{\cdot}04\cos x + 0{\cdot}53\cos 2x - 0{\cdot}1\cos 3x$
$- 0{\cdot}63\sin x - 0{\cdot}23\sin 2x + 0{\cdot}085\sin 3x$.

5. $7850\sin\theta + 1500\sin 2\theta;\ T = 8332$ 6. $1{\cdot}071$.

8. $y = 2{\cdot}10 - 0{\cdot}283\cos x - 0{\cdot}18\cos 2x + 1{\cdot}6\sin x - 0{\cdot}49\sin 2x$

10. $y = 6{\cdot}44 - 1{\cdot}65\cos x - 0{\cdot}34\cos 2x + 0{\cdot}71\sin x - 0{\cdot}69\sin 2x$

Exercises 1(i) Page 189

1. $e^{ax} = \frac{\sinh a\pi}{a\pi} + \frac{\sinh a\pi}{\pi^2}\sum_{1}^{\infty}\frac{(-)^n\,2a\cos nx}{a^2+n^2} - \frac{2\sinh a\pi}{\pi^2}\sum_{1}^{\infty}\frac{(-1)^n\,n\sin nx}{a^2+n^2}$.

2. (i) $\frac{\sin\alpha\pi}{\pi}\left\{\frac{1}{\alpha}+\sum_{1}^{\infty}\frac{(-1)^n\,2\alpha\cos nx}{\alpha^2-n^2}\right\}$

(ii) $\frac{2\sin\alpha\pi}{\pi}\sum_{1}^{\infty}\frac{(-1)^n\,n\sin nx}{\alpha^2-n^2}$.

3. $e^{-x} = \sum_{n=-\infty}^{\infty}(-1)^n\frac{(1-in\pi)\sinh 1}{1+n^2\pi^2}e^{in\pi x}$.

$= \sinh 1 - 2\sinh 1\left\{\frac{\cos\pi x}{1+\pi^2}-\frac{\cos 2\pi x}{1+4\pi^2}+\frac{\cos 3\pi x}{1+9\pi^2}-\dots\right\}$

$-2\pi\sinh 1\left\{\frac{\sin\pi x}{1+\pi^2}-\frac{2\sin 2\pi x}{1+4\pi^2}+\frac{3\sin 3\pi x}{1+9\pi^2}-\dots\right\}$

4. $C_0 = \frac{1}{6}$; $C_n = \frac{3}{4n^2\pi^2}\left\{\left(1+\frac{2}{3}in\pi\right)e^{-\frac{2}{3}in\pi} - 1\right\}$.

8. (*i*) $\sin ax = \frac{\sin a\pi}{\pi} \sum_{-\infty}^{\infty} (-1)^n \frac{n}{a^2 - n^2} e^{inx}$

(*ii*) $e^x = \frac{\sinh \pi}{\pi} \sum_{-\infty}^{\infty} (-1)^n \frac{1 + in}{1 + n^2} e^{inx}$.

CHAPTER 2

FOURIER TRANSFORMS

Definition. If $f(x)$ is defined in (a, b), the integral transform of $f(x)$ with the Kernel $K(s, x)$ is defined by

$$F(s) = \bar{f}(s) = \int_a^b f(x)\, K(s, x)\, dx, \text{ if the integral exists.}$$

Here, $K(s, x)$ is called the Kernel of the transform while a, b are fixed limits. If a, b are finite, the transform is finite and if a, b are infinite, it is an infinite transform.

$$\text{If } \quad K(s, x) = e^{-sx} \qquad \text{for } x \geq 0$$
$$= 0 \qquad \text{for } x < 0,$$

the above infinite transform becomes the well known Laplace transform.

$$\text{If } \quad F(s) = \int_a^b f(x)\, K(s, x)\, dx \qquad \text{...(1)}$$

it may be possible to get $f(x)$ as

$$f(x) = \int_c^d F(s)\, H(s, x)\, ds \qquad \text{...(2)}$$

In such a case, (2) is called the inversion formula for (1). In (1), if $F(s)$ is known while $f(x)$ is unknown, it may be regarded as an integral equation.

From the general integral transform definition, we can get various integral transforms by properly defining the kernel.

Some of the well known integral transforms are listed below together with the corresponding inversion formula.

1. Infinite Fourier Transform (complex form)

$$F(s) = \frac{1}{\sqrt{2\pi}} \int_{-\infty}^{\infty} f(x)\, e^{isx}\, dx$$

$$f(x) = \frac{1}{\sqrt{2\pi}} \int_{-\infty}^{\infty} F(s)\, e^{-isx}\, ds \quad \text{(inversion formula)}$$

2. Infinite Fourier Cosine Transform

$$F_c(s) = \bar{f}_c(x) = \sqrt{\frac{2}{\pi}} \int_0^\infty f(x) \cos sx \, dx$$

$$f(x) = \sqrt{\frac{2}{\pi}} \int_0^\infty F_c(s) \cos sx \, ds$$

3. Infinite Fourier Sine Transform

$$F_s(s) = \sqrt{\frac{2}{\pi}} \int_0^\infty f(x) \sin sx \, dx$$

$$f(x) = \sqrt{\frac{2}{\pi}} \int_0^\infty F_s(s) \sin sx \, ds$$

4. Laplace Transform

$$F(s) = \int_0^\infty e^{-sx} f(x) \, dx$$

$$f(x) = \frac{1}{2\pi i} \int_{c-i\infty}^{c+i\infty} F(s) \, e^{sx} \, ds$$

5. Hankel Transform

$$F_n(s) = \int_0^\infty x f(x) \, J_n(sx) \, dx$$

$$f(x) = \int_0^\infty s \, F_n(s) \, J_n(sx) \, ds$$

where $J_n(x)$ is the Bessel function of the first kind of order n.

6. Mellin Transform

$$F(s) = \int_0^\infty f(x) \, x^{s-1} \, dx$$

$$f(x) = \frac{1}{2\pi i} \int_{c-i\infty}^{c+i\infty} F(s) \, x^{-s} \, ds$$

Theorem 1. *If $f(x)$ satisfies Dirichlet's conditions in $(-l, l)$, then the complex Fourier series of $f(x)$ is*

$$f(x) = \sum_{n=-\infty}^{\infty} C_n e^{\frac{in\pi x}{l}} \text{ where} \quad \text{...(1)}$$

$$C_n = \frac{1}{2l}\int_{-l}^{l} f(x)\, e^{\frac{-in\pi x}{l}}\, dx \quad \text{...(2)}$$

(For proof, refer to chapter 'Fourier series', Page 184, Note 3)

Theorem 2. (FOURIER INTEGRAL THEOREM)

If f (x) is piece-wise continuously differentiable and absolutely integrable in (− ∞, ∞), then

$$f(x) = \frac{1}{2\pi}\int_{-\infty}^{\infty}\int_{-\infty}^{\infty} f(t)\, e^{i(x-t)s}\, dt\, ds.$$

Proof. In theorem 1, substituting the value of C_n from (2) in (1), and changing the variable of integration in (2) as t, we get,

$$f(x) = \sum_{n=-\infty}^{\infty}\left(\frac{1}{2l}\int_{-l}^{l} f(t)\, e^{\frac{in\pi(x-t)}{l}}\, dt\right)$$

Let $\frac{\pi}{l} = \delta s.$ $\quad (-l \le x \le l)$

$$\therefore \quad f(x) = \sum_{n=-\infty}^{\infty}\left(\frac{1}{2l}\int_{-l}^{l} f(t)\, e^{i(x-t)n\delta s}\, dt\right)$$

$$= \frac{1}{2\pi}\sum_{n=-\infty}^{\infty} \delta s\left(\int_{-l}^{l} f(t)\, e^{i(x-t)n\delta s}\, dt\right) \quad \text{since } l = \frac{\pi}{\delta s}$$

Let $l \to \infty$, so that the range $(-l, l)$ becomes $(-\infty, \infty)$. As $l \to \infty$, $\delta s \to 0$. Changing the summation to definite integral, we get,

$$f(x) = \frac{1}{2\pi}\int_{-\infty}^{\infty}\int_{-\infty}^{\infty} f(t)\, e^{i(x-t)s}\, dt\, ds.$$

Note 1. Using Euler's theorem, $e^{i\theta} = \cos\theta + i\sin\theta$, we get

$$f(x) = \frac{1}{2\pi}\int_{-\infty}^{\infty}\int_{-\infty}^{\infty} f(t)\,[\cos(x-t)s + i\sin(x-t)s]\, dt\, ds.$$

Equating real and imaginary parts on both sides ($f(x)$ real) we get,

$$f(x) = \frac{1}{2\pi} \int_{-\infty}^{\infty} \int_{-\infty}^{\infty} f(t) \cos (x - t) s \, dt \, ds \, .$$

and
$$= \frac{1}{2\pi} \int_{-\infty}^{\infty} \int_{-\infty}^{\infty} f(t) \sin (x - t) s \, dt \, ds = 0.$$

Complex Fourier Transform (Infinite)

Let $f(x)$ be a function defined in $(-\infty, \infty)$ and be piece-wise continuous in each finite partial interval and absolutely integrable in $(-\infty, \infty)$. Then the complex Fourier Transform of $f(x)$ is defined by

$$\mathcal{F}(s) = F(s) = \frac{1}{\sqrt{2\pi}} \int_{-\infty}^{\infty} f(x) \, e^{isx} \, dx.$$

Theorem 3. Inversion Theorem for Complex Fourier Transform

If $f(x)$ satisfies the Dirichlet's conditions in every finite interval $(-l, l)$ and if it is absolutely integrable in the range, and if

$F(s)$ denotes the complex Fourier transform of $f(x)$, then at every point of continuity of $f(x)$, we have

$$f(x) = \frac{1}{\sqrt{2\pi}} \int_{-\infty}^{\infty} F(s) \, e^{-isx} \, ds$$

Proof. From Fourier integral theorem

$$f(x) = \frac{1}{2\pi} \int_{-\infty}^{\infty} \int_{-\infty}^{\infty} f(t) \, e^{i(x-t)s} \, dt \, ds$$

$$= \frac{1}{2\pi} \int_{-\infty}^{\infty} e^{ixs} \left(\int_{-\infty}^{\infty} f(t) \, e^{-its} \, dt \right) ds$$

$$= \frac{1}{2\pi} \int_{-\infty}^{\infty} e^{-ixv} \left(\int_{-\infty}^{\infty} f(t) \, e^{itv} \, dt \right) dv \qquad \text{putting } s = -v$$

$$= \frac{1}{\sqrt{2\pi}} \int_{-\infty}^{\infty} e^{-ixv} \left(\frac{1}{\sqrt{2\pi}} \int_{-\infty}^{\infty} f(t) \, e^{itv} \, dt \right) dv$$

$$= \frac{1}{\sqrt{2\pi}} \int_{-\infty}^{\infty} e^{-ixv} \, (F(v)) \, dv$$

$$= \frac{1}{\sqrt{2\pi}} \int_{-\infty}^{\infty} F(s) e^{-isx}\, ds \qquad \text{changing } v \text{ as } s.$$

Note 1. The parameter s is taken as p by some authors.

Note 2. Some authors define the Complex Fourier transforms in different forms, changing the Kernels of the transform. Various forms are listed below.

1. $$\bar{f}(s) = \int_{-\infty}^{\infty} e^{-isx} f(x)\, dx$$

$$f(x) = \frac{1}{2\pi} \int_{-\infty}^{\infty} e^{isx} \bar{f}(s)\, ds$$

2. $$F(s) = \int_{-\infty}^{\infty} e^{isx} f(x)\, dx$$

$$f(x) = \frac{1}{2\pi} \int_{-\infty}^{\infty} e^{-isx} F(s)\, ds$$

3. $$F(s) = \frac{1}{\sqrt{2\pi}} \int_{-\infty}^{\infty} e^{-isx} f(x)\, dx$$

$$f(x) = \frac{1}{\sqrt{2\pi}} \int_{-\infty}^{\infty} e^{isx} F(s)\, ds$$

Properties of Fourier Transforms

Theorem 1. *Fourier transform is linear.*

i.e. $F[af(x) + bg(x)] = a[Ff(x)] + bF[g(x)]$

where F stands for Fourier transform.

$$F[af(x) + bg(x)] = \frac{1}{\sqrt{2\pi}} \int_{-\infty}^{\infty} (af(x) + bg(x))\, e^{isx}\, dx$$

$$= a \cdot \frac{1}{\sqrt{2\pi}} \int_{-\infty}^{\infty} f(x)\, e^{isx}\, dx + b \cdot \frac{1}{\sqrt{2\pi}} \int_{-\infty}^{\infty} g(x)\, e^{isx}\, dx$$

$$= a\, F[f(x)] + b\, [Fg(x)].$$

Theorem 2. Shifting theorem

If $F\{f(x)\} = F(s)$, *then* $F\{f(x-a)\} = e^{isa} F(s)$

$$F\{f(x-a)\} = \frac{1}{\sqrt{2\pi}} \int_{-\infty}^{\infty} f(x-a)\, e^{isx}\, dx$$

$$= \frac{1}{\sqrt{2\pi}} \int_{-\infty}^{\infty} f(t)\, e^{i(a+t)s}\, dt \qquad \text{putting } x - a = t$$

$$= e^{ias} \cdot \frac{1}{\sqrt{2\pi}} \int_{-\infty}^{\infty} f(t)\, e^{its}\, dt$$

$$= e^{ias} F(s).$$

Theorem 3. Change of scale property

If $F\{f(x)\} = F(s)$, *then* $F\{f(ax)\} = \frac{1}{|a|} F\left(\frac{s}{a}\right)$ *where* $a \neq 0$

$$F\{f(ax)\} = \frac{1}{\sqrt{2\pi}} \int_{-\infty}^{\infty} e^{isx} f(ax)\, dx$$

$$= \frac{1}{\sqrt{2\pi}} \int_{-\infty}^{\infty} e^{i\left(\frac{s}{a}\right)t} f(t) \frac{dt}{a} \qquad \text{putting } ax = t \text{ and } a > 0.$$

$$= \frac{1}{a} F\left(\frac{s}{a}\right) \quad \text{if } a > 0.$$

$$F\{f(ax)\} = \frac{1}{\sqrt{2\pi}} \int_{+\infty}^{-\infty} e^{i\left(\frac{s}{a}\right)t} f(t) \frac{dt}{a} \text{ if } a < 0$$

$$= -\frac{1}{a} F\left(\frac{s}{a}\right) \text{ if } a < 0$$

Hence $F\{f(ax)\} = \frac{1}{|a|} F\left(\frac{s}{a}\right)$

Theorem 4. $F\{e^{iax} f(x)\} = F(s+a)$ ***(BR 1995 Ap.)***

$$F\{e^{iax} f(x)\} = \frac{1}{\sqrt{2\pi}} \int_{-\infty}^{\infty} e^{iax} f(x)\, e^{isx}\, dx$$

$$= \frac{1}{\sqrt{2\pi}} \int_{-\infty}^{\infty} e^{i(s+a)x} f(x)\, dx$$

$$= F(s+a).$$

Theorem 5. Modulation theorem

If $\quad F\{f(x)\} = F(s)$, then

$$F\{f(x)\cos ax\} = \tfrac{1}{2}[F(s-a) + F(s+a)]$$

Proof :

$$F\{f(x)\cos ax\} = \frac{1}{\sqrt{2\pi}}\int_{-\infty}^{\infty} e^{isx} f(x)\cos ax\, dx$$

$$= \frac{1}{\sqrt{2\pi}}\int_{-\infty}^{\infty} e^{isx} f(x) \cdot \frac{e^{iax} + e^{-iax}}{2}\, dx$$

$$= \frac{1}{2}\left[\frac{1}{\sqrt{2\pi}}\int_{-\infty}^{\infty} e^{i(s+a)x} f(x)\, dx + \frac{1}{\sqrt{2\pi}}\int_{-\infty}^{\infty} e^{i(s-a)x} f(x)\, dx\right]$$

$$= \tfrac{1}{2}[F(s+a) + F(s-a)]$$

Theorem 6. *If* $F\{f(x)\} = F(s)$, *then* $F\{x^n f(x)\} = (-i)^n \dfrac{d^n}{ds^n} F(s)$

$$F(s) = \frac{1}{\sqrt{2\pi}}\int_{-\infty}^{\infty} e^{isx} f(x)\, dx$$

Differentiating w.r.t. s both sides, n times,

$$\frac{d^n F(s)}{ds^n} = \frac{1}{\sqrt{2\pi}}\int_{-\infty}^{\infty} (ix)^n e^{isx} f(x)\, dx$$

$$= (i)^n \frac{1}{\sqrt{2\pi}}\int_{-\infty}^{\infty} x^n e^{isx} f(x)\, dx$$

$$= (i)^n F\{x^n f(x)\}$$

$$F\{x^n f(x)\} = (-i)^n \frac{d^n}{ds^n}\{F(s)\}$$

Theorem 7. $F\{f'(x)\} = -is\, F(s)$ if $f(x) \to 0$ as $x \to \pm\infty$

$$F\{f'(x)\} = \frac{1}{\sqrt{2\pi}}\int_{-\infty}^{\infty} e^{isx} f'(x)\, dx$$

$$= \frac{1}{\sqrt{2\pi}}\int_{-\infty}^{\infty} e^{isx}\, d\{f(x)\}$$

$$= \frac{1}{\sqrt{2\pi}} \left[\left\{ e^{isx} f(x) \right\}_{-\infty}^{\infty} - is \int_{-\infty}^{\infty} f(x)\, e^{isx}\, dx \right]$$

$$= - is\, F(s) \text{ if } f(x) \to 0 \text{ as } x \to \pm \infty$$

Cor.: $F\{f^n(x)\} = (-is)^n F(s)$ if $f, f', f'', \ldots f^{(n-1)} \to 0$ as $x \to \pm\infty$

Theorem 8. $F\left\{ \int_a^x f(x)\, dx \right\} = \dfrac{F(s)}{(-is)}$

Let $\qquad \phi(x) = \int_a^x f(x)\, dx$

Then, $\qquad \phi'(x) = f(x)$

$$F\{\phi'(x)\} = (-is)\, \bar{\phi}(s)$$

$$= (-is)\, F(\phi(x))$$

$$= (-is)\, F \int_a^x f(x)\, dx$$

$$F\left(\int_a^x f(x)\, dx \right) = \frac{1}{(-is)} F\{\phi'(x)\}$$

$$= \frac{1}{(-is)} F(f(x)) = \frac{F(s)}{(-is)}$$

Example 1. *Find the Complex Fourier transform of*

$$f(x) = x \text{ for } |x| \le a$$
$$= 0 \text{ for } |x| > a$$

$$F\{f(x)\} = \frac{1}{\sqrt{2\pi}} \int_{-\infty}^{\infty} e^{isx} f(x)\, dx$$

$$= \frac{1}{\sqrt{2\pi}} \int_{-a}^{a} x\, e^{isx}\, dx, \quad \text{the other integrals vanish.}$$

$$= \frac{1}{\sqrt{2\pi}} \left[\frac{x\, e^{isx}}{is} - \frac{e^{isx}}{(is)^2} \right]_{-a}^{a}$$

$$= \frac{1}{\sqrt{2\pi}} \left[\frac{1}{is} \left\{ a\, e^{isa} + a\, e^{-isa} \right\} + \frac{1}{s^2} \left\{ e^{isa} - e^{-isa} \right\} \right]$$

$$= \frac{1}{\sqrt{2\pi}} \left[\frac{-2ai}{s} \cos sa + \frac{2i}{s^2} (\sin sa) \right]$$

$$= \frac{2i}{s^2} \cdot \frac{1}{\sqrt{2\pi}} [\, \sin sa - as \cos sa \,]$$

Example 2. *Find the Fourier transform (complex) of*

$$f(x) = e^{ikx},\ a < x < b$$

$$= 0,\ x < a \text{ and } x > b$$

$$F\{f(x)\} = \frac{1}{\sqrt{2\pi}} \int_a^b e^{ikx} e^{isx}\, dx$$

$$= \frac{1}{\sqrt{2\pi}} \left[\frac{e^{i(k+s)x}}{i(k+s)} \right]_a^b$$

$$= \frac{i}{\sqrt{2\pi}\,(k+s)} \left[e^{i(k+s)a} - e^{i(k+s)b} \right]$$

Example 3. *Find the Fourier transform of*

$$f(x) = \begin{cases} 1 - x^2 & \text{if } |x| < | \\ 0 & \text{if } |x| > | \end{cases}$$ **(Madras 1986, 1987.)**

Hence evaluate $\int_0^\infty \left(\frac{x \cos x - \sin x}{x^3} \right) \cos \frac{x}{2}\, dx$

$$F\{f(x)\} = \frac{1}{\sqrt{2\pi}} \int_{-\infty}^{\infty} e^{isx} f(x)\, dx$$

$$= \frac{1}{\sqrt{2\pi}} \int_{-1}^{1} e^{isx} (1 - x^2)\, dx$$

$$= \frac{1}{\sqrt{2\pi}} \left[(1 - x^2) \left(\frac{e^{isx}}{is} \right) - (-2x) \left(\frac{e^{isx}}{i^2 s^2} \right) + (-2) \left(\frac{e^{isx}}{i^3 s^3} \right) \right]_{-1}^{1}$$

$$= \frac{1}{\sqrt{2\pi}} \left[\frac{-2}{s^2} \{ e^{is} + e^{-is} \} - \frac{2i}{s^3} \{ e^{is} - e^{-is} \} \right]$$

$$= -\frac{2}{s^3}\frac{1}{\sqrt{2\pi}}[2\, s\cos s - 2\sin s]$$

$$= \frac{-4}{\sqrt{2\pi}}\left[\frac{s\cos s - \sin s}{s^3}\right]$$

Using inversion formula

$$f(x) = \frac{1}{\sqrt{2\pi}}\int_{-\infty}^{\infty}\left[\frac{-4}{\sqrt{2\pi}}\left(\frac{s\cos s - \sin s}{s^3}\right)e^{-isx}\, ds\right.$$

$$= -\frac{2}{\pi}\int_{-\infty}^{\infty}\left(\frac{s\cos s - \sin s}{s^3}\right)e^{-isx}\, ds$$

$$\int_{-\infty}^{\infty}\left(\frac{s\cos s - \sin s}{s^3}\right)(\cos sx - i\sin sx)\, ds$$

$$= -\frac{\pi}{2}(1-x^2) \text{ if } |x| < |$$

$$= 0 \text{ if } |x| > 1$$

Equating real parts,

$$\int_{-\infty}^{\infty}\left(\frac{s\cos s - \sin s}{s^3}\right)\cos sx\, ds = -\frac{\pi}{2}(1-x^2) \text{ if } |x| < 1$$

$$= 0 \text{ if } |x| > 1$$

Set $x = \frac{1}{2}$

$$\therefore \quad \int_{-\infty}^{\infty}\left(\frac{s\cos s - \sin s}{s^3}\right)\cos\frac{s}{2}\, ds = -\frac{\pi}{2}\left(1 - \frac{1}{4}\right) = -\frac{3\pi}{8}$$

$$2\int_{0}^{\infty}\left(\frac{s\cos s - \sin s}{s^3}\right)\cos\frac{s}{2}\, ds = -\frac{3\pi}{8}$$

$$\therefore \quad \int_{0}^{\infty}\left(\frac{s\cos s - \sin s}{s^3}\right)\cos\frac{s}{2}\, ds = -\frac{3\pi}{16}$$

Example 4. *Show that the transform of* $e^{-\frac{x^2}{2}}$ *is* $e^{-\frac{s^2}{2}}$ *by finding the Fourier transform of* $e^{-a^2x^2}$, $a > 0$

$$F\{e^{-a^2x^2}\} - \frac{1}{\sqrt{2\pi}}\int_{-\infty}^{\infty} e^{-a^2x^2}e^{isx}\, dx$$

$$= \frac{1}{\sqrt{2\pi}} \int_{-\infty}^{\infty} e^{-\left(ax - \frac{is}{2a}\right)^2} e^{-\frac{s^2}{4a^2}} dx$$

$$= e^{-\frac{s^2}{4a^2}} \cdot \frac{1}{a\sqrt{2\pi}} \int_{-\infty}^{\infty} e^{-t^2} dt, \text{ putting } ax - \frac{is}{2a} = t \text{ and } a > 0$$

$$= e^{-\frac{s^2}{4a^2}} \cdot \frac{1}{a\sqrt{2\pi}} \times \sqrt{\pi} \text{ since } \int_{-\infty}^{\infty} e^{-t^2} dt = \sqrt{\pi}$$

$$= \frac{1}{\sqrt{2}\, a} e^{-\frac{s^2}{4a^2}}$$

Setting $a = \frac{1}{\sqrt{2}}$,

$$F\left\{e^{-\frac{x^2}{2}}\right\} = e^{-\frac{s^2}{2}}$$

That is, $e^{-\frac{x^2}{2}}$ is self-reciprocal.

Example 5. *Find the Fourier transform of* $f(x)$ given by

$$f(x) = \begin{cases} 1 \text{ for } |x| < a \\ 0 \text{ for } |x| > a > 0 \end{cases}$$

and hence evaluate $\int_0^{\infty} \frac{\sin x}{x} dx$ **(*Anna Univ. 2002*)**

and $\int_{-\infty}^{\infty} \frac{\sin as \cos sx}{s} ds$ **(*1987 Ap-Barathidasan*)**

$$F\{f(x)\} = \frac{1}{\sqrt{2\pi}} \int_{-\infty}^{\infty} f(x)\, e^{ixs} dx$$

$$= \frac{1}{\sqrt{2\pi}} \int_{-a}^{a} e^{ixs} dx$$

$$= \frac{1}{\sqrt{2\pi}} \left(\frac{e^{isx}}{is}\right)_{-a}^{a}$$

$$= \frac{1}{\sqrt{2\pi}} \cdot \frac{1}{is} \left[e^{ias} - e^{-ias} \right]$$

$$= \sqrt{\frac{2}{\pi}} \frac{\sin as}{s}$$

Using inversion formula, we get,

$$\frac{1}{\sqrt{2\pi}} \int_{-\infty}^{\infty} \left(\sqrt{\frac{2}{\pi}} \frac{\sin as}{s} \right) e^{-ixs} \, ds = f(x) = \begin{cases} 1 \text{ for } |x| < a \\ 0 \text{ for } |x| > a \end{cases}$$

$$\frac{1}{\sqrt{2\pi}} \int_{-\infty}^{\infty} \left(\sqrt{\frac{2}{\pi}} \frac{\sin as}{s} \right) (\cos xs - i \sin xs) \, ds = f(x)$$

Equating real parts,

$$\frac{1}{\pi} \int_{-\infty}^{\infty} \frac{\sin as}{s} \cos xs \, ds = f(x) = \begin{cases} 1 \text{ for } |x| < a \\ 0 \text{ for } |x| > a \end{cases}$$

$$\frac{2}{\pi} \int_{0}^{\infty} \frac{\sin as}{s} \cos xs \, ds = f(x)$$

$$\int_{0}^{\infty} \frac{\sin as}{s} \cos xs \, ds = \frac{\pi}{2} \text{ for } |x| < a$$

$$= 0 \text{ for } |x| > a$$

$$= \frac{1}{2}\left(\frac{\pi}{2} + 0\right) = \frac{\pi}{4} \text{ for } |x| = a$$

Setting $x = 0$,

$$\int_{0}^{\infty} \frac{\sin as}{s} \, ds = \frac{\pi}{2}$$

$$\therefore \int_{0}^{\infty} \frac{\sin ax}{x} \, dx = \frac{\pi}{2}$$

Setting $ax = \theta$, we get,

$$\int_{0}^{\infty} \frac{\sin \theta}{\theta} \, d\theta = \frac{\pi}{2}$$

Convolution Theorem or Faltung Theorem

Def. The convolution of two functions $f(x)$ and $g(x)$ is defined as

$$f * g = \frac{1}{\sqrt{2\pi}} \int_{-\infty}^{\infty} f(t)\, g(x-t)\, dt$$

Theorem. *The Fourier transform of the convolution of f (x) and g(x) is the product of their Fourier transforms.*

That is, $F\{f(x) * g(x)\} = F(s) \,.\, G(s) = F\{f(x)\} \,.\, F\{g(x)\}$

$$F\{f * g\} = \frac{1}{\sqrt{2\pi}} \int_{-\infty}^{\infty} (f * g)\, e^{ixs}\, dx$$

$$= \frac{1}{\sqrt{2\pi}} \int_{-\infty}^{\infty} \left(\frac{1}{\sqrt{2\pi}} \int_{-\infty}^{\infty} f(t)\, g(x-t)\, dt \right) e^{ixs}\, dx$$

$$= \frac{1}{2\pi} \int_{-\infty}^{\infty} f(t) \left(\int_{-\infty}^{\infty} g(x-t)\, e^{ixs}\, dx \right) dt$$

by changing the order of integration

$$= \frac{1}{\sqrt{2\pi}} \int_{-\infty}^{\infty} f(t)\, F\{g(x-t)\}\, dt$$

$$= \frac{1}{\sqrt{2\pi}} \int_{-\infty}^{\infty} f(t)\, e^{its}\, G(s)\, dt, \text{ using shifting theorem}$$

$$= G(s) \,.\, \frac{1}{\sqrt{2\pi}} \int_{-\infty}^{\infty} f(t)\, e^{its}\, dt$$

$$= G(s) \,.\, F(s)$$

$$= F(s) \,.\, G(s)$$

By inversion,

$F^{-1}\{F(s)\, G(s)\} = f * g = F^{-1}\{F(s)\} * F^{-1}\{G(s)\}$

Parseval's identity. If $F(s)$ is the Fourier transform of $f(x)$,

then $$\int_{-\infty}^{\infty} |f(x)|^2\, dx = \int_{-\infty}^{\infty} |F(s)|^2\, ds$$

Proof. Firstly let us prove

$\overline{F\{f(-x)\}} = \overline{F(s)}$ where $\overline{F(s)}$ indicates the complex conjugate of $F(s)$

$$\overline{F(s)} = \frac{1}{\sqrt{2\pi}} \int_{-\infty}^{\infty} \overline{f(x)}\, e^{-isx}\, dx$$

$$= \frac{1}{\sqrt{2\pi}} \int_{-\infty}^{\infty} \overline{f(-v)}\, e^{isv}\, dv \quad \text{putting } x = -v$$

$$= F\,\overline{\{f(-v)\}}$$

$$= F\,\overline{\{f(-x)\}}, \text{ changing the dummy variable} \qquad \ldots(1)$$

By convolution Theorem,

$$F\{f(x) * g(x)\} = F(s)\, G(s)$$

$$f * g = F^{-1}\{F(s)\, G(s)\}$$

$$\frac{1}{\sqrt{2\pi}} \int_{-\infty}^{\infty} f(t)\, g(x-t)\, dt = \frac{1}{\sqrt{2\pi}} \int_{-\infty}^{\infty} F(s)\, G(s)\, e^{-isx}\, ds$$

Putting $x = 0$, we get

$$\int_{-\infty}^{\infty} f(t)\, g(-t)\, dt = \int_{-\infty}^{\infty} F(s)\, G(s)\, ds \qquad \ldots(2)$$

Since it is true for all $g(t)$, take $g(t) = \overline{f(-t)}$

$\therefore \quad g(-t) = \overline{f(t)}$

$G(s) = F\{g(t)\} = F\,\{\overline{f(-t)}\} = \overline{F(s)}$ using (1). Use this in (2).

$$\int_{-\infty}^{\infty} f(t)\, \overline{f(t)}\, dt = \int_{-\infty}^{\infty} F(s)\, \overline{F(s)}\, ds$$

$$\therefore \int_{-\infty}^{\infty} |f(t)|^2\, dt = \int_{-\infty}^{\infty} |F(s)|^2\, ds.$$

Example 6. *Using Parseval's identity, prove* $\int_0^{\infty} \left(\frac{\sin t}{t}\right)^2 dt = \frac{\pi}{2}$.

In example 5, if $f(x) = \begin{cases} 1 \text{ for } |x| < a \\ 0 \text{ for } |x| > a > 0 \end{cases}$

We have proved $F(s) = \sqrt{\frac{2}{\pi}}\, \frac{\sin as}{s}$.

Using Parseval's identity.

$$\int_{-\infty}^{\infty} |f(t)|^2\, dt = \int_{-\infty}^{\infty} |F(s)|^2\, ds$$

$$\therefore \qquad \int_{-a}^{a} 1\, dt = \int_{-\infty}^{\infty} \frac{2}{\pi}\left(\frac{\sin as}{s}\right)^2 ds$$

$$2a = \frac{2}{\pi}\int_{-\infty}^{\infty}\left(\frac{\sin as}{s}\right)^2 ds.$$

Setting $as = t$, we get,

$$\int_{-\infty}^{\infty}\left(\frac{\sin t}{t}\right)^2 dt = \pi$$

$$\therefore \qquad 2\int_{-\infty}^{\infty}\left(\frac{\sin t}{t}\right)^2 dt = \pi$$

$$\therefore \qquad \int_{0}^{\infty}\left(\frac{\sin t}{t}\right)^2 dt = \pi/2$$

Example 7. *Find the Fourier transform of* $f(x) = 1 - |x|$ *if* $|x| < 1$

and hence find the value $\int_0^\infty \frac{\sin^4 t}{t^4}\, dt.$ $\qquad = 0$ *for* $|x| > 1$

$$F\{f(x)\} = \frac{1}{\sqrt{2\pi}}\int_{-\infty}^{\infty}(1 - |x|)\, e^{isx}\, dx$$

$$= \frac{1}{\sqrt{2\pi}}\int_{-1}^{1}(1 - |x|)(\cos sx + i\sin sx)\, dx$$

$$= \frac{1}{\sqrt{2\pi}}\cdot 2\int_{0}^{1}(1 - x)\cos sx\, dx \quad \text{since } (1 - |x|) \text{ is even}$$

$$= \sqrt{\frac{2}{\pi}}\left(\frac{1 - \cos s}{s^2}\right)$$

Using Parseval's identity,

$$\frac{2}{\pi}\int_{-\infty}^{\infty}\frac{(1 - \cos s)^2}{s^4}\, ds = \int_{-1}^{1}(1 - |x|)^2\, dx$$

$$\frac{4}{\pi}\int_0^\infty \frac{(1-\cos s)^2}{s^4}\, ds = 2\int_0^1 (1-x)^2\, dx = 2/3$$

$$\frac{16}{\pi}\int_0^\infty \frac{\sin^4 s/2}{s^4}\, ds = 2/3$$

Setting $\frac{s}{2} = x$, we get

$$\int_0^\infty \frac{\sin^4 x}{x^4}\, dx = \frac{\pi}{3}$$

Infinite Fourier cosine transform and sine transform:

Infinite Fourier cosine transform.

Let $f(x)$ be defined for all $x \geq 0$. Now we shall extend the function $f(x)$ to the negative side of x axis. That is, let the extended function $F(x)$ be defined as

$$F(x) = f(x) \text{ for } x \geq 0$$
$$= f(-x) \text{ for } x < 0 \quad \text{so that } F(x) \text{ is even in } (-\infty, \infty). \text{ Now,}$$

$$F\{F(x)\} = \frac{1}{\sqrt{2\pi}}\int_{-\infty}^{\infty} F(x)\, e^{isx}\, dx$$

$$= \frac{1}{\sqrt{2\pi}}\int_{-\infty}^{\infty} F(x)\, [\cos sx + i \sin sx]\, dx$$

$$= \frac{1}{\sqrt{2\pi}} \times 2\int_0^\infty F(x)\cos sx\, dx \qquad \text{since } F(x) \text{ is even}$$

$$= \sqrt{\frac{2}{\pi}}\int_0^\infty f(x)\cos sx\, dx \text{ since } F(x) = f(x) \text{ for } x > 0$$

The R.H.S. is defined as the infinite Fourier cosine transform of $f(x)$ denote by $F_c(s)$

$\therefore$ Infinite Fourier cosine transform of $f(x)$ is defined to be

$$F_c(s) = \sqrt{\frac{2}{\pi}}\int_0^\infty f(x)\cos sx\, dx \qquad ...(1)$$

which is an even function of s

Then by inversion theorem,

$$F(x) = \frac{1}{\sqrt{2\pi}} \int_{-\infty}^{\infty} F_c(s)\, e^{-isx}\, ds$$

$$= \frac{1}{\sqrt{2\pi}} \int_{-\infty}^{\infty} F_c(s)\{\cos sx - i \sin sx\} ds$$

$$= \frac{1}{\sqrt{2\pi}} \times 2 \int_{0}^{\infty} F_c(s) \cos sx\, dx \text{ since } F_c(s) \text{ is even}$$

Since $F(x) = f(x)$ *in* $(0, \infty)$,

$$f(x) = \sqrt{\frac{2}{\pi}} \int_{0}^{\infty} F_c(s) \cos sx\, dx \qquad ...(2)$$

Equation (1) defines the Fourier cosine transform of $f(x)$ and (2) gives the Inversion Theorem for Fourier cosine transform.

Infinite Fourier Sine Transform

Let $f(x)$ be defined for $x \geq 0$. We extend $f(x)$ to the negative side of x axis. Thus, let,

$$F(x) = f(x) \text{ for } x \geq 0$$

$$= -f(-x) \text{ for } x < 0$$

so that $F(x)$ is odd in $(-\infty, \infty)$

$$F\{F(x)\} = \frac{1}{\sqrt{2\pi}} \int_{-\infty}^{\infty} F(x)\, e^{isx}\, dx$$

$$= \frac{1}{\sqrt{2\pi}} \int_{-\infty}^{\infty} F(x)\, [\cos sx + i \sin sx]\, dx$$

$$= \frac{i}{\sqrt{2\pi}} \int_{-\infty}^{\infty} F(x) \sin sx\, dx \text{ since } F(x) \text{ is odd}$$

$$= i\sqrt{\frac{2}{\pi}} \int_{0}^{\infty} f(x) \sin sx\, dx$$

Now, we define Fourier sine transform of $f(x)$ as

$$F_s(s) = \text{ imaginary point of } F\{F(x)\}$$

$$F_s\{f(x)\} = F_s(s) = \sqrt{\frac{2}{\pi}} \int_0^{\infty} f(x) \sin sx \, dx \quad \text{...(3)}$$

By inversion theorem,

$$F(x) = F^{-1}\{i F_s(s)\}$$

$$= i \frac{1}{\sqrt{2\pi}} \int_{-\infty}^{\infty} F_s(s) e^{-isx} ds$$

$$f(x) = \sqrt{\frac{2}{\pi}} \int_0^{\infty} F_s(s) \sin sx \, dx \quad \text{...(4)}$$

since $F_s(s)$ is odd function of s and $F(x) = f(x)$ in $(0, \infty)$

Equation (3) defines infinite Fourier sine transform of $f(x)$ and (4) gives the inversion theorem for Fourier sine transform.

Hence the transform pairs are

$$F_c\{f(x)\} = \sqrt{\frac{2}{\pi}} \int_0^{\infty} f(x) \cos sx \, dx \, ; f(x) = \sqrt{\frac{2}{\pi}} \int_0^{\infty} F_c(s) \cos sx \, ds$$

$$F_s\{f(x)\} = \sqrt{\frac{2}{\pi}} \int_0^{\infty} f(x) \sin sx \, dx \, ; f(x) = \sqrt{\frac{2}{\pi}} \int_0^{\infty} F_s(s) \sin sx \, ds$$

Properties regarding cosine and sine transforms

1. Cosine and sine transforms are linear.

   That is,

   $F_c\{af(x) + bg(x)\} = a F_c\{f(x)\} + b F_c[g(x)]$

   The proof is obvious.

2. $F_s[af(x) + bg(x)] = a F_s\{f(x)\} + b F_s\{g(x)\}$

3. $F_s[f(x) \sin ax] = \frac{1}{2}[F_c(s-a) - F_c(s+a)]$

4. $F_s[f(x) \cos ax] = \frac{1}{2}[F_s(s+a) + F_s(s-a)]$

5. $F_c[f(x) \sin ax] = \frac{1}{2}[F_s(a+s) + F_s(a-s)]$

6. $F_c[f(x) \cos ax] = \frac{1}{2}[F_c(s+a) + F_c(s-a)]$

Proof. $F_s\,[f(x)\sin ax] = \sqrt{\frac{2}{\pi}}\int_0^\infty f(x)\sin ax \sin sx\,dx$

$$= \frac{1}{2}\sqrt{\frac{2}{\pi}}\int_0^\infty f(x)\,.\,[\cos(s-a)x - \cos(a+s)x]\,dx$$

$$= \frac{1}{2}[F_c(s-a) - F_c(s+a)]$$

$$F_s\,[f(x)\cos ax] = \sqrt{\frac{2}{\pi}}\int_0^\infty f(x)\cos ax \sin sx\,dx$$

$$= \frac{1}{2}\sqrt{\frac{2}{\pi}}\int_0^\infty f(x)\,.\,[\sin(s+a)x + \sin(s-a)x]\,dx$$

$$= \frac{1}{2}[F_s(s+a) + F_s(s-a)]$$

Similarly, we can prove the results (5) and (6).

7. $F_c\,\{f(ax)\} = \frac{1}{a}F_c\left(\frac{s}{a}\right)$

8. $F_s\,\{f(ax)\} = \frac{1}{a}F_s\left(\frac{s}{a}\right)$

Hint. put $ax = u$ and proceed.

Identities

If $F_c(s)$, $G_c(s)$ are the Fourier cosine transforms and $F_s(s)$, $G_s(s)$ are the Fourier sine transforms of $f(x)$ and $g(x)$ respectively, then

1. $\int_0^\infty f(x)\,g(x)\,dx = \int_0^\infty F_c(s)\,G_c(s)\,ds$

2. $\int_0^\infty f(x)\,g(x)\,dx = \int_0^\infty F_s(s)\,G_s(s)\,ds$

3. $\int_0^\infty |f(x)|^2\,dx = \int_0^\infty |F_c(s)|^2\,ds = \int_0^\infty |F_s(s)|^2\,ds.$

Proof. $\int_0^\infty F_s(s)\,G_s(s)\,ds = \int_0^\infty F_s(s)\left\{\sqrt{\frac{2}{\pi}}\int_0^\infty g(x)\sin sx\,dx\right\}ds$

$$= \int_0^\infty g(x) \left\{ \sqrt{\frac{2}{\pi}} \int_0^\infty F_s(s) \sin sx \, ds \right\} dx$$

$$= \int_0^\infty g(x) f(x) \, dx$$

Similarly, we can prove the first identity and the third follows by setting $g(x) = f(x)$.

Example 8. *Find Fourier cosine and sine transforms of e^{-ax}, $a > 0$ and hence deduce the inversion formula.*

$$F_s(e^{-ax}) = \sqrt{\frac{2}{\pi}} \int_0^\infty e^{-ax} \sin sx \, dx$$

$$= \sqrt{\frac{2}{\pi}} \left\{ \frac{e^{-ax}}{a^2 + s^2} (-a \sin sx - s \cos sx) \right\}_0^\infty$$

$$= \sqrt{\frac{2}{\pi}} \frac{s}{a^2 + s^2}, \text{ if } a > 0 \qquad \text{...(1)}$$

$$F_c(e^{-ax}) = \sqrt{\frac{2}{\pi}} \int_0^\infty e^{-ax} \cos sx \, dx$$

$$= \sqrt{\frac{2}{\pi}} \left[\frac{e^{-ax}}{a^2 + s^2} (-a \cos sx + s \sin sx) \right]_0^\infty$$

$$= \sqrt{\frac{2}{\pi}} \cdot \frac{a}{a^2 + s^2} \text{ if } a > 0 \qquad \text{...(2)}$$

By inversion formula of (1),

$$e^{-ax} = \sqrt{\frac{2}{\pi}} \int_0^\infty \left(\sqrt{\frac{2}{\pi}} \frac{s}{a^2 + s^2} \cdot \sin sx \right) ds$$

$$\therefore \int_0^\infty \frac{s}{a^2 + s^2} \sin sx \, ds = \frac{\pi}{2} e^{-ax}, a > 0$$

Changing the variables,

$$\int_0^\infty \frac{x \sin \alpha x}{a^2 + x^2}\, dx = \frac{\pi}{2} e^{-\alpha a}, a > 0 \qquad ...(3)$$

Again, by inversion formula of (2)

$$e^{-ax} = \sqrt{\frac{2}{\pi}} \int_0^\infty \left(\sqrt{\frac{2}{\pi}} \frac{a}{a^2 + s^2} \right) \cos sx\, ds$$

$$\therefore \quad \int_0^\infty \frac{\cos sx}{a^2 + s^2}\, ds = \frac{\pi}{2a} e^{-ax}$$

Changing the variables,

$$\int_0^\infty \frac{\cos \alpha x}{a^2 + x^2}\, dx = \frac{\pi}{2a} e^{-\alpha a}, a > 0 \qquad ...(4)$$

Example 9. *Find Fourier sine transform of* $\frac{x}{a^2 + x^2}$ *and Fourier cosine transform of* $\frac{1}{a^2 + x^2}$.

$$F_s\left(\frac{x}{a^2 + x^2}\right) = \sqrt{\frac{2}{\pi}} \int_0^\infty \frac{x}{a^2 + x^2} \sin sx\, dx$$

$$= \sqrt{\frac{2}{\pi}} \left[\frac{\pi}{2} e^{-as} \right] \text{ using (3) of Example 8.}$$

$$= \sqrt{\frac{\pi}{2}}\, e^{-as}$$

$$F_c\left(\frac{1}{a^2 + x^2}\right) = \sqrt{\frac{2}{\pi}} \int_0^\infty \frac{1}{a^2 + x^2} \cos sx\, dx$$

$$= \sqrt{\frac{2}{\pi}} \left[\frac{\pi}{2a} e^{-as} \right] \text{ using (4) of Example 8.}$$

$$= \sqrt{\frac{\pi}{2}} \cdot \frac{1}{a} e^{-as}.$$

Example 10. *Using Parseval's identity evaluate*

$$\int_0^\infty \frac{dx}{(a^2 + x^2)^2} \text{ and } \int_0^\infty \frac{x^2}{(a^2 + x^2)^2}\, dx \text{ if } a > 0.$$

By example 8, If $f(x) = e^{-ax}$, $F_s(s) = \sqrt{\frac{2}{\pi}} \frac{s}{a^2 + s^2}$

and $F_c(s) = \sqrt{\frac{2}{\pi}} \frac{a}{a^2 + s^2}$

Using Parseval's identity,

$$\int_0^\infty \frac{2}{\pi} \frac{s^2}{(a^2 + s^2)^2} ds = \int_0^\infty e^{-2ax} dx$$

$$= \left(\frac{e^{-2ax}}{-2a} \right)_0^\infty$$

$$= \frac{1}{2a}$$

$$\therefore \quad \int_0^\infty \frac{x^2}{(a^2 + x^2)^2} dx = \frac{\pi}{4a}. \text{ if } a > 0$$

Also $\int_0^\infty \frac{2}{\pi} \frac{a^2}{(a^2 + s^2)^2} ds = \int_0^\infty e^{-2ax} dx = \frac{1}{2a}$

$$\int_0^\infty \frac{dx}{(a^2 + x^2)^2} = \frac{\pi}{4a^3} \text{ if } a > 0$$

Example 11. *Evaluate* $\int_0^\infty \frac{dx}{(a^2 + x^2)(b^2 + x^2)}$ *using transform methods.*

Let $f(x) = e^{-ax}$, $g(x) = e^{-bx}$

$$F_c(s) = \sqrt{\frac{2}{\pi}} \int_0^\infty e^{-ax} \cos sx dx$$

$$= \sqrt{\frac{2}{\pi}} \frac{a}{a^2 + s^2} \text{ by example 8}$$

Similarly,

$$G_c(s) = \sqrt{\frac{2}{\pi}} \frac{b}{b^2 + s^2}$$

$\therefore$ Using, $\int_0^\infty F_c(s)\, G_c(s)\, ds = \int_0^\infty f(x)\, g(x)\, dx$

$$\frac{2}{\pi}\int_0^\infty \frac{ab}{(a^2+s^2)(b^2+s^2)}\,ds = \int_0^\infty e^{-(a+b)x}\,dx$$

$$= \frac{1}{a+b}$$

$$\int_0^\infty \frac{dx}{(a^2+x^2)(b^2+x^2)} = \frac{\pi}{2\,ab\,(a+b)}, \text{ if } a, b > 0.$$

Example 12. *Find Fourier cosine transform of*

$$f(x) = \begin{cases} \cos x \text{ in } 0 < x < a \\ 0 \qquad x \geq a \end{cases}$$

$$F_c(s) = \sqrt{\frac{2}{\pi}}\int_0^\infty f(x)\cos sx\,dx$$

$$= \sqrt{\frac{2}{\pi}}\int_0^a \cos x \cos sx\,dx$$

$$= \sqrt{\frac{2}{\pi}}\,\frac{1}{2}\int_0^a [\cos(s+1)x + \cos(s-1)x]\,dx$$

$$= \frac{1}{\sqrt{2\pi}}\left[\frac{\sin(s+1)a}{s+1} + \frac{\sin(s-1)a}{s-1}\right] \text{if } s \neq 1, -1$$

Example 13. *Find Fourier sine transform of* $\frac{1}{x}$

$$F_s\left(\frac{1}{x}\right) = \sqrt{\frac{2}{\pi}}\int_0^\infty \frac{\sin sx}{x}\,dx$$

$$= \sqrt{\frac{2}{\pi}}\int_0^\infty \frac{\sin\theta}{\theta}\,d\theta, \text{ putting } sx = \theta$$

$$= \sqrt{\frac{2}{\pi}} \times \frac{\pi}{2}$$

$$= \sqrt{\frac{\pi}{2}}$$

Example 14. *Find Fourier sine and cosine transform of* x^{n-1}.

We know, $\Gamma(n) = \int_0^\infty e^{-x} x^{n-1}\,dx, n > 0$

$$= \int_0^\infty e^{-ax} (ax)^{n-1}\, a dx \text{ setting } x \text{ as } ax, a > 0$$

$$\therefore \quad \int_0^\infty e^{-ax} x^{n-1}\, dx = \frac{\Gamma(n)}{a^n}, n > 0, a > 0$$

We can prove the above result even if a is complex.

Setting $a = is$,

$$\int_0^\infty e^{-isx} x^{n-1}\, dx = \frac{\Gamma(n)}{(is)^n}$$

$$= \frac{(-i)^a\, \Gamma(n)}{s^n}$$

$$= \frac{e^{-\frac{\pi}{2} ni}}{s^n} \Gamma(n) \text{ since } -i = e^{-\frac{\pi}{2} i}$$

Equating real and imaginary parts on both sides, we get

$$\int_0^\infty x^{n-1} \cos sx\, dx = \frac{\Gamma(n)}{s^n} \cos \frac{n\pi}{2}$$

$$\int_0^\infty x^{n-1} \sin sx\, dx = \frac{\Gamma(n)}{s^n} \sin \frac{n\pi}{2}$$

$$\therefore \quad F_c(x^{n-1}) = \sqrt{\frac{2}{\pi}} \frac{\Gamma(n)}{s^n} \cos \frac{n\pi}{2}$$

$$F_s(x^{n-1}) = \sqrt{\frac{2}{\pi}} \frac{\Gamma(n)}{s^n} \sin \frac{n\pi}{2}$$

Taking $n = 1/2$,

$$F_c\left(\frac{1}{\sqrt{x}}\right) = \frac{1}{\sqrt{s}} \text{ since } \Gamma\left(\frac{1}{2}\right) = \sqrt{\pi}$$

$$F_s\left(\frac{1}{\sqrt{x}}\right) = \frac{1}{\sqrt{s}}$$

Note. : $\frac{1}{\sqrt{x}}$ is self reciprocal under Fourier sine and cosine transforms.

Example 15. *Show that*

(*i*) $F_s[xf(x)] = -\frac{d}{ds} F_c(s)$ (*ii*) $F_c[xf(x)] = \frac{d}{ds} F_s(s)$

and hence find Fourier cosine and sine transform of xe^{-ax}.

$$F_c(s) = \sqrt{\frac{2}{\pi}} \int_0^\infty f(x) \cos sx \, dx$$

$$\frac{d}{ds} F_c(s) = -\sqrt{\frac{2}{\pi}} \int_0^\infty x f(x) \sin sx \, dx$$

$$= -F_s\{xf(x)\}$$

Similarly,

$$F_s(s) = \sqrt{\frac{2}{\pi}} \int_0^\infty f(x) \sin sx \, dx$$

$$\frac{d}{ds} F_s(s) = \sqrt{\frac{2}{\pi}} \int_0^\infty x f(x) \cos sx \, dx$$

$$= F_c\{xf(x)\}$$

$$F_c(xe^{-ax}) = \frac{d}{ds} F_s(e^{-ax})$$

$$= \frac{d}{ds}\left(\sqrt{\frac{2}{\pi}} \frac{s}{a^2+s^2}\right) \text{ using example 8}$$

$$= \sqrt{\frac{2}{\pi}} \frac{a^2-s^2}{(a^2+s^2)^2}$$

$$F_s(xe^{-ax}) = -\frac{d}{ds} F_c(e^{-ax})$$

$$= -\frac{d}{ds}\left(\sqrt{\frac{2}{\pi}} \frac{a}{a^2+s^2}\right) \text{ (Refer example 8)}$$

$$= \sqrt{\frac{2}{\pi}} \frac{2as}{(a^2+s^2)^2}$$

Example 16. *Find Fourier cosine transform of $e^{-a^2x^2}$ and hence evaluate Fourier sine transform of $xe^{-a^2x^2}$*

$$F_c(e^{-a^2x^2}) = \sqrt{\frac{2}{\pi}} \int_0^\infty e^{-a^2x^2} . \cos sx \, dx$$

$$= \text{Real part of } \sqrt{\frac{2}{\pi}} \int_0^\infty e^{-a^2x^2+isx}\, dx$$

$$= \frac{1}{a\sqrt{2}} e^{-\frac{s^2}{4a^2}} \quad \text{(Refer example 4)}$$

$$F_s(xe^{-a^2x^2}) = -\frac{d}{ds} F_c(e^{-a^2x^2})$$

$$= \frac{1}{a\sqrt{2}} e^{-\frac{s^2}{4a^2}} \frac{s}{2a^2}$$

$$= \frac{1}{2a^3\sqrt{2}} e^{-\frac{s^2}{4a^2}}$$

Example 17. *Solve for f (x) from the integral equation*

$$\int_0^\infty f(x) \cos \alpha x\, dx = e^{-\alpha}$$

Proof. Multiplying by $\sqrt{\frac{2}{\pi}}$, we get

$$\sqrt{\frac{2}{\pi}} \int_0^\infty f(x) \cos \alpha x\, dx = \sqrt{\frac{2}{\pi}} e^{-\alpha}$$

$$F_c\{f(x)\} = \sqrt{\frac{2}{\pi}} e^{-\alpha}, \ \alpha \text{ parameter}$$

$$\therefore \quad f(x) = F_c^{-1}\left(\sqrt{\frac{2}{\pi}} e^{-\alpha}\right)$$

$$= \frac{2}{\pi} \int_0^\infty e^{-\alpha} \cos \alpha x\, d\alpha$$

$$= \frac{2}{\pi} \cdot \frac{1}{1+x^2} \text{ on integration}$$

Example 18. *Solve for f(x) from the integral equation*

$$\int_0^\infty f(x) \sin sx\, dx = \begin{cases} 1 \text{ for } 0 \le s < 1 \\ 2 \text{ for } 1 \le s < 2 \\ 0 \text{ for } s \ge 2 \end{cases}$$

Proof. Multiplying by $\sqrt{\frac{2}{\pi}}$, both sides,

$$F_s(f(x)) = \begin{cases} \sqrt{\dfrac{2}{\pi}} & \text{for } 0 \le s < 1 \\ 2\sqrt{\dfrac{2}{\pi}} & \text{for } 1 \le s < 2 \\ 0 & \text{for } s \ge 2 \end{cases}$$

$$\therefore \quad f(x) = F_s^{-1}\,(\text{R.H.S.})$$

$$= \frac{2}{\pi}\int_0^1 \sin sx\, ds + \frac{4}{\pi}\int_1^2 \sin sx\, ds$$

$$= \frac{2}{\pi}\left(\frac{1-\cos x}{x}\right) + \frac{4}{\pi}\left(\frac{\cos x - \cos 2x}{x}\right)$$

$$f(x) = \frac{2}{\pi x}(1 + \cos x - 2\cos 2x)$$

Example 19. *Find the Fourier cosine transform of e^{-x^2} (second method).*

$$I = F_c(e^{-x^2}) = \sqrt{\frac{2}{\pi}}\int_0^\infty e^{-x^2}\cos sx\, dx$$

$$\frac{dI}{ds} = -\sqrt{\frac{2}{\pi}}\int_0^\infty xe^{-x^2}\sin sx\, dx$$

$$= +\sqrt{\frac{2}{\pi}}\int_0^\infty \sin sx\, d\left(\frac{e^{-x^2}}{2}\right)$$

$$= \sqrt{\frac{2}{\pi}}\left[\left(\frac{e^{-x^2}}{2}\sin sx\right)_0^\infty - \frac{1}{2}\int_0^\infty se^{-x^2}\cos sx\, dx\right]$$

$$= -\frac{s}{2}I$$

$$\therefore \quad \frac{dI}{I} = -\frac{s}{2}\,ds$$

$$\therefore \quad \log I = -\frac{s^2}{4} + \log c$$

$$I = ce^{-\frac{s^2}{4}} \qquad \text{...(1)}$$

when $s = 0$,
$$I = \sqrt{\frac{2}{\pi}}\int_0^\infty e^{-x^2}\, dx = \frac{1}{\sqrt{2}}$$

Using in (1),

$$\frac{1}{\sqrt{2}} = c$$

$$\therefore \qquad I = F_c\,(e^{-x^2}) = \frac{1}{\sqrt{2}}\, e^{-\frac{s^2}{4}}$$

Example 20. *Find the complex Fourier transform of dirac delta function* $\delta\,(t-a)$.

$$F\{\delta\,(t-a)\} = \frac{1}{\sqrt{2\pi}} \int_{-\infty}^{\infty} e^{ist}\,\delta\,(t-a)\,dt$$

$$= \frac{1}{\sqrt{2\pi}} \operatorname*{Lt}_{h\to 0} \int_{a}^{a+h} \frac{1}{h}\, e^{ist}\,dt$$

$$= \frac{1}{\sqrt{2\pi}} \operatorname*{Lt}_{h\to 0} \frac{1}{h} \left(\frac{e^{ist}}{is}\right)_a^{a+h}$$

$$= \frac{1}{\sqrt{2\pi}} \operatorname*{Lt}_{h\to 0} e^{isa} \left(\frac{e^{ish}-1}{ish}\right)$$

$$= \frac{e^{isa}}{\sqrt{2\pi}} \text{ since } \operatorname*{Lt}_{\theta\to 0} \frac{e^{\theta}-1}{\theta} = 1$$

Note. Dirac delta function $\delta\,(t-a)$ is defined as

$$\delta\,(t-a) = \operatorname*{Lt}_{h\to 0} I(h, t-a) \text{ where}$$

$$I\,(h, t-a) = \frac{1}{h} \text{ for } a < t < a+h$$

$$= 0 \text{ for } t < a \text{ and } t > a+h$$

Example 21. *Find the function if its sine transform is* $\dfrac{e^{-as}}{s}$.

Let $F_s\,(f\,(x)) = \dfrac{e^{-as}}{s}$

$$\text{Then, } f\,(x) = \sqrt{\frac{2}{\pi}} \int_0^{\infty} \frac{e^{-as}}{s} \sin sx\,ds \qquad ...(1)$$

$$\therefore \qquad \frac{df}{dx} = \sqrt{\frac{2}{\pi}} \int_0^{\infty} e^{-as} \cos sx\,ds$$

$$= \sqrt{\frac{2}{\pi}} \cdot \frac{a}{a^2+x^2}$$

$$\therefore \qquad f(x) = \sqrt{\frac{2}{\pi}} \cdot a \int \frac{dx}{a^2+x^2}$$

$$= \sqrt{\frac{2}{\pi}} \tan^{-1}\frac{x}{a} + c \qquad \text{...(2)}$$

At $x = 0, f(0) = 0$ using (1)

Using this in (2), $c = 0$

Hence, $\qquad f(x) = \sqrt{\frac{2}{\pi}} \tan^{-1}\frac{x}{a}$.

Setting $a = 0$,

$$F_s^{-1}\left(\frac{1}{s}\right) = \sqrt{\frac{2}{\pi}} \cdot \frac{\pi}{2} = \sqrt{\frac{\pi}{2}}.$$

Example 22. *Prove* (*i*) $F\{x^n f(x)\} = (-i)^n \dfrac{d^n F(s)}{ds^n}$

and (*ii*) $F\{f^n(x)\} = (-is)^n F(s)$ (*iii*) *Hence solve for* $f(x)$ *if*

$$\int_{-\infty}^{\infty} f(t)\, e^{-|x-t|}\, dt = \phi(x) \text{ where } \phi(x) \text{ is known.}$$

Proof. (*i*) $F(s) = \dfrac{1}{\sqrt{2\pi}} \displaystyle\int_{-\infty}^{\infty} e^{isx} f(x)\, dx$

$$\frac{d^n}{ds^n} F(s) = \frac{1}{\sqrt{2\pi}} \int_{-\infty}^{\infty} (ix)^n e^{isx} f(x)\, dx$$

$$(-i)^n \frac{d^n}{ds^n} F(s) = \frac{1}{\sqrt{2\pi}} \int_{-\infty}^{\infty} x^n e^{isx} f(x)\, dx$$

$$= F\{x^n f(x)\}$$

(*ii*) Similarly,

$$F\{f^n(x)\} = \frac{1}{\sqrt{2\pi}} \int_{-\infty}^{\infty} e^{isx} \frac{d^n}{dx^n} f(x)\, .\, dx$$

$$= (is)^n F(s),$$

Using integration by parts successively and making assumptions that $f, f', ..., f^{(n-1)} \to 0$ as $x \to \pm\infty$.

(*iii*) $\frac{1}{\sqrt{2\pi}}\phi(x) = \frac{1}{\sqrt{2\pi}}\int_{-\infty}^{\infty} f(t)\, e^{-|x-t|}\, dt$, from the given equation

$$= f(x) * e^{-|x|}$$

By convolution theorem,

$$\frac{1}{\sqrt{2\pi}}\bar{\phi}(s) = F(s) \cdot \sqrt{\frac{2}{\pi}}\frac{1}{1+s^2}$$

$$F(s) = \tfrac{1}{2}(1+s^2)\,\bar{\phi}(s)$$

$$= \tfrac{1}{2}[\bar{\phi}(s) - (-is)^2\,\bar{\phi}(s)]$$

$\therefore\ f(x) = \frac{1}{2}\phi(x) - \frac{1}{2}\phi''(x)$ using the result derived in (*ii*)

Exercises 1 (a)

1. Show that the Fourier transform of

$$f(x) = a - |x| \text{ for } |x| < a$$
$$= 0 \quad \text{for } |x| > a > 0$$

is $\sqrt{\frac{2}{\pi}}\frac{1-\cos as}{s^2}$. Hence show that $\int_0^{\infty}\left(\frac{\sin t}{t}\right)^2 dt = \pi/2$

2. Show that the Fourier transform of

$$f(x) = 0 \text{ for } x < \alpha$$
$$= 1 \text{ for } \alpha < x < \beta$$
$$= 0 \text{ for } x > \beta$$

is $\frac{1}{\sqrt{2\pi}}\left(\frac{e^{i\beta s} - e^{i\alpha s}}{is}\right)$

3. Show that the Fourier transform of

$$f(x) = \begin{cases} \frac{\sqrt{2\pi}}{2a}, & \text{for } |x| \le a \\ 0 & \text{for } |x| > a \end{cases}$$

is $\frac{\sin sa}{sa}$

4. Find the Fourier sine and cosine transforms of e^{-x} and hence using the inversion formulae, show that

$$\int_0^{\infty}\frac{x\sin\alpha x}{1+x^2}\,dx = \frac{\pi}{2}e^{-\alpha} = \int_0^{\infty}\frac{\cos\alpha x}{1+x^2}\,dx$$

5. Show that the Fourier sine transform of

$$f(x) = \begin{cases} x & \text{for } 0 < x < 1 \\ 2-x & \text{for } 1 < x < 2 \\ 0 & \text{for } x > 2 \end{cases}$$

is $2 \sin s\,(1 - \cos s)/s^2$. **(1987 Ap. Barathadasan)**

6. Find the Fourier sine and cosine transform of $\cosh x - \sinh x$ (same as question 4).

7. Find the Fourier sine and cosine transform of $ae^{-\alpha x} + be^{-\beta x}$, $\alpha, \beta > 0$.

8. Find the Fourier transform of $f(x) = \begin{cases} 1 \text{ for } |x| \le a \\ 0 \text{ for } |x| > a \end{cases}$

9. Show that the Fourier cosine transform of $\dfrac{1}{1+x^2}$ is $\sqrt{\dfrac{\pi}{2}}\, e^{-s}$

10. Show that the Fourier sine transform of $\dfrac{x}{1+x^2}$ is $\sqrt{\dfrac{\pi}{2}}\, e^{-s}$

11. Show that the Fourier transform of $e^{\frac{-x^2}{2}}$ is self-reciprocal.

12. Find Fourier transform of $e^{-a|x|}$ if $a > 0$.

13. Find Fourier transform of $\dfrac{1}{\sqrt{|x|}}$.

Fourier Transform of Derivatives

We have already seen (refer Example 22) that,

$$F\{f^{n}(x)\} = (-is)^n F(s)$$

(*i*) $\therefore F\left(\dfrac{\partial^2 u}{\partial x^2}\right) = (-is)^2 F\{u(x)\} = -s^2 \bar{u}$ where $\bar{u}$ is Fourier transform of u w.r.t. x

(*ii*) $F_c\{f'(x)\} = -\sqrt{\dfrac{2}{\pi}}\, f(o) + s\,F_s(s)$

$$\text{L.H.S.} = \sqrt{\frac{2}{\pi}} \int_0^\infty f'(x) \,.\, \cos sx\, dx$$

$$= \sqrt{\frac{2}{\pi}} \int_0^\infty \cos sx\, d\{f(x)\}$$

$$= \sqrt{\frac{2}{\pi}} \left[\{f(x) \cos sx\}_0^\infty + s \int_0^\infty f(x) \sin sx\, dx \right]$$

$$= s\,F_s(s) - \sqrt{\frac{2}{\pi}}\, f(0) \text{ assuming } f(x) \to 0 \text{ as } x \to \infty$$

$$(iii)\quad F_s\{f'(x)\} = \sqrt{\frac{2}{\pi}}\int_0^{\infty} \sin sx \, d\,[f(x)]$$

$$= \sqrt{\frac{2}{\pi}}\left[(f(x)\sin sx)_0^{\infty} - s\int_0^{\infty} f(x)\cos sx\,dx \right]$$

$$= -s\,F_c(s)$$

$$(iv)\quad F_c\{f''(x)\} = \sqrt{\frac{2}{\pi}}\int_0^{\infty} \cos sx\, d\,[f'(x)]$$

$$= \sqrt{\frac{2}{\pi}}\left[\{f'(x)\cos sx\}_0^{\infty} + s\int_0^{\infty} f'(x)\sin sx\,dx \right]$$

$$= -\sqrt{\frac{2}{\pi}}\,f'(0) + s\,F_s\{f'(x)\}$$

$$= -s^2 F_c(s) - \sqrt{\frac{2}{\pi}}\,f'(0)$$

assuming $f(x), f'(x) \to 0$ as $x \to \infty$

$$(v)\quad F_s\{f''(x)\} = \sqrt{\frac{2}{\pi}}\left[\int_0^{\infty} \sin sx\, d\,[f'(x)]\right]$$

$$= \sqrt{\frac{2}{\pi}}\left[\left(f'(x)\sin sx\right)_0^{\infty} - s\int_0^{\infty} f'(x)\cos sx\,dx \right]$$

$$= -s\,F_c\{f'(x)\}$$

$$= -s\left[s\,F_s(s) - \sqrt{\frac{2}{\pi}}\,f(0) \right]$$

$$= -s^2 F_s(s) + \sqrt{\frac{2}{\pi}}\,sf(0).$$

assuming $f(x), f'(x) \to 0$ as $x \to \infty$.

Relationship between Fourier and Laplace Transforms

Consider

$$f(t) = \begin{cases} e^{-xt}\,g(t) & \text{for } t > 0 \\ 0 & \text{for } t < 0 \end{cases} \qquad \text{...(1)}$$

Then the Fourier transform of $f(t)$ is given by

$$F\{f(t)\} = \frac{1}{\sqrt{2\pi}} \int_{-\infty}^{\infty} e^{ist} f(t)\, dt$$

$$= \frac{1}{\sqrt{2\pi}} \int_{0}^{\infty} e^{ist} f(t)\, dt$$

$$= \frac{1}{\sqrt{2\pi}} \int_{0}^{\infty} e^{(is-x)t}\, g(t)\, dt$$

$$= \frac{1}{\sqrt{2\pi}} \int_{0}^{\infty} e^{-pt}\, g(t)\, dt \text{ where } p = x - is$$

$$= \frac{1}{\sqrt{2\pi}} L\{g(t)\}$$

$\therefore$ Fourier transform of $f(t) = \frac{1}{\sqrt{2\pi}} \times$ Laplace transform of $g(t)$ defined by (1).

*APPLICATIONS TO BOUNDARY VALUE PROBLEMS

Example 1. *Solve the diffusion equation*

$$\frac{\partial u}{\partial t} = K \frac{\partial^2 u}{\partial x^2},\ -\infty < x < \infty,\ t > 0 \text{ with the conditions,}$$

$u(x, 0) = f(x)$, *and* $\frac{\partial u}{\partial x}$, *u tend to zero as x tend to* $\pm\infty$.

Proof. Here, $u = u(x, t),\ -\infty < x < \infty,\ t > 0$

Defining Fourier transform of $u(x, t)$ as

$\bar{u}(s, t) = \frac{1}{\sqrt{2\pi}} \int_{-\infty}^{\infty} u(x, t)\, e^{isx}\, dx$, take Fourier transform of the given differential equation.

$$KF\left\{\frac{\partial^2 u}{\partial x^2}\right\} = F\left\{\frac{\partial u}{\partial t}\right\}$$

i.e., $$K(-s^2 \bar{u}(s, t)) = \frac{1}{\sqrt{2\pi}} \int_{0}^{\infty} \frac{\partial u}{\partial t} e^{isx}\, dx$$

$$= \frac{d}{dt}\{F(u)\}$$

$$= \frac{d\bar{u}}{dt}$$

$\therefore \quad \frac{d\bar{u}}{dt} + s^2 . k\bar{u} = 0$ where $\bar{u}$ = Fourier transform of $\bar{u}$ ...(1)

Solving (1), $\bar{u}(s, t) = ce^{-s^2 kt}$ ...(2)

Since $u(x, 0) = f(x)$, taking transform w.r.t. x,

$$\bar{u}(s, 0) = F(s)$$

Using (3) in (2), we get, $c = F(s)$

$$\bar{u}(s, t) = F(s)\, e^{-s^2 kt}$$

Taking inverse transform,

$$u(x, t) = F^{-1}\{F(s)\, e^{-s^2 kt}\}$$

$$= f(x) * F^{-1}(e^{-s^2 kt})$$

$$= f(x) * \frac{e^{-\frac{x^2}{4kt}}}{\sqrt{2kt}}$$

(since $F^{-1}\left(e^{-\frac{s^2}{4a^2}}\right) = \sqrt{2}a\, e^{-a^2x^2}$ by example 4)

$$= \frac{1}{\sqrt{4\pi kt}} \int_{-\infty}^{\infty} f(\theta)\, e^{-\frac{(x-\theta)^2}{4kt}}\, d\theta$$

Putting, $\frac{x-\theta}{2\sqrt{kt}} = \phi$, we get $\theta = x - 2\sqrt{kt}\,\phi$

$$\therefore \quad u(x, t) = \frac{1}{\sqrt{\pi}} \int_{-\infty}^{\infty} f(x - 2\sqrt{kt}\,\phi)\, e^{-\phi^2}\, d\phi$$

Example 2. *Solve* $\frac{\partial^2 u}{\partial t^2} = \alpha^2 \frac{\partial^2 u}{\partial x^2}, -\infty < x < \infty, t \geq 0$ *with conditions* $u(x, 0) = f(x), \frac{\partial u}{\partial t}(x, 0) = g(x)$ *and assuming* $u, \frac{\partial u}{\partial x} \to 0$ *as* $x \to \pm\infty$.

Taking Fourier transform on both sides of the differential equation,

$\frac{d^2\bar{u}}{dt^2} = \alpha^2(-s^2\bar{u})$ where $\bar{u}$ is Fourier transform of u with respect to x.

$$\frac{d^2\bar{u}}{dt^2} + \alpha^2 s^2 \bar{u} = 0$$

Auxiliary equation is $m^2 + \alpha^2 s^2 = 0$

$$m = \pm i\alpha s$$

$$\therefore \quad \bar{u}(s, t) = Ae^{i\alpha st} + Be^{-i\alpha st} \quad ...(1)$$

Since $u(x, 0) = f(x)$ and $\frac{\partial u}{\partial t}(x, 0) = g(x)$,

$$\bar{u}(s, 0) = F(s) \text{ and } \frac{d\bar{u}}{dt}(s, 0) = G(s) \text{ on taking transforms.}$$

Using these conditions in (1),

$$\bar{u}(s, 0) = A + B = F(s) \quad ...(2)$$

$$\frac{d\bar{u}}{dt}(s, 0) = i\,\alpha s(A - B) = G(s) \quad ...(3)$$

$$\text{Solving } A = \frac{1}{2}\left[F(s) + \frac{G(s)}{i\alpha s}\right]$$

$$B = \frac{1}{2}\left[F(s) - \frac{G(s)}{i\alpha s}\right]$$

Using these values in (1),

$$\bar{u}(s, t) = \frac{1}{2}\left[F(s) + \frac{G(s)}{i\alpha s}\right]e^{i\alpha st} + \frac{1}{2}\left[F(s) - \frac{G(s)}{i\alpha s}\right]e^{-i\alpha st} \quad ...(4)$$

By inversion theorem, (4) reduces to,

$$u(x, t) = \frac{1}{2}\left[f(x - \alpha t) - \frac{1}{\alpha}\int_{\alpha}^{x-\alpha t} g(\theta)\, d\theta\right]$$

$$+ \frac{1}{2}\left[f(x + \alpha t) + \frac{1}{\alpha}\int_{\alpha}^{x+\alpha t} g(\theta)\, d\theta\right]$$

Using the result

$$F\left(\int_{\alpha}^{x} f(t)\, dt\right) = \frac{F(s)}{(-is)}$$

Boundary Value Problems using sine and cosine Transforms

Example 3. *Solve* $\frac{\partial u}{\partial t} = \frac{\partial^2 u}{\partial x^2}$ *for* $x > 0, t > 0$ *given that*

(*i*) $u(0, t) = 0$ *for* $t > 0$

(*ii*) $u(x, 0) = \begin{cases} 1 \text{ for } 0 < x < 1 \\ 0 \text{ for } x \geq 1 \end{cases}$

and (*iii*) $u(x, t)$ *is bounded.*

(**Note.** If u at $x = 0$ is given, take Fourier sine transform and if $\frac{\partial u}{\partial x}$ at $x = 0$ is given, take Fourier cosine transform,)

In this problem, u at $x = 0$ is given. Therefore, take fourier (infinite) sine transform.

$$F_s\left(\frac{\partial u}{\partial t}\right) = F_s\left(\frac{\partial^2 u}{\partial x^2}\right)$$

$$\sqrt{\frac{2}{\pi}}\int_0^\infty \frac{\partial u}{\partial t}\sin sx\,dx = \sqrt{\frac{2}{\pi}}\int_0^\infty \frac{\partial^2 u}{\partial x^2}\sin sx\,dx$$

i.e., $\frac{d\bar{u}}{dt} = -s^2\bar{u} + \sqrt{\frac{2}{\pi}}\,s\,u\,(0, t)$ where $\bar{u}$ stands for Fourier sine transform of u.

Using boundary condition (*i*), we get

$$\frac{d\bar{u}}{dt} + s^2\bar{u} = 0$$

Solving

$$\bar{u}\,(s, t) = ce^{-s^2 t} \qquad \text{...}(iv)$$

Now, $u\,(x, 0) = \begin{cases} 1 \text{ for } 0 < x < 1 \\ 0 \text{ for } x \geq 1 \end{cases}$

Taking Fourier sine transform of this initial condition,

$$\bar{u}\,(s, 0) = \sqrt{\frac{2}{\pi}}\int_0^1 \sin sx\,dx$$

$$= \sqrt{\frac{2}{\pi}}\left(\frac{1-\cos s}{s}\right) \qquad \text{...}(v)$$

Using (*v*) in (*iv*),

$$c = \sqrt{\frac{2}{\pi}}\left(\frac{1-\cos s}{s}\right)$$

Hence (*iv*) reduces to,

$$\bar{u}\,(s, t) = \sqrt{\frac{2}{\pi}}\left(\frac{1-\cos s}{s}\right)\cdot e^{-s^2 t}$$

By inversion theorem, this becomes,

$$u\,(x, t) = \frac{2}{\pi}\int_0^\infty \left(\frac{1-\cos s}{s}\right) e^{-s^2 t}.\sin sx\,ds.$$

Example 4. *Solve* $\frac{\partial u}{\partial t} = K\frac{\partial^2 u}{\partial x^2}$ *for* $0 \leq x < \infty$, $t > 0$ *given the conditions*

(*i*) $u\,(x, 0) = 0$ for $x \geq 0$

(ii) $\dfrac{\partial u}{\partial x}(0, t) = -a$ (constant)

(iii) $u(x, t)$ is bounded.

In this problem, $\dfrac{\partial u}{\partial x}$ at $x = 0$ is given. Hence, take Fourier cosine transform on both sides of the given equation.

$$F_c\left(\frac{\partial u}{\partial t}\right) = F_c\left(K\frac{\partial^2 u}{\partial u^2}\right)$$

$$\frac{d\bar{u}}{dt} = K\left(-s^2\bar{u} - \sqrt{\frac{2}{\pi}}\cdot\frac{\partial u}{\partial x}(0, t)\right)$$

$$= -ks^2\bar{u} + \sqrt{\frac{2}{\pi}}\,ka \quad \text{using condition} \qquad ...(ii)$$

$$\frac{d\bar{u}}{dt} + ks^2\bar{u} = \sqrt{\frac{2}{\pi}}\,ka$$

This is linear in $\bar{u}$. Therefore, solving

$$\bar{u}e^{ks^2t} = \int\sqrt{\frac{2}{\pi}}\,ka\,e^{ks^2t}\,dt$$

$$= \sqrt{\frac{2}{\pi}}\,ka\,\frac{e^{ks^2t}}{ks^2} + c$$

$$\bar{u}(s, t) = \sqrt{\frac{2}{\pi}}\,\frac{a}{s^2} + c\,e^{-ks^2t} \qquad ...(iv)$$

Since $u(x, 0) = 0$ for $x \geq 0$,

$\bar{u}(s, 0) = 0.$

Using this in (*iv*), we get

$$\bar{u}(s, 0) = c + \sqrt{\frac{2}{\pi}}\,\frac{a}{s^2} = 0$$

$$\therefore \qquad c = -\sqrt{\frac{2}{\pi}}\,\frac{a}{s^2}$$

Substituting this in (*iv*)

$$\bar{u}(s, t) = \sqrt{\frac{2}{\pi}}\,\frac{a}{s^2}\left(1 - e^{-ks^2t}\right)$$

By inversion theorem,

$$u(x, t) = \frac{2}{\pi} \cdot a \int_0^\infty \frac{1 - e^{-ks^2 t}}{s^2} \cos sx \, ds.$$

Example 5. *Solve* $\frac{\partial u}{dt} = k \frac{\partial^2 u}{\partial x^2}$ *for* $x \geq 0, t \geq 0$ *under the given conditions* $u = u_0$ *at* $x = 0, t > 0$ *with initial condition* $u(x, 0) = 0, x \geq 0$

Taking Fourier sine transforms

$$F_s\left(\frac{\partial u}{\partial t}\right) = F_s\left(k \frac{\partial^2 u}{\partial x^2}\right)$$

$$\frac{d}{dt}\bar{u} = k\left[-s^2\bar{u} + \sqrt{\frac{2}{\pi}} \, su(0, t)\right]$$

$$= -ks^2\bar{u} + \sqrt{\frac{2}{\pi}} \, ksu_0 \quad \text{where } \bar{u} \text{ is the Fourier sine transform of } u.$$

$$\frac{d\bar{u}}{dt} + ks^2\bar{u} = \sqrt{\frac{2}{\pi}} \, ksu_0$$

This is linear in $\bar{u}$.

$$\therefore \quad \bar{u} \, e^{ks^2 t} = \sqrt{\frac{2}{\pi}} \, ku_0 \int s \, e^{ks^2 t} \, dt$$

$$= \sqrt{\frac{2}{\pi}} \frac{u_0}{s} e^{ks^2 t} + c \qquad (1)$$

Since, $u(x, 0) = 0$, $\bar{u}(s, 0) = 0$. Using this in (1)

$$0 = \sqrt{\frac{2}{\pi}} \frac{u_0}{s} + c$$

$$\therefore \quad c = -\sqrt{\frac{2}{\pi}} \frac{u_0}{s}$$

$$e^{ks^2 t}\bar{u} = \sqrt{\frac{2}{\pi}} \frac{u_0}{s}\left(e^{ks^2 t} - 1\right)$$

$$\therefore \quad \bar{u} = \sqrt{\frac{2}{\pi}} \frac{u_0}{s}\left(1 - e^{ks^2 t}\right)$$

By inversion theorem,

$$u(x, t) = \frac{2 u_0}{\pi} \int_0^\infty \left(\frac{1 - e^{-ks^2 t}}{s}\right) \sin sx \, ds.$$

Exercises 1 (b)

1. Show that the solution of $\dfrac{\partial u}{\partial t} = 2\dfrac{\partial^2 u}{\partial x^2}$,

subject to $u(0, t) = 0$, for $t > 0$ and $u(x, 0) = e^{-x}$ for $x > 0$ and $u(x, t)$ is bounded,

is $\dfrac{2}{\pi}\displaystyle\int_0^{\infty} \dfrac{s\, e^{-2s^2 t}}{1+s^2} \sin sx\, ds$

2. Solve $\dfrac{\partial u}{\partial t} = \dfrac{\partial^2 u}{\partial x^2}$ if

(*i*) $\dfrac{\partial u}{\partial x}(0, t) = 0$ for $t > 0$.

(*ii*) $u(x, 0) = \begin{cases} x \text{ for } 0 \le x \le 1 \\ 0 \text{ for } x > 1 \end{cases}$

(*iii*) and $u(x, t)$ is bounded for $x > 0, t > 0$.

$$\left[\textbf{Ans. } u(x, t) = \frac{2}{\pi}\int_0^{\infty}\left(\frac{\sin s}{s} + \frac{\cos s - 1}{s^2}\right) e^{-s^2 t} \cos sx\, ds \right]$$

*FINITE FOURIER TRANSFORMS

Let $f(x)$ denote a function which is sectionally continuous over the range $(0, l)$. Then the **finite Fourier sine transform** of $f(x)$ on this interval is defined as

$$F_s(p) = \bar{f}_s(p) = \int_0^l f(x) \sin \frac{p\pi x}{l}\, dx$$

where p is an integer (Instead of s, we take p as a parameter).

Inversion formula for sine transform

If $\bar{f}_s(p) = F_s(p)$ is the finite Fourier sine transform of $f(x)$ in $(0, l)$ then the inversion formula for sine transform is

$$f(x) = \frac{2}{l}\sum_{p=1}^{\infty} \bar{f}_s(p) \sin \frac{p\pi x}{l}$$

Proof. For the given function $f(x)$ in $(0, l)$, if we find the half range Fourier sine series, we get,

$$f(x) = \sum_{n=1}^{\infty} b_n \sin \frac{n\pi x}{l} \qquad ...(1)$$

where $$b_n = \frac{2}{l} \int_0^l f(x) \sin \frac{n\pi x}{l} \, dx$$

$$\therefore \quad b_p = \frac{2}{l} \int_0^l f(x) \sin \frac{p\pi x}{l} \, dx$$

$$= \frac{2}{l} \bar{f}_s(p) \text{ by definition}$$

Substituting in (1),

$$\therefore \quad f(x) = \frac{2}{l} \sum_{p=1}^{\infty} \bar{f}_s(p) \sin \frac{p\pi x}{l}$$

Finite Fourier Cosine Transform

Let $f(x)$ denote a sectionally continuous function in $(0, l)$.

Then the Finite Fourier cosine transform of $f(x)$ over $(0, l)$ is defined as

$$f_c(p) = \bar{f}_c(p) = \int_0^l f(x) \cos \frac{p\pi x}{l} \, dx \text{ where } p \text{ is an integer.}$$

Inversion formula for cosine transform

If $\bar{f}_c(p)$ is the finite Fourier cosine transform of $f(x)$ in $(0, l)$, then the inversion formula for cosine transform is

$$f(x) = \frac{1}{l} \bar{f}_c(0) + \frac{2}{l} \sum_{p=l}^{\infty} \bar{f}_c(p) \cos \frac{p\pi x}{l}$$

where $$\bar{f}_c(0) = \int_0^l f(x) \, dx.$$

Proof. If we find half range Fourier cosine series for $f(x)$ in $(0, l)$, we obtain,

$$f(x) = \frac{a_0}{2} + \sum_{n=1}^{\infty} a_n \cos \frac{n\pi x}{l} \quad ...(2) \text{ where}$$

$$a_n = \frac{2}{l} \int_0^l f(x) \cos \frac{n\pi x}{l} \, dx$$

$$\therefore \quad a_p = \frac{2}{l} \bar{f}_c(p)$$

$$a_0 = \frac{2}{l}\int_0^l f(x)\,dx$$

$$= \frac{2}{l}\bar{f}_c(0).$$

Substituting in (2), we get,

$$f(x) = \frac{1}{l}\bar{f}_c(0) + \frac{2}{l}\sum_{p=1}^{\infty}\bar{f}_c(p)\cos\frac{p\pi x}{l}$$

Example 1. *Find the finite Fourier sine and cosine transforms of*

(*i*) $f(x) = 1$ in $(0, \pi)$

(*ii*) $f(x) = x$ in $(0, l)$

(*iii*) $f(x) = x^2$ in $(0, l)$

(*iv*) $f(x) = 1$ in $0, < x < \pi/2$

$= -1$ in $\pi/2 < x < \pi$

(*v*) $f(x) = x^3$ in $(0, l)$

(*vi*) $f(x) = e^{ax}$ in $(0, l)$

(*i*) $\bar{f}_s(p) = F_s(1) = \int_0^{\pi} 1 \cdot \sin\frac{p\pi x}{\pi}\,dx$

$$= \left(-\frac{\cos px}{p}\right)_0^{\pi}$$

$$= \frac{1-\cos p\pi}{p} \qquad \text{if } p \neq 0$$

$$\bar{f}_c(p) = \int_0^{\pi} 1 \cdot \cos px\,dx$$

$$= \left(\frac{\sin px}{p}\right)_0^{\pi} = \frac{1}{p}(0-0) = 0$$

(*ii*) $\bar{f}_s(p) = F_s(p) = \int_0^l x\sin\frac{p\pi x}{l}\,dx$

$$= \left[(x)\left(\frac{-\cos\frac{p\pi x}{l}}{\frac{p\pi}{l}}\right) - (1)\left(-\frac{\sin\frac{p\pi x}{l}}{\frac{p^2\pi^2}{l^2}}\right)\right]_0^l$$

$$= \frac{-l}{p\pi}(l\cos p\pi)$$

$$= \frac{-l^2}{p\pi}(-1)^p \qquad \text{if } p \neq 0$$

$$\mathcal{F}_c(p) = F_c(x) = \int_0^1 x\cos\frac{p\pi x}{l}\,dx$$

$$= \left[(x)\left(\frac{\sin\frac{p\pi x}{l}}{\frac{p\pi}{l}}\right) - (1)\left(-\frac{\cos\frac{p\pi x}{l}}{\frac{p^2\pi^2}{l^2}}\right)\right]_0^l$$

$$= \frac{l^2}{p^2\pi^2}[(-1)^p - 1] \qquad \text{if } p \neq 0$$

(*iii*) $$\mathcal{F}_s(p) = F_s(x^2) = \int_0^l x^2\sin\frac{p\pi x}{l}\,dx$$

$$= \left[(x^2)\left(-\frac{\cos\frac{p\pi x}{l}}{\frac{p\pi}{l}}\right) - (2x)\left(-\frac{\sin\frac{p\pi x}{l}}{\frac{p^2\pi^2}{l^2}}\right) + (2)\left(\frac{\cos\frac{p\pi x}{l}}{\frac{p^3\pi^3}{l^3}}\right)\right]_0^l$$

$$= \frac{-l^3}{p\pi}[(-1)^p] + \frac{2l^3}{p^3\pi^3}[(-1)^p - 1] \qquad \text{if } p \neq 0$$

$$\mathcal{F}_c(p) = \int_0^l (x^2)\cos\frac{p\pi x}{l}\,dx$$

$$= \left[(x^2)\left(\frac{\sin\frac{p\pi x}{l}}{\frac{p\pi}{l}}\right) - (2x)\left(-\frac{\cos\frac{p\pi x}{l}}{\frac{p^2\pi^2}{l^2}}\right) + (2)\left(\frac{\sin\frac{p\pi x}{l}}{\frac{p^3\pi^3}{l^3}}\right)\right]_0^l$$

$$= \frac{2l^3}{p^2\pi^2}[(-1)^p] \qquad \text{if } p \neq 0$$

(iv) $$F_s\{f(x)\} = \int_0^{\pi/2} \sin px\, dx + \int_{\pi/2}^{\pi} (-1) \sin px\, dx$$

$$= \left(-\frac{\cos px}{p}\right)_0^{\pi/2} + \left(\frac{\cos px}{p}\right)_{\pi/2}^{\pi}$$

$$= -\frac{1}{p}\left(\cos\frac{p\pi}{2} - 1\right) + \frac{1}{p}(\cos p\pi - \cos p\pi/2)$$

$$= \frac{1}{p}\left(\cos p\pi - -2\cos\frac{p\pi}{2} + 1\right) \qquad \text{if } p \neq 0$$

$$F_c(f(x)) = \int_0^{\pi/2} \cos px\, dx - \int_{\pi/2}^{\pi} \cos px\, dx$$

$$= \left(\frac{\sin px}{p}\right)_0^{\pi/2} - \left(\frac{\sin px}{p}\right)_{\pi/2}^{\pi} = \frac{2}{p}\sin\frac{p\pi}{2} \qquad \text{If } p \neq 0$$

(v) $$F_s(x^3) = \int_0^l x^3 \sin\frac{p\pi x}{l}\, dx$$

$$= \left[(x^3)\left(-\frac{\cos\frac{p\pi x}{l}}{\frac{p\pi}{l}}\right) - (3x^2)\left(-\frac{\sin\frac{p\pi x}{l}}{\frac{p^2\pi^2}{l^2}}\right) + (6x)\left(\frac{\cos\frac{p\pi x}{l}}{\frac{p^3\pi^3}{l^3}}\right) - (6)\left(-\frac{\sin\frac{p\pi x}{l}}{\frac{p^4\pi^4}{l^4}}\right)\right]_0^l$$

$$= -\frac{l^4}{p\pi}\left[(-1)^p\right] + \frac{6l^4}{p^3\pi^3}(-1)^p \qquad \text{if } p \neq 0$$

$$F_c(x^3) = \int_0^l x^3 \cos\frac{p\pi x}{l}\,dx$$

$$= \left[(x^3)\left(\frac{\sin\frac{p\pi x}{l}}{\frac{p\pi}{l}}\right) - (3x^2)\left(-\frac{\cos\frac{p\pi x}{l}}{\frac{p^2\pi^2}{l^2}}\right) + (6x)\left(\frac{\sin\frac{p\pi x}{l}}{\frac{p^3\pi^3}{l^3}}\right) - (6)\left(\frac{\cos\frac{p\pi x}{l}}{\frac{p^4\pi^4}{l^4}}\right)\right]_0^l$$

$$= \frac{3l^4}{\pi^2 p^2}(-1)^p - \frac{6l^4}{p^4\pi^4}\left[(-1)^p - 1\right] \qquad \text{if } p \neq 0$$

(vi) $$F_s(e^{ax}) = \int_0^l e^{ax}\sin\frac{p\pi x}{l}\,dx$$

$$= \left\{\frac{e^{ax}}{a^2 + \frac{p^2\pi^2}{l^2}}\left[a\sin\frac{p\pi x}{l} - \frac{p\pi}{l}\cos\frac{p\pi x}{l}\right]\right\}_0^l$$

$$= \frac{e^{ax}}{a^2 + \frac{p^2a^2}{l^2}}\left(-\frac{p\pi}{l}(-1)^p\right) + \frac{1}{a^2 + \frac{p^2\pi^2}{l^2}}\left(\frac{p\pi}{l}\right)$$

$$F_c(e^{ax}) = \left\{\frac{e^{ax}}{a^2 + \frac{p^2\pi^2}{l^2}}\left[a\cos\frac{p\pi x}{l} + \frac{p\pi}{l}\sin\frac{p\pi x}{l}\right]\right\}_0^l$$

$$= \frac{e^{al}}{a^2 + \frac{p^2\pi^2}{l^2}}a(-1)^p - \frac{1}{a^2 + \frac{p^2\pi^2}{l^2}}\left(a\right)$$

Note. In all the above problems, if $p = 0$, do the integration separately.

Example 2. *Find $f(x)$ if its finite Fourier sine transform is $\frac{2\pi}{p^3}(-1)^{p-1}$ for $p = 1, 2, \ldots\ldots, 0 < x < \pi$.*

By inversion Theorem,

$$f(x) = \frac{2}{\pi} \sum_{p=1}^{\infty} \frac{2\pi}{p^3} (-1)^{p-1} \sin px$$

$$= 4 \sum_{p=1}^{\infty} \frac{(-1)^{p-1}}{p^3} \sin px$$

Example 3. *Find f(x) if its finite Fourier sine transform is given by*

(i) $F_s(p) = \dfrac{1 - \cos p\pi}{p^2 \pi^2}$ for $p = 1, 2, 3, \ldots\ldots$ and $0 < x < \pi$

(ii) $F_s(p) = \dfrac{16(-1)^{p-1}}{p^3}$ for $p = 1, 2, 3, \ldots.$ where $0 < x < 8$

(iii) $F_s(p) = \dfrac{\cos \frac{2\pi p}{3}}{(2p+1)^2}$ for $p = 1, 2, 3, \ldots.$ and $0 < x < 1$.

Solution :

By inversion theorem

(i) $$f(x) = \frac{2}{\pi} \sum_{p=1}^{\infty} \left(\frac{1 - \cos p\pi}{p^2 \pi^2} \right) . \sin px$$

$$= \frac{2}{\pi^3} \sum_{p=1}^{\infty} \left(\frac{1 - \cos p\pi}{p^2} \right) . \sin px$$

(ii) $$f(x) = \frac{2}{l} \sum_{p=1}^{\infty} F_s(p) \sin \left(\frac{p\pi x}{l} \right)$$

$$= \frac{2}{8} \sum_{p=1}^{\infty} \frac{16(-1)^{p-1}}{p^3} \sin \left(\frac{p\pi x}{8} \right) \quad \text{since } l = 8$$

$$= 4 \sum_{p=1}^{\infty} \frac{(-1)^{p-1}}{p^3} \sin \left(\frac{p\pi x}{8} \right)$$

(iii) $$f(x) = \frac{2}{l} \sum_{p=1}^{\infty} F_s(p) \sin \left(\frac{p\pi x}{l} \right)$$

$$= 2 \sum_{p=1}^{\infty} \frac{\cos\left(\frac{2\pi p}{3}\right)}{(2p+1)^2} \sin (p\pi x) \quad \text{since } l = 1$$

Example 4. *Find f(x) if its finite Fourier cosine transform is* (*i*) $F_c(p) = \dfrac{1}{2p} \sin\left(\dfrac{p\pi}{2}\right)$ for $p = 1, 2, 3, ...$

$= \dfrac{\pi}{4}$ for $p = 0$ given $0 < x < 2\pi$

(*ii*) $F_c(p) = \dfrac{6 \sin \dfrac{p\pi}{2} - \cos p\pi}{(2p + 1)\pi}$ for $p = 1, 2, 3, ...$

$= \dfrac{2}{\pi}$ for $p = 0$ given $0 < x < 4$

(*iii*) $F_c(p) = \dfrac{\cos\left(\dfrac{2p\pi}{3}\right)}{(2p + 1)^2}$ for $p = 1, 2, 3, ...$

$= 1$ for $p = 0$ given $0 < x < 1$

Solution :

By inversion theorem,

$$f(x) = \frac{1}{l} F_c(0) + \frac{2}{l} \sum_{p=1}^{\infty} F_c(p) . \cos \frac{p\pi x}{l}.$$

(*i*) Here $F_c(0) = \pi/4$ and $l = 2\pi$

$$\therefore f(x) = \frac{1}{2\pi}\left(\frac{\pi}{4}\right) + \frac{2}{2\pi} \sum_{p=1}^{\infty} \frac{1}{2p} \sin\left(\frac{p\pi}{2}\right) \cos\left(\frac{p\pi x}{2\pi}\right)$$

$$= \frac{1}{8} + \frac{1}{2\pi} \sum_{p=1}^{\infty} \frac{1}{p} \sin\left(\frac{p\pi}{2}\right) \cos\left(\frac{px}{2}\right).$$

(*ii*) Here, $F_c(0) = \dfrac{2}{\pi}$ and $l = 4$

$$f(x) = \frac{1}{4}\left(\frac{2}{\pi}\right) + \frac{2}{4} \sum_{p=1}^{\infty} \left(\frac{6 \sin \dfrac{p\pi}{2} - \cos p\pi}{(2p + 1)\pi}\right) \cos\left(\frac{p\pi x}{4}\right)$$

$$= \frac{1}{2\pi} + \frac{1}{2\pi} \sum_{p=1}^{\infty} \left(\frac{6 \sin \dfrac{p\pi}{2} - \cos p\pi}{(2p + 1)}\right) . \cos\left(\frac{p\pi x}{4}\right)$$

(*iii*) Here $F_c(0) = 1$, $l = 1$

$$\therefore f(x) = \frac{1}{1} + \frac{2}{1} \sum_{p=1}^{\infty} \frac{1}{(2p + 1)^2} \cos\left(\frac{2p\pi}{3}\right) . \cos(p\pi x)$$

$$= 1 + 2 \sum_{p=1}^{\infty} \frac{\cos\left(\frac{2p\pi}{3}\right)}{(2p+1)^2} \cos\ (p\pi x).$$

Example 5. *Find the finite Fourier sine transform of $f(x) = 1$ in $(0, \pi)$. Use the inversion theorem and find Fourier sine series for $f(x) = 1$ in $(0, \pi)$ Hence prove*

(i) $1 - \frac{1}{3} + \frac{1}{5} - \frac{1}{7} + ... = \pi/4$

(ii) $\frac{1}{1^2} + \frac{1}{3^2} + \frac{1}{5^2} + ... = \pi^2/8$

Solution.

$$F_s(1) = \int_0^{\pi} 1.\ \sin\left(\frac{p\pi x}{\pi}\right) dx$$

$$\bar{f}_s(p) = \frac{1 - \cos p\pi}{p} \quad \text{if } p \neq 0$$

By inversion theorem,

$$f(x) = \frac{2}{l} \sum_{p=1}^{\infty} F_s\ (p)\ \sin \frac{p\pi x}{l}$$

$$1 = \frac{2}{\pi} \sum_{p=1}^{\infty} \frac{1-(-1)^p}{p}.\ \sin px \ \text{ since } \ l = \pi$$

$$1 = \frac{4}{\pi}\left[\sin x + \frac{1}{3} \sin 3x + \frac{1}{5} \sin 5x + ...\right]$$

This is the half range Fourier sine series for $f(x) = 1$ in $(0, \pi)$ getting $x = \pi/2$,

$$\frac{4}{\pi}\left[1 - \frac{1}{3} + \frac{1}{5} - \frac{1}{7} + ...\right] = 1$$

$$\therefore \qquad 1 - \frac{1}{3} + \frac{1}{5} - \frac{1}{7} + ... = \pi/4$$

In the half Fourier sine series $l_n = \frac{4}{\pi} \cdot \frac{1}{n}$ for n odd

By using Parseval's Theorem.

(range) $$\left[\frac{1}{2}\sum b_n^2\right] = \int_0^{\pi} (1)^2\, dx$$

$$\pi\left[\frac{1}{2} \cdot \frac{16}{\pi^2} \sum_{n=1,3,5....}^{\infty} \frac{1}{n^2}\right] = \pi$$

i.e., $$\sum_{n=1,3,5....}^{\infty} \frac{1}{n^2} = \frac{\pi^2}{8}$$

Example 6. *Find f (x) if its finite Fourier cosine transform is* $\frac{2l^3}{p^2\pi^2}(-1)^p$ *for* $p = 1, 2, 3, ...$ *and is* $\frac{l^3}{3}$ *for* $p = 0; 0 < x < l$.

Proof: By inversion formula,

$$f(x) = \frac{1}{l} \mathcal{F}_c(0) + \frac{2}{l} \sum_{p=1}^{\infty} \mathcal{F}_c(p) \cos\left(\frac{p\pi x}{l}\right)$$

$$= \frac{1}{l} \cdot \frac{l^3}{3} + \frac{2}{l} \sum_{p=1}^{\infty} \frac{2l^3(-1)^p}{p^2\pi^2} \cos\frac{p\pi x}{l}$$

$$= \frac{l^2}{3} + \frac{4l^2}{\pi^2} \sum_{p=1}^{\infty} \frac{(-1)^p}{p^2} \cos\left(\frac{p\pi x}{l}\right)$$

Exercises 1 (c)

Find the finite Fourier sine and cosine transforms of

1. $f(x) = 2x$ in $(0, 4)$
2. $f(x) = x$ in $(0, \pi)$ *(BR 1995 April)*
3. $f(x) = \cos ax$ in $(0, \pi)$
4. $f(x) = 1 - \frac{x}{\pi}$ in $(0, \pi)$
5. $f(x) = \begin{cases} x \text{ in } (0, \pi/2) \\ \pi - x \text{ in } (\pi/2, \pi) \end{cases}$
6. $f(x) = e^{-ax}$ in $(0, l)$
7. Find finite Fourier cosine transform of $\left(1 - \frac{x}{\pi}\right)^2$, $0 < x < \pi$
8. Find $f(x)$ if $\mathcal{F}_c(p) = \dfrac{\sin\left(\frac{p\pi}{2}\right)}{2p}$, $p = 1, 2, 3....$

   and $= \frac{\pi}{4}$ if $p = 0$ given $0 < x < 2\pi$.
9. Find finite Fourier cosine and sine transform of

   $f(x) = \frac{\pi}{3} - x + \frac{x^2}{2\pi}$. *(BR 1995 April)*

*Finite Fourier sine and cosine transform of derivatives

Using the definition and the integration by parts, we can easily prove the following results. For $0 \le x \le l$,

$$F_s\{f''(x)\} = -\frac{p^2\pi^2}{l^2} \mathcal{F}_s(p) + \frac{p\pi}{l}\left[f(0) - (-1)^p f(l)\right]$$

$$F_c\ \{f''(x)\} = -\frac{p^2\pi^2}{l^2}\bar{f}_c\ (p) + f'\ (l)\ (-1)^p - f'\ (0)$$

$$F_s\ \{f'(x)\} = -\frac{p\pi}{l}\bar{f}_c\ (p)$$

$$F_c\ \{f'(x)\} = f(l)\ (-1)^p - f(0) + \frac{p\pi}{l}\bar{f}_s\ (p)$$

Proof: $(i) F_s\ \{f'(x)\} = \int_0^l f'(x) \sin\frac{p\pi x}{l}\,dx$

$$= \int_0^l \sin\frac{p\pi x}{l}\cdot d[f(x)]$$

$$= \left(f(x)\sin\frac{p\pi x}{l}\right)_0^l - \int_0^l f(x)\cdot\cos\frac{p\pi x}{l}\cdot\frac{p\pi}{l}\,dx$$

$$= -\frac{p\pi}{l}\bar{f}_c\ (p)$$

$(ii) F_c\ \{f'(x)\} = \int_0^l f'(x)\cos\frac{p\pi x}{l}\,dx$

$$= \left(f(x)\cos\frac{p\pi x}{l}\right)_0^l - \int_0^l f(x)\cdot\frac{p\pi}{l}\sin\frac{p\pi x}{l}\,dx$$

$$= (-1)^p\ f(l) - f(0) + \frac{p\pi}{l}\bar{f}_s\ (p)$$

$(iii)\ F_s\ \{f''(x)\} = \int_0^l \sin\frac{p\pi x}{l}\,d\,[f'\ (x)]$

$$= \left(f'\ (x)\sin\frac{p\pi x}{l}\right)_0^l - \frac{p\pi}{l}\int_0^l f'\ (x)\cos\frac{p\pi x}{l}\,dx$$

$$= -\frac{p\pi}{l}\left[(-1)^p f(l) - f(0) + \frac{p\pi}{l}\bar{f}_s\ (p)\right]$$

$$= -\frac{p^2\pi^2}{l^2}\bar{f}_s\ (p) + \frac{p\pi}{l}\left[f(0) - (-1)^p f(l)\right]$$

(iv) $F_c\{f''(x)\} = \int_0^l \cos\frac{p\pi x}{l}\, d[f'(x)]$

$$= \left[f'(x)\cos\frac{p\pi x}{l}\right]_0^l + \frac{p\pi}{l}\int_0^l f'(x)\sin\frac{p\pi x}{l}\,dx$$

$$= (-1)^p f'(l) - f'(0) + \frac{p\pi}{l}\left[-\frac{p\pi}{l}\bar{f}_c(p)\right]$$

$$= -\frac{p^2\pi^2}{l^2}\bar{f}_c(p) + f'(l)(-1)^p - f'(0)$$

Note. If $u = u(x, t)$, then

$$F_s\left[\frac{\partial u}{\partial x}\right] = -\frac{p\pi}{l}F_c(u)$$

$$F_c\left[\frac{\partial u}{\partial x}\right] = \frac{p\pi}{l}F_s(u) - u(0, t) + (-1)^p u(l, t)$$

$$F_s\left[\frac{\partial^2 u}{\partial x^2}\right] = -\frac{p^2\pi^2}{l^2}F_s(u) + \frac{p\pi}{l}[u(0, t) - (-1)^p u(l, t)]$$

$$F_c\left[\frac{\partial^2 u}{\partial x^2}\right] = -\frac{p^2\pi^2}{l^2}F_c(u) + \frac{\partial u}{\partial x}(l, t)\cos p\pi - \frac{\partial u}{\partial x}(0, t)$$

Example 1. *Using finite Fourier transform, solve*

$$\frac{\partial u}{\partial t} = \frac{\partial^2 u}{\partial x^2} \text{ given } u(0, t) = 0 \text{ and } u(4, t) = 0$$

and $u(x,0) = 2x$ *where* $0 < x < 4,\ t > 0$

Proof. Since $u(0, t)$ is given, take finite Fourier sine transform.

$$\int_0^4 \frac{\partial u}{\partial t}\sin\frac{p\pi x}{4}\,dx = \int_0^4 \frac{\partial^2 u}{\partial x^2}\sin\frac{p\pi x}{4}\,dx$$

$$\frac{d}{dt}\bar{u}_s = F_s\left(\frac{\partial^2 u}{\partial x^2}\right)$$

$$= -\frac{p^2\pi^2}{16}\bar{u}_s + \frac{p\pi}{4}[u(0, t) - (-1)^p u(4, t)]$$

$$= -\frac{p^2\pi^2}{16}\bar{u}_s \quad \text{using } u(0, t) = 0,\ u(4, t) = 0$$

$$\frac{d\bar{u}_s}{\bar{u}_s} = -\frac{p^2\pi^2}{16}\,dt$$

Integrating

$$\log \bar{u}_s = -\frac{p^2\pi^2}{16}t + c$$

$$\bar{u}_s = A\,e^{-\frac{\pi^2 p^2}{16}t} \qquad \ldots(1)$$

Since $u(x, 0) = 2x$

$$\bar{u}_s(p, 0) = \int_0^4 (2x)\sin\left(\frac{p\pi x}{4}\right)dx$$

$$= -\frac{32}{p\pi}\cos p\pi \qquad \ldots(2)$$

Using (2) in (1),

$$\bar{u}_s(p, 0) = A = -\frac{32}{p\pi}\cos p\pi.$$

Substituting in (1),

$$\therefore \qquad \bar{u}_s = -\frac{32}{p\pi}(-1)^p\,e^{-\frac{p^2\pi^2}{16}t}$$

By inversion theorem,

$$u(x, t) = \frac{2}{4}\sum_{p=1}^{\infty}\frac{32}{p\pi}(-1)^{p+1}\,e^{-\frac{p^2\pi^2}{16}t}\sin\left(\frac{p\pi x}{4}\right)$$

Example 2. *Solve* $\frac{\partial u}{\partial t} = \frac{\partial^2 u}{\partial x^2}$, $0 < x < 6, t > 0,$

given $\frac{\partial u}{\partial x}(0, t) = 0$, $\frac{\partial u}{\partial x}(6, t) = 0$ and $u(x, 0) = 2x$

Proof. Since $\frac{\partial u}{\partial x}(0, t)$ is given, use finite Fourier cosine transform.

$$\int_0^6 \frac{\partial u}{\partial t}\cos\frac{p\pi x}{6}\,dx = \int_0^6 \frac{\partial^2 u}{\partial x^2}\cos\frac{p\pi x}{6}\,dx$$

$$\frac{d}{dt}\bar{u}_c = -\frac{p^2\pi^2}{36}\bar{u}_c + \frac{\partial u}{\partial x}(6, t)\cos p\pi - \frac{\partial u}{\partial x}(0, t)$$

$$= -\frac{p^2\pi^2}{36}\bar{u}_c$$

$$\frac{d\bar{u}_c}{\bar{u}_c} = -\frac{p^2\pi^2}{36}\,dt$$

$$\log \bar{u}_c = -\frac{p^2\pi^2}{36}t + c$$

$$\bar{u}_c = A\,e^{-\frac{p^2\pi^2}{36}t} \qquad \text{...(1)}$$

$u\,(x, 0) = 2x.$

$\therefore$ At $t = 0$

$$\bar{u}_c\,(p, 0) = \int_0^6 (2x)\cos\frac{p\pi x}{6}\,dx$$

$$= \frac{72}{p^2\pi^2}(\cos p\pi - 1) \qquad \text{...(2)}$$

Using this in (1),

$$\bar{u}_c\,(p, 0) = A = \frac{72}{p^2\pi^2}(\cos p\pi - 1)$$

Substituting in (1),

$$\bar{u}_c\,(p, t) = \frac{72}{p^2\pi^2}(\cos p\pi - 1)\,e^{-\frac{p^2\pi^2}{36}t}$$

By inversion theorem,

$$u\,(x, t) = \frac{1}{l}\bar{f}_c\,(0) + \frac{2}{l}\sum_{p=1}^{\infty}\bar{f}_c\,(p)\cos\left(\frac{p\pi x}{l}\right)$$

$$= \frac{1}{6}\int_0^6 (2x)\,dx + \frac{2}{6}\sum_{p=1}^{\infty}\frac{72}{p^2\pi^2}(\cos p\pi - 1)\,e^{-\frac{p^2\pi^2}{36}t}\,.\cos\left(\frac{p\pi x}{6}\right)$$

$$= 6 + \frac{24}{\pi^2}\sum_{p=1}^{\infty}\frac{(\cos p\pi - 1)}{p^2}\,e^{-\frac{p^2\pi^2}{36}t}\cos\left(\frac{p\pi x}{6}\right).$$

Example 3. *Solve* $\frac{\partial u}{\partial t} = 2\frac{\partial^2 u}{\partial x^2},\ 0 < x < 4,\ t > 0$

given $u\,(0, t) = 0,\ u\,(4, t) = 0\ ;\ u\,(x, 0) = 3\sin \pi x - 2\sin 5\pi x.$

Proof. Since $u\,(0, t)$ is given, take finite Fourier sine transform. The equation becomes (as in example 1)

$$\frac{d}{dt}\bar{u}_c = 2\left[-\frac{p^2\pi^2}{16}\bar{u}_s + \frac{p\pi}{4}\{u\,(0, t) - (-1)^p\,u\,(4, t)\}\right]$$

$$= -\frac{p^2\pi^2}{8}\bar{u}_s$$

Solving we get,

$$\bar{u}_s = A\, e^{-\frac{p^2\pi^2}{8}t} \qquad \text{...(1)}$$

$$u(x, 0) = 3 \sin \pi x - 2 \sin 5\pi x$$

Taking sine transform,

$$\bar{u}_s(p, 0) = \int_0^4 (3 \sin \pi x - 2 \sin 5\pi x) \sin \frac{p\pi x}{4}\, dx$$

$$= 0 \text{ if } p \neq 4, p \neq 20.$$

If $p = 4, \bar{u}_s(4, 0) = 6$

If $p = 20, \bar{u}_s(20, 0) = -4$

$$u(x, t) = \frac{2}{4}\sum_{p=1}^{\infty} \bar{u}_s(p, t) \sin\left(\frac{p\pi x}{4}\right)$$

$$= \tfrac{1}{2}\left[6\, e^{-\frac{p^2\pi^2}{8}t} \sin \pi x - 4\, e^{-\frac{p^2\pi^2}{8}t} \sin 5\pi x\right]$$

where p in the first term is 4 and p in the second term is 20

$$= 3\, e^{-2\pi^2 t} \sin \pi x - 2\, e^{-50\pi^2 t} \sin 5\pi x.$$

Example 4. *Solve* $\frac{\partial u}{\partial t} = \frac{\partial^2 u}{\partial x^2}$, $0 < x < \pi$, $t > 0$ *given* $u(0, t) = u(\pi, t) = 0$ *for* $t > 0$ *and* $u(x, 0) = \sin^3 x$.

Solution. Since $u(0, t)$ and $u(\pi, t)$ are given, take finite Fourier sine transforms on both sides of the given equation with respect to x in $(0, \pi)$.

$$\int_0^{\pi} \frac{\partial u}{\partial t} \sin\frac{p\pi x}{\pi}\, dx = \int_0^{\pi} \frac{\partial^2 u}{\partial x^2} \sin\frac{p\pi x}{\pi}\, dx$$

$$\frac{d}{dt}\bar{u}_s = F_s\left(\frac{\partial^2 u}{\partial x^2}\right)$$

$$= -p^2\bar{u}_s + p\,[u(0, t) - (-1)^p\, u(\pi, t)]$$

$$= -p^2\bar{u}_s \text{ using } u(0, t) = u(\pi, t) = 0.$$

$$\frac{d\,\overline{u}_s}{\overline{u}_s} = -p^2 dt$$

Integrating

$$\log(\overline{u}_s) = -p^2 t + k$$

$\therefore$ $$\overline{u}_s(p, t) = c\, e^{-p^2 t} \quad \text{... (1)}$$

Since, $$u(x, 0) = \sin^3 x$$

$$\overline{u}_s(p, 0) = \int_0^\pi \sin^3 x \sin px\, dx$$

$$= \int_0^\pi \left(\frac{3}{4}\sin x - \frac{1}{4}\sin 3x\right) \sin px\, dx$$

$$= 0 \text{ for } p \neq 1,\ p \neq 3$$

when $$p = 1, \overline{u}_s(1,0) = \int_0^\pi \sin^4 x dx$$

$$= 2\int_0^{\pi/2} \sin^4 x\, dx$$

$$= 2\left(\frac{3}{4} \cdot \frac{1}{2} \cdot \pi/2\right)$$

$$= \frac{3}{8}\pi \quad \text{.... (2)}$$

when $$p = 3, \overline{u}_s(3, 0) = \int_0^\pi \left(-\frac{1}{4}\sin^2 3x\right) dx$$

$$= -\frac{1}{4}\int_0^\pi \frac{1-\cos 6x}{2}\, dx$$

$$= -\frac{1}{8}\left[x - \frac{\sin 6x}{6}\right]_0^\pi$$

$$= -\pi/8 \quad \text{... (3)}$$

Using (2) and (3) in (1),

when $p = 1, \overline{u}_s(1, 0) = C = \dfrac{3\pi}{8}$

when $p = 3, \overline{u}_s(3, 0) = C = -\pi/8$

For all other values of p, $c = 0$

By inversion formula, $u(x, t) = \frac{2}{\pi} \sum_{p=1}^{\infty} \bar{u}_s(p, t) \sin nx$

i.e.
$$u(x, t) = \frac{2}{\pi}\left[\frac{3\pi}{8} e^{-t} \sin x - \frac{\pi}{8} e^{-9t} \sin 3x\right]$$
$$= \frac{3}{4} e^{-t} \sin x - \frac{1}{4} e^{-9t} \sin 3x$$

Example 5. *Solve* $\frac{\partial u}{\partial t} = \frac{\partial^2 u}{\partial x^2}$, $0 < x < 10$ *given* $u(0, t) = u(10, t) = 0$ *for* $t > 0$ *and* $u(x, 0) = 10x - x^2$ *for* $0 < x < 10$.

Solution. As the previous problem, take finite Fourier sine transform on both sides of the heat equation with respect to x in (0, 10).

$$\therefore \quad \frac{d}{dt}\bar{u}_s = -\frac{p^2\pi^2}{100}\bar{u}_s + \frac{p\pi}{10}[u(0,t) - (-1)^p u(10, t)]$$
$$= -\frac{p^2\pi^2}{100}\bar{u}_s \text{ using } u(0, t) = u(10, t) = 0$$

Solving for $\bar{u}_s$, we get

$$\bar{u}_s = C e^{\frac{-p^2\pi^2}{100}t} \quad \ldots (1)$$

Since $u(x, 0) = 10x - x^2$,

$$\bar{u}_s(p, 0) = \int_0^{10} (10x - x^2) \sin\frac{p\pi x}{10}\, dx$$

$$= \left\{(10x - x^2)\left[\frac{-\cos\frac{p\pi x}{10}}{\frac{p\pi}{10}}\right] - (10 - 2x)\left[-\frac{\sin\frac{p\pi x}{10}}{\frac{p^2\pi^2}{100}}\right] + (-2)\left[\frac{\cos\frac{p\pi x}{10}}{\frac{p^3\pi^3}{1000}}\right]\right\}_0^{10}$$

$$= -\frac{2000}{p^3\pi^3}\left[(-1)^p - 1\right]$$

$$= \begin{cases} \frac{4000}{p^3\pi^3} & \text{if p is odd} \\ 0 & \text{if p is even} \end{cases}$$

Using (1), $\bar{u}_s(p, 0) = C$.. using this again in (1), C is eliminated. Hence, taking inversion formula, (use (1))

$$u(x, t) = \frac{2}{10} \sum \bar{u}_s \,.\, \sin\frac{p\pi x}{10}$$

$$= \frac{800}{\pi^3} \sum_{p=1,3,5....}^{\infty} \frac{1}{p^3} \sin \frac{p\pi x}{10} e^{\frac{-p^2\pi^2}{100}t}$$

Exercises 1 (d)

1. Solve $\frac{\partial u}{\partial t} = \frac{\partial^2 u}{\partial x^2}$, $0 < x < 6$, $t > 0$ given that

$u(0, t) = 0 = u(6, t)$ and $u(x, 0) = \begin{cases} 1 \text{ for } 0 < x < 3 \\ 0 \text{ for } 3 < x < 6 \end{cases}$

$$\left[\textbf{Ans. } u(x, t) = \frac{2}{\pi} \sum_{p=1}^{\pi} \left(\frac{1 - \cos \frac{p\pi}{2}}{p} \right) e^{-\frac{p^2\pi^2 t}{36}} \sin\left(\frac{p\pi x}{6}\right)\right]$$

2. Solve $\frac{\partial v}{\partial t} = \frac{\partial^2 v}{\partial x^2}$ subject to conditions $v(0, t) = 1$, $v(\pi, t) = 3$

$v(x, 0) = 1$ for $0 < x < \pi$, $t > 0$.

$$\left[\textbf{Ans. } v(x, t) = \frac{4}{\pi} \sum_{p=1}^{\infty} \frac{\cos p\pi}{p} e^{-p^2 t} \sin px + 1 + \frac{2x}{\pi}.\right]$$

3. Solve $\frac{\partial \theta}{\partial t} = \frac{\partial^2 \theta}{\partial x^2}$ given

$\theta(0, t) = 0$, $\theta(\pi, t) = 0$, $\theta(x, 0) = 2x$ for $0 < x < \pi$, $t > 0$

[Refer example 3]

4. Solve $\frac{\partial u}{\partial t} = \frac{\partial^2 u}{\partial x^2}$ for $0 < x < l$, $t > 0$ given $\frac{\partial u}{\partial x}(0, t) = 0$, $\frac{\partial u}{\partial x}(l, t) = 0$ for $t > 0$

and $u(x, 0) = lx - x^2$ for $0 < x < l$

$$\left[\textbf{Ans. } u(x, t) = \frac{l^2}{6} - \frac{l^2}{\pi^2} \sum \frac{1}{p^2} \cos \frac{2p\pi x}{l} e^{\frac{-4p^2\pi^2 t}{l^2}}\right.$$

5. Solve $\frac{\partial u}{\partial t} = 2\frac{\partial^2 u}{\partial x^2}$ for $0 < x < 1$, $t > 0$ given $u(0, t) = u(1, t) = 0$ for $t > 0$ and

$u(x, 0) = \sin 3\pi x + \sin \pi x$

$$\left[\textbf{Ans: } u(x, t) = e^{-2\pi^2 t} \sin \pi x + e^{-18\pi^2 t} \sin 3\pi x\right]$$

SHORT ANSWER QUESTIONS

1. Define Periodic function.
2. If $f(x)$ is expressed in Fourier series of period 2π in $(c, c+2\pi)$ write down the Fourier series and formulae for Euler's coefficients.
3. If $f(x)$ is expressed in Fourier series of period $2l$ in $(c, c+2l)$ write down the Fourier series and formulae for Euler's coefficients.
4. If $f(x)$ is expressed in half range Fourier cosine series or sine series in $(0, l)$ of periodicity $2l$, write down two series and the corresponding integrals for coefficients.
5. State Dirichlet's conditions
6. Draw the graph of $f(x) = |x|$
7. Define $|x|$ in $(-\infty, \infty)$
8. Find half range Fourier Cosine series or sine series in $(0, \pi)$ of periodicity 2π if $f(x) = k$.
9. Define odd and even functions.
10. State any property you know regarding $\int_{-l}^{l} f(x)\, dx$ if $f(x)$ is odd or even.
11. Evaluate $\int_{-1}^{1} (1 - |x|) \cos n\pi x\, dx$, n any integer
12. Find the half range sine series in $0 < x < \pi$ if $f(x) = x^2$.
13. Find the half range cosine series in $0 < x < \pi$ if $f(x) = x + 1$
14. Write down the analytic expression for $f(x)$ if $y = f(x)$ has the graph

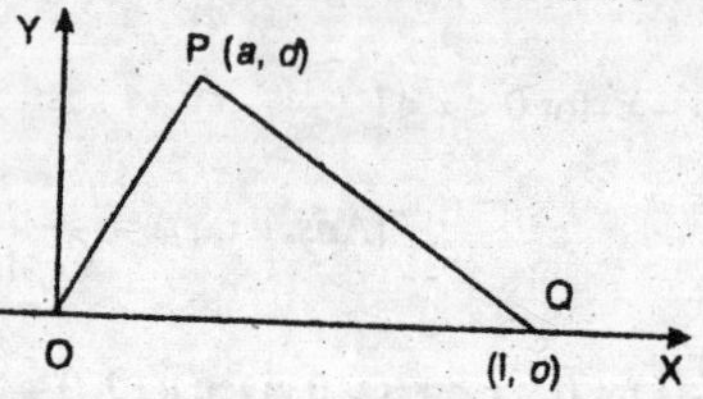

15. Define root mean square value of $f(x)$ over the range (a, b)
16. Using the R.M.S. value fill up the blanks in terms of Fourier coefficients

$$\int_{c}^{c+2\pi} [f(x)]^2\, dx - \ldots\ldots$$

17. Write down the complex Fourier series, stating the formula for coefficient C_n.

Write true or false, with reason, if possible, (18 to 19)

18. In $(-l, l)$, $f(x) = (1 - |x|) \sin nx$ is odd
19. $\int_{c}^{c+2\pi} \sin mx \cos nx\, dx = 0$, for all integral values of m, n.

20. Find the Fourier series of periodicity 2π for $f(x) = x$ in $-\pi < x < \pi$

21. $f(x) = 1 + \frac{2x}{\pi}$ in $-\pi < x < 0$

$= 1 - \frac{2x}{\pi}$ in $0 < x < \pi$ is even. State true of false

22. If Fourier series is found out for $f(x) = x^2$ in $-1 < x < 1$, the series will contain only cosine terms. State true or false.

23. If $f(x) = x^3$ in $-\pi < x < \pi$, find the constant term of its Fourier series.

24. If $f(x) = \sin \alpha x$, *in* $-\pi < x < \pi$, α not an integer, find the constant term of its Fourier expansion.

25. If $f(x) = x^2$ in $(-l, l)$ is expressed as a Fourier series of periodicity $2l$, find the constant term of the series.

26. In the definition of Fourier transform, $F(s) = \frac{1}{\sqrt{2\pi}} \int_{-\infty}^{\infty} f(x)\, e^{isx}\, dx$ what is the Kernel?

27. Define infinite complex Fourier transform of $f(x)$ and write its inversion formula also.

28. State Fourier integral theorem.

29. State Inversion theorem for complex Fourier Transform.

30. Prove Fourier Transform is linear.

31. If $F\{f(x)\} = F(s)$, $F\{f(x-a)\} =$ .....

32. If $F\{f(x)\} = F(s)$, $F\{f(ax)\} =$ ......

33. $F\{e^{iax} f(x)\} =$ ....

34. $F\{f(x) \cos ax\} = \frac{1}{2}\left[\quad + \quad\right]$ (modulation theorem).

35. $F\{x^n f(x)\} =$ ....

36. $F\{f'(x)\} =$ .... if $f(x) \to 0$ as $x \to \pm\infty$

37. Find Fourier transform of

$$f(x) = \begin{cases} 1 & \text{if} \quad |x| < a \\ 0 & \text{if} \quad |x| > a > 0. \end{cases}$$

38. Define convolution of two functions of $f(x)$ and $g(x)$.

39. Prove convolution of two functions is commutative.

40. $F^{-1}\{F(s)\, G(s)\} =$ ......

41. State convolution Theorem.

42. State parseval's identity.

43. Define infinite Fourier sine and cosine Transform.

44. Write the corresponding inversion forms.

45. Find Fourier sine and cosine transform of e^{-x}, $x \geq 0$,

46. Find Fourier sine transform of $\frac{1}{x}$.

47. Solve $f(x)$ if $\int_0^\infty f(x) \cos \alpha x = e^{-\alpha}$.

48. The Fourier transform of $e^{-\frac{1}{2}x^2}$ is ....

49. Define Finite Fourier sine and cosine transform of $f(x)$.

50. Write down the corresponding inversion results.

51. Find Fourier sine and cosine transforms of $f'(x)$.

52. Find Fourier sine and cosine transforms of $f''(x)$.

53. Find finite Fourier sine and cosine transform of

(*i*) $f(x) = x$ in $(0, \pi)$

(*ii*) $f(x) = x^2$ in $(0, l)$

(*iii*) $f(x) = 1 - \frac{x}{\pi}$ in $(0, \pi)$

(*iv*) $f(x) = 1$ in $(0, \pi)$

54. If $F_s(p) = \frac{2\pi}{p^3}(-1)^{p-1}$ for $p = 1, 2, 3....$ $0 \le x \le \pi$ find $f(x)$

55. Find $f(x)$, if $F_c(p) = \dfrac{\sin \frac{p\pi}{2}}{2p}$ if $p = 1, 2, 3....$

$= \frac{\pi}{4}$ if $p = 0$ in $0 < x < 2\pi$

56. Prove $F_s\{f'(x)\} = -\frac{p\pi}{l}\bar{f}_c(p)$ (finite)

57. Prove $F_c(f''(x)) = -\frac{p^2\pi^2}{l^2}\bar{f}_c(p) + f'(l)(-1)^p - f'(0)$

58. Prove $F_c\{f'(x)\} = f(l)(-1)^p - f(0) + \frac{p\pi}{l}\bar{f}_s(p)$ (finite)

59. Prove $F_s(f''(x)) = \frac{-p^2\pi^2}{l^2}\bar{f}_s(p) + \frac{p\pi}{l}\left[f(0) - (-1)^p f(l)\right]$

60. Find Fourier cosine and sine transform of e^{-x}

61. $F_c\{f(ax)\} =$ ...... **61 (*a*).** $F_s\{f(ax)\} =$ ....

62. If $F\{f(x)\} = F(s)$ then $F[f(x-a)] =$ .....

63. $F\left[e^{-|x|}\right]$ (*i*) does not exist **(*BR 1995 Ap.*)**

(*ii*). is $\frac{2}{1+s^2}$

(*iii*) is $\frac{1}{1+s^2}$

(*iv*) is $\frac{s}{1+s^2}$

(*v*) is $-\frac{2}{1+s^2}$

State true or false:

64. Dirichlet's conditions are necessary for the uniform convergence of Fourier series.

65. If the complex form of the Fourier series of $f(x)$ is

$\sum_{-\infty}^{\infty} C_n e^{inx}$ in $(0, 2\pi)$, then C_n is (*a*) $\frac{1}{\pi}\int_0^{2\pi} f(x)\, e^{-inx}\, dx.$

(*b*) $\frac{2}{\pi}\int_0^{2\pi} f(x)\, e^{-inx}\, dx.$ (*c*) $\frac{1}{2\pi}\int_0^{2\pi} f(x)\, e^{-inx}\, dx.$

(*d*) $\frac{2}{\pi}\int_0^{\pi} f(x)\, e^{-inx}\, dx.$ (*e*) $\frac{1}{2\pi}\int_0^{2\pi} f(x)\, e^{inx}\, dx.$

66. If the Fourier series of $f(x)$ and $g(x)$ are

$\frac{a_0}{2} + \Sigma\,[a_n \cos nx + b_n \sin nx]$ and $A_0 + \Sigma\,[2A_n \cos nx + 2B_n \sin nx]$ over $[-\pi, \pi]$, write down the Fourier series of $2f(x) + g(x)$ over $[-\pi, \pi]$.

67. Sum of the Fourier series for

$\left\{\begin{array}{l} f(x) = x,\ 0 < x < 1 \\ \quad\ \ = 2\ 1 < x < 2 \end{array}\right\}$ at $x = 1$ is

(*a*) 0 (*b*) 1 (*c*) 2 (*d*) 1/2 (*e*) $\frac{3}{2}$.